THE URBAN TRANSFORMATION OF THE DEVELOPING WORLD

The Urban Transformation of the Developing World

Edited by

JOSEF GUGLER

OXFORD UNIVERSITY PRESS

1996

Oxford University Press, Walton Street, Oxford OX2 6DP
Oxford New York
Athens Auckland Bangkok Bombay
Calcutta Cape Town Dar es Salaam Delhi
Florence Hong Kong Istanbul Karachi
Kuala Lumpur Madras Madrid Melbourne
Mexico City Nairobi Paris Singapore
Taipei Tokyo Toronto
and associated companies in
Berlin Ibadan

Oxford is a trade mark of Oxford University Press

Published in the United States
by Oxford University Press Inc., New York

British Library Cataloguing in Publication Data
Data available

Library of Congress Cataloging in Publication Data
The urban transformation of the developing world : regional trajectories / edited by Josef Gugler.
1. Urbanization—Developing countries. 2. Cities and towns—Developing countries. I. Gugler, Josef.
HT149.5.U7337 1996 95–34814
307.76'09172'4—dc20
ISBN 0–19–874159–6 (Pbk)
ISBN 0–19–874158–8

Typeset by Graphicraft Typesetters Ltd., Hong Kong

Printed in Great Britain
on acid-free paper by
Bookcraft (Bath) Ltd., Midsomer Norton, Avon.

To celebrate diversity as well as
our shared humanity on this
our planet

Preface

THE urban transformation of the globe may come to be seen as the lasting legacy of the twentieth century. Already close to half the world's population lives in urban settlements. Now the last phase of this profound human transformation is playing itself out in the less developed countries of Asia, Oceania, Africa, and Latin America and the Caribbean: we are witnessing the urbanization of the globe. At this time, nearly two-thirds of the world's urban population, more than one and a half billion people, live in the cities of the South. Within little more than a generation their number will triple.

There has been a tendency to generalize about urbanization in the 'Third World'. But, of course, the countries thus commonly lumped together span three continents and are home to more than three-quarters of humanity. Not surprisingly then, there are important differences in the urban transformation across 'the South'. The first essay in this volume discusses these divergences as well as the convergences. The second essay presents the urban history of Asia, a region that holds more than half the world's population today, an urban history unparalleled in its time span, geographical spread, and cultural riches.

The next three essays discuss urbanization in the Third World's three largest countries, each the largest country within one of the three regions of Asia: China in East Asia, India in South Asia, and Indonesia in South-East Asia. Each of these three countries may be considered a region in its own right in terms of both the size of its population and its diversity. They thus present units of analysis that bear comparison with the other three major less developed World regions which are discussed in the last three essays: the Arab states, Africa South of the Sahara, and Latin America. Indeed, the population of China is about the same size as that of the Arab States, Africa, and Latin America combined. The three regions and three countries together represent more than 80 per cent of the population living in less developed countries.

I was fortunate in securing the collaboration of leading experts on these countries/regions. All but one contributed original essays. Our contributions reflect to some extent our diverse disciplinary orientations: six sociologists, an economist, a geographer, and an historian.

There are notable differences as well in the perspectives we privilege: modernization, dependency, political economy, class, urban bias, each is emphasized by one or the other. And, of course, the specific foci of our individual research agendas inform our contributions. Our joint efforts thus provide an opportunity to compare the entire range of social science approaches to urbanization.

The different emphases given to various topics in each contribution, and the diverse analytical perspective privileged, are not simply idiosyncratic. Rather they correspond, at least in part, to significant regional differences. Indeed, the specific features of each region played a major role in shaping the research agenda of each contributor.

China has gone through a series of major transformations over the last half century as successive regimes experimented with a variety of approaches to the political, social, and economic transformation of a very poor country. The World's most populous country thus presents an extraordinary case-study in the modelling of an alternative socialist city, the adoption of a range of policies to realize the model, and the intended and unintended consequences of the implementation of these policies.

India has experienced an exceptional degree of continuity in government and political orientation since independence in 1947. The country thus provides an opportunity to explore the difficulties of urban governance over several decades.

In Indonesia, the distinction between urban and rural is blurring. The population movement between village and city strengthens linkages between urban and rural areas with an increasing flow of information, goods, money, and people. Indeed, short-term migrants and commuters constitute a large part of the urban work-force.

Several Arab states have been profoundly transformed through oil wealth in barely a generation. The stark contrasts thus created within the region invite an analysis that focuses on the political economy of oil.

In Africa South of the Sahara, the contrast between urban and rural living conditions continues to be stark. At the same time, a large proportion of the urban population remains deeply involved in the affairs of their rural community of origin. This involvement reinforces ties among people of common origin in the urban setting and affects alignments in urban conflict.

Latin America is furthest along in the urban transition among all the less developed regions. Socio-economic and political changes have drastically modified social structures. Still, the region's cities continue to be characterized by severe inequality. At the same time, there is considerable heterogeneity in urban social stratification across countries as well as among regions within countries.

This volume brings together authors from the far corners of the globe. I wish to thank them for taking on a task rather undervalued in academia: contributing to an edited collection. Appreciation is due to some for bearing with me through the successive drafts I urged on them and for the hard work they invested in revising, and revising yet again, to others who turned in dazzling first drafts. I am grateful to Loretta Bass, for most efficient and invariably cheerful assistance, and to staff at the Homer Babbidge Library, University of Connecticut, who time and again went well beyond the call of duty.

The entire globe is caught up in the great human transformation to an urban world, but the trajectories vary. With this volume we celebrate diversity as well as shared humanity on this our planet.

Josef Gugler

Storrs
March 1995

Contents

List of Tables xiii

List of Figures xv

Notes on Contributors xvii

1. Regional Trajectories in the Urban Transformation: Convergences and Divergences 1

   Josef Gugler

2. A History of the City in Monsoon Asia 18

   Rhoads Murphey

3. Urbanization in China: Reassessing an Evolving Model 61

   Xiangming Chen and William L. Parish

4. Urbanization in India: Patterns and Emerging Policy Issues 93

   Rakesh Mohan

5. Urbanization in Indonesia: City and Countryside Linked 133

   Graeme Hugo

6. Urbanization in the Arab World and the International System 185

   Janet Abu-Lughod

7. Urbanization in Africa South of the Sahara: New Identities in Conflict 211

   Josef Gugler

8. Urban Development and Social Inequality in Latin America 253

   Orlandina de Oliveira and Bryan Roberts

Name Index 315

Subject Index 320

List of Tables

1.1 Demographic, Economic, and Human Development Indicators for Major Developing Countries and Regions, about 1990 3

3.1 Labour-Force in Chinese Cities: Size Category by Sector, 1989 and 1993 71

3.2 New Entrants to China's Urban Labour-Force, 1978–1993 73

3.3 Quality of Life Indicators for Chinese Cities, 1984 and 1991 78

4.1 Growth of Urban Population in India, 1901–1991 94

4.2 Employment Structure in India, 1961–1991 101

4.3 Distribution and Growth of Urban Population by Size Classes in India, 1961–1991 104

4.4 Annual Growth Rate of Urban Population by Size of Town in India, 1971–1981 105

4.5 Annual Growth Rate of Urban Population by Size of Town in India, 1981–1991 106

4.6 Growth of Urban and Rural Population in India by State, 1951–1991 108

4.7 Level of Urbanization in India by State, 1951–1991 109

4.8 Selected Indicators of Economic Development in India by State, 1961–1991 110

5.1 Basic Economic and Social Indicators for Indonesia, 1970–1991 135

5.2 Employed Persons by Industry in Indonesia, 1971–1990 135

5.3 Population of Batavia by Origin, 1673 and 1815 141

5.4 Population in Major Indonesian Cities, 1855–1930 143

5.5 Four-City Primacy Index for Indonesia, 1890–1990 144

5.6 Indigenous and Non-Indigenous Populations of Indonesia, 1860–1930 146

5.7 Distribution of Indigenous and Non-Indigenous Population between Urban and Rural Areas in Indonesia, 1930 146

5.8 Distribution of Population between Urban and Rural Areas in Indonesia, 1920–1990 150

5.9 Population in Jakarta Metropolitan Area According to Different Definitions, 1971–1990 155

5.10 Distribution of Population between Urban and Rural Areas in Indonesia by Province, 1980 and 1990 157

5.11 Migration Into and Out of Java, 1971, 1980, and 1990 161

5.12 Migration Into and Out of Java by Urban/Rural Destination, 1990 161

5.13 Inmigrants in the Capital District of Jakarta, 1961–1990 166

5.14 Population and Population Growth Rate of Urban Centres with more than 100,000 Inhabitants in 1980 for 1920–1990 169

6.1 Relationship Between Economy and Urbanization in Arab States, 1950, 1980, and 1990 192

7.1 Demographic, Economic, and Human Development Indicators for Larger Countries in Africa South of the Sahara, about 1990 222

8.1 Distribution of Population and its Growth in Six Latin American Countries, 1940–1980 258

8.2 Growth of Mexican Cities, 1940–1990 267

8.3 Urban Occupational Stratification in Six Latin American Countries, 1940–1980 279

8.4 Educational Levels of Economically Active Population in Six Latin American Countries, 1960, 1970, 1980 295

List of Figures

1.1 World Urban Agglomerations with more than 4 Million Inhabitants in 1990 xx

2.1 Major Ports and Chief Commercially Productive Areas in Monsoon Asia, 1600–1940 16

3.1 Urban Agglomerations in China with more than 1 Million Inhabitants in 1990 60

3.2 China's Urban and Labour-Force Trends, 1949–1993 62

3.3 Chinese Cities by Size Categories and Entitled Population, 1982–1993 69

4.1 Urban Agglomerations in India with more than 1 Million Inhabitants in 1990 92

5.1 Urban Agglomerations in Indonesia with more than 1 Million Inhabitants in 1990 132

5.2 Indonesia: Population Density, 1990 136

5.3 Sacred and Market Cities in Indonesia, 750–1400 140

5.4 Actual and Idealized Rank-Size Distributions of Indonesian Cities in 1850, 1930, 1980, and 1990 153

5.5 The Functioning Urban Region of Greater Jakarta 156

5.6 Indonesian Provinces: Relationship between Percentage of Population Living in Urban Areas and Gross Domestic Product Per Capita, 1990 160

6.1 Urban Agglomerations in the Arab World with more than 1 Million Inhabitants in 1990 184

7.1 Urban Agglomerations in Africa South of the Sahara with more than 1 Million Inhabitants in 1990 210

7.2 African Urban Centres, 1500 BCE–CE 1850 213

8.1 Urban Agglomerations in Latin America with more than 1 Million Inhabitants in 1990 252

Notes on Contributors

JANET ABU-LUGHOD is Professor of Sociology and Historical Studies in the Graduate Faculty of the New School for Social Research. She has carried out research on Cairo and Rabat, as well as the Lower East Side in New York, and worked on the world system in medieval and modern times. At this time she is engaged in a study comparing New York, Chicago, and Los Angeles as world cities.

XIANGMING CHEN is Assistant Professor of Sociology at the University of Illinois at Chicago and Research Fellow at the IC^2 Institute at the University of Texas at Austin. His recent research addresses the interface between urban and economic development from a comparative perspective, with a central focus on China.

JOSEF GUGLER is Professor of Sociology at the University of Connecticut. Previously he served as Director of Sociological Research at the Makerere Institute of Social Research, Uganda. Research and teaching have taken him to Cuba, India, Kenya, Nigeria, Senegal, Tanzania, and Zaïre. At this time his research focuses on the role of gender in migration and on the relationship between literature and politics in Africa.

GRAEME HUGO is Professor and Head of the Department of Geography at the University of Adelaide. Previously he taught at Flinders University and held visiting appointments at the University of Iowa, the University of Hawaii, Hasanuddin University in Indonesia, and the Australian National University. Much of his research deals with population issues in South-east Asia, especially Indonesia, and Australia.

RAKESH MOHAN is Economic Adviser to the Government of India in the Ministry of Industry. He obtained his doctoral degree at Princeton University, then joined the World Bank in 1976 to work on the City Study of Bogotá and Cali. He has pursued his scholarly interests along with an active life in policy-making in government. His current work focuses on industry and technology policy. Most recently he has been Distinguished Visiting Fellow at the United Nations University, Institute for New Technologies at Maastricht.

RHOADS MURPHEY is Professor Emeritus of History and Director of the Program in Asian Studies at the University of Michigan. He edits the Monograph Series of the Association for Asian Studies, of which he is past-president. He has lived and worked in India, China, and Japan, and travelled widely in the rest of Asia. He has taught at several Asian universities as a visitor.

ORLANDINA DE OLIVEIRA is a senior researcher at the Center for Sociological Studies at El Colegio de México. She was born in Brazil and completed her undergraduate studies there, before studying at FLACSO in Chile and at the University of Texas at Austin where she obtained her doctoral degree. She has served as Director of the Center for Sociological Studies at El Colegio de México. Her extensive research in Latin America focuses on Mexico and Brazil in particular.

WILLIAM L. PARISH is Professor of Sociology at the University of Chicago. He has done research on family, community, and development issues in rural China, urban China, and Taiwan. His current work on China includes studies of labour transition, private enterprises, women's education, family assistance networks, and sexual behaviour.

BRYAN ROBERTS is Professor of Sociology at the University of Texas at Austin where he holds the C. B. Smith Sr. Chair in US–Mexico Relations. He obtained his doctoral degree from the University of Chicago, then returned to the United Kingdom to teach at the University of Manchester from 1964 to 1986. He has carried out research in Central America, Peru, and Mexico.

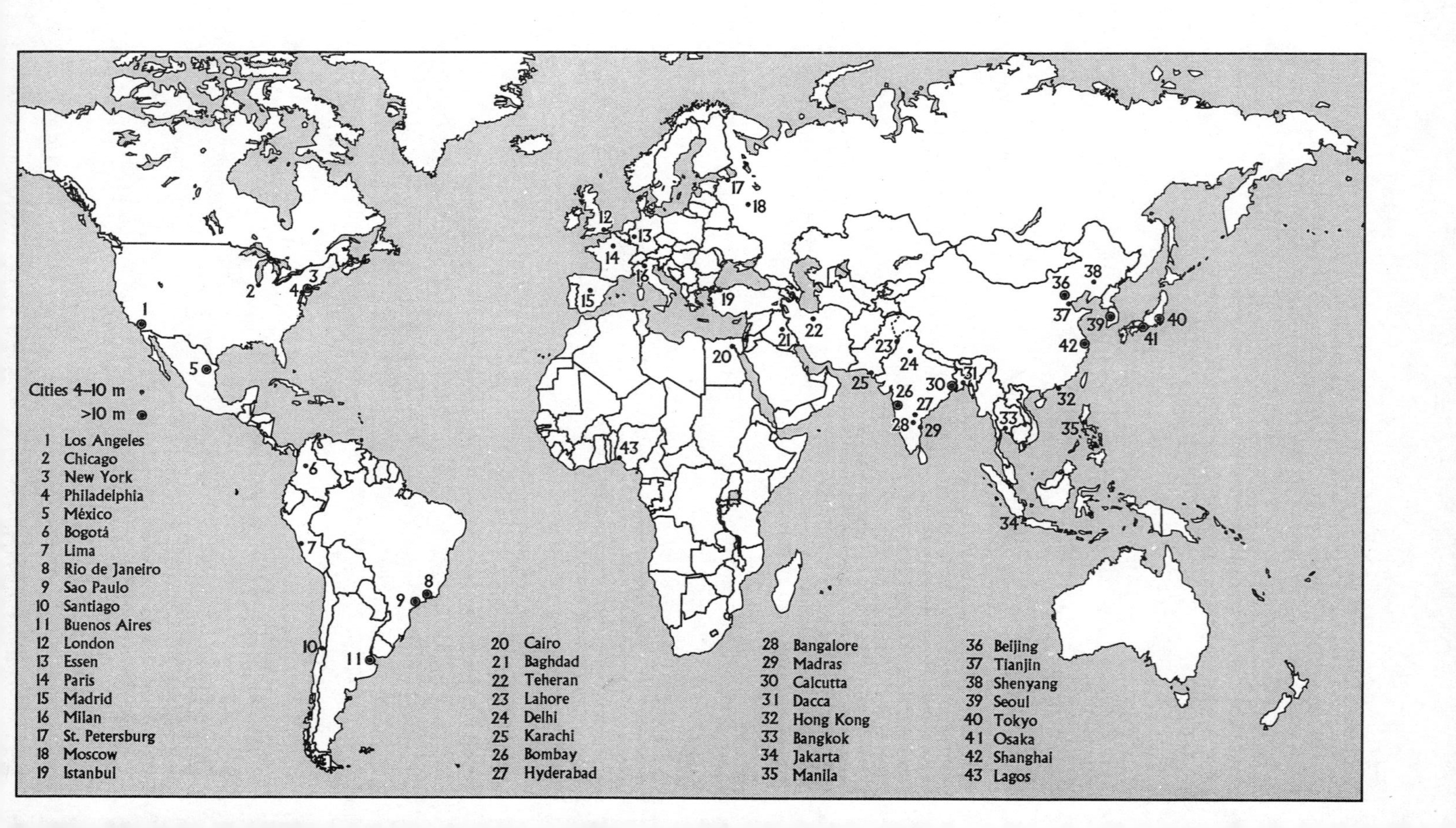

Cities 4–10 m
>10 m
1 Los Angeles
2 Chicago
3 New York
4 Philadelphia
5 México
6 Bogotá
7 Lima
8 Rio de Janeiro
9 Sao Paulo
10 Santiago
11 Buenos Aires
12 London
13 Essen
14 Paris
15 Madrid
16 Milan
17 St. Petersburg
18 Moscow
19 Istanbul
20 Cairo
21 Baghdad
22 Teheran
23 Lahore
24 Delhi
25 Karachi
26 Bombay
27 Hyderabad
28 Bangalore
29 Madras
30 Calcutta
31 Dacca
32 Hong Kong
33 Bangkok
34 Jakarta
35 Manila
36 Beijing
37 Tianjin
38 Shenyang
39 Seoul
40 Tokyo
41 Osaka
42 Shanghai
43 Lagos

1

Regional Trajectories in the Urban Transformation: Convergences and Divergences

JOSEF GUGLER

THE last phase of the urban transformation is now unfolding in the less developed countries of the world. We used to lump most of them together as the 'Third World', a term born in the 1950s from the desire of leaders from countries newly independent or soon to become independent, to establish a third force in world affairs, aligned with neither the capitalist 'First World' nor the socialist 'Second World'. Only a handful managed to remain non-aligned. 'Third World' continued to be used, its original intent largely forgotten, to designate the less developed countries of 'the South', outside the Eastern Bloc. With the dissolution of the 'Second World' the term has become obsolete altogether.

'Third World' is a designation with clearer historical than current applicability. Yet, the tradition of scholarship that focuses on the less developed countries in the 'Third World', to the exclusion of others, retains some unity and momentum. This collection remains within this tradition in delineating its purview. It covers much of Asia, Africa, and Latin America, but leaves aside not only highly industrialized Japan but also those Asian countries that were once part of the Soviet Union. Nor does it include in its purview those European countries that may be considered less developed. More than four billion people, more than three-quarters of the world's population live in the South thus defined. More than a third of them, more than one and a half billion people, live in urban areas.[1]

The less developed countries of Asia, Africa, and Latin America share, by definition, one characteristic: they are poor compared to most of the rest of the world. In many other respects there are significant

I wish to thank, without implicating, William G. Flanagan and Bryan Roberts for helpful comments on an earlier version.

differences. Here we intend to show that some of these differences can be delineated as distinct regional patterns.[2] We will focus on the urban dimension of these regional patterns and relate them to major differences in level of economic development, political economy, and cultural heritage.

The distinct patterns of urbanization in different regions are shaped by the legacy of their urban history as well. This history differs markedly across the South. The roots of urbanization reach farther back in Asia than anywhere else (see Chapter 2). Subsequent urban developments in Africa and the Americas remained more limited until well after the imposition of colonial rule. The impact of colonialism varied with different imperial powers. To take only the most visible legacy: the distinct approaches of the Spanish, Portuguese, British, French, and Dutch affect the morphology and architecture of many cities in their former colonies to this day.

The Urban Transition

The South is urbanizing rapidly, but its various regions differ markedly in the level of urbanization they have attained at this time (Table 1.1). At one extreme, nearly three-quarters of the population of Latin America and the Caribbean live in urban areas, just about the same level of urbanization as in developed countries. At the other extreme, two-thirds or more of the population remain rural in China, India, Indonesia, and Africa South of the Sahara. The Arab states fall in between. Of course, these averages hide major variations among countries, e.g. between the highly urbanized Southern Cone of South America and Central America, and within countries, e.g. between the coastal provinces of China and inland regions.

The differences in level of urbanization among the regions correspond roughly to the differences in per capita income.[3] This is the case whether the conventional measure of GNP per capita is used or real GDP per capita, i.e. domestic product figures converted into US dollars in terms of purchasing power parity.[4] The causality in the relationship between levels of urbanization and of income is easily assumed to go from a larger, more productive urban population to higher incomes. The opposite causal relationship is, however, more plausible in a world where industrialization no longer provides the

TABLE 1.1 Demographic, Economic, and Human Development Indicators for Major Developing Countries and Regions, about 1990

Country/region	Total population (m)	Annual population growth-rates (%)		Total fertility[a] rate (%)	Urban population (% of total population)		Urban population annual growth-rates (%)		Per cent of urban population in largest city	Urban sex ratio (males per 1000 females)	GNP per capita (US$)	GNP per capita annual growth (%)		Real GDP[b] per capita (US$)	Infant mortality (per 1000 live births)
	1992	1960–92	1992–2000	1992	1960	1992	1960–92	1992–2000	1990	1990–1	1992	1965–80	1980–92	1992	1992
China	1184	1.9	1.0	2.0	19	28	3.1	3.8	4	1083	480	4.1	7.6	1950	44
India	884	2.2	1.8	3.8	18	26	3.4	3.0	6	1119	310	1.5	3.1	1230	82
Indonesia	189	2.1	1.5	2.9	15	33	4.7	4.3	17	999	680	5.2	4.0	2950	58
Arab states[c]	230	2.6	2.9	4.8	30	50	4.5	3.5	31	—	—	—	—	4452	67
Africa South of the Sahara[c]	510	2.8	2.9	6.3	15	30	5.0	4.5	—	—	559	1.4	-1.8	1346	97
Latin America and the Caribbean[c]	450	2.4	1.8	3.1	50	73	3.6	2.4	24	—	2791	2.7	1.0	5730	45
Developing[c] countries	4220	2.3	1.8	3.5	22	36	3.8	3.2	—	—	982	4.6	4.0	2595	70
WORLD[c]	5420	1.9	1.5	3.1	34	44	2.7	2.6	—	—	4534	—	—	5430	—

[a] The total fertility rate is the average number of children born to a woman during her lifetime.
[b] Real GDP is based on conversion in terms of purchasing power parity.
[c] The values for regions, developing countries, and the world are appropriately weighted, except for total population.

Sources: United Nations Development Programme (1995), tables 2, 4, 15, 16, 20, 39 (corrected by UNDP); urban sex ratios from United Nations (1994), table 6.

major thrust to urbanization. Higher incomes allow the concentration of resources in urban areas in terms of public bureaucracies, of élite and middle-class standards of living that support a large service sector, and of public works. To take the most dramatic example, the high levels of urbanization of oil-rich countries are a function not of the requirements of producing these riches but of the urban consumption they finance.

The very same distribution in terms of level of urbanization—China, India, Indonesia, and Africa South of the Sahara low; the Arab states intermediate; Latin America and the Caribbean high—obtained three decades ago. However, the differences in the rate of urban growth among the regions over the last three decades are quite striking (Table 1.1). China, India, and Latin America and the Caribbean report less than 4 per cent annual growth in their urban populations, Indonesia and the Arab states 4.7 and 4.5 per cent annual growth respectively, and Africa South of the Sahara 5 per cent, i.e. the population of African cities and towns doubled every fourteen years.

Natural population growth is a major element in urban growth, but rural–urban migration makes an even larger contribution in many less developed countries.[5] Thus the populations of India and Indonesia grew at about the same rate over the last three decades, but the cities and towns of Indonesia grew substantially faster than those of India.[6] And the Arab states and Africa South of the Sahara had similar population growth-rates, but urban growth was faster in Africa. Noteworthy are the small differences between China and India because they run counter to the common assumption that China effectively restricted rural–urban migration.

A consideration of economic growth can serve to explain some of these differences. Thus GNP per capita in India grew at a markedly slower pace than in Indonesia between 1960 and 1992. The case of Africa South of the Sahara, however, calls for another explanation. The region experienced urban growth much faster than any other, while its economic performance was the worst among the regions. This peculiar constellation may be accounted for by two factors: the fastest rate of natural population growth and particularly pronounced urban bias (see Chapter 7).

Current urban-growth trends indicate distinct changes from the patterns of the last three decades (Table 1.1). In China, urban growth has accelerated even while population growth has declined. A booming economy has stimulated large-scale rural–urban migration that is no longer fettered by migration controls. Elsewhere urban growth is slowing down along with population growth. Africa South of the Sahara continues to have the most rapid urban growth: its urban population is expected to increase by more than half in the 1990s.

Urban concentration may be measured in terms of the proportion of a country's population living in its largest city (Table 1.1). China and India show a pattern typical of very large countries where several large urban centres dominate different regions. Thus Shanghai has only 4 per cent of China's, Bombay only 6 per cent of India's population, while Beijing and Tianjin, Calcutta and Delhi, with populations around ten million, are not much smaller. The Arab states, Africa South of the Sahara, and Latin America and the Caribbean present a sharp contrast. In each region a quarter or more of the average country's population lives in its largest city. The similarity among these three regions is all the more striking as their level of urbanization varies so widely.

Urban sex ratios vary considerably across the Third World (Table 1.1). In China and India men outnumber women by a substantial margin. In Indonesia the urban sex ratio is balanced. Male-dominated urban populations are characteristic of most countries in the Arab World and Africa South of the Sahara. In Latin America, however, women outnumber men in every country, frequently by a substantial margin that cannot be explained by sex differentials in mortality.[7]

Rural–Urban Migration

Distinct patterns of rural–urban migration account for most of the differences in urban sex ratios across the Third World. Where men outnumber women in the urban population this is a function of men predominating in net rural–urban migration.[8] Some of these men are young and single, some stay in the city only for a short while. However, given widespread urban unemployment, the more common pattern is for these men to become long-term urban workers while leaving wives and children in their rural area of origin. As Weisner (1972) put it, they have 'one family, two households'.

The Industrial Revolution engendered the distinction of workplace and home, but the separation of men from their wives and children has been drastically magnified in less developed countries. Many men spend long years, commonly their entire working life, in the city and visit their village-based families as employment conditions and transport costs permit. But they are not the only ones to visit their rural homes. Many urban families maintain strong ties with a rural community which they continue to consider their home and where they anticipate retiring eventually. Such a pattern of 'life in a dual system' can be quite enduring (Gugler, 1996).

In India, Indonesia, the Arab states, and Africa South of the Sahara large numbers of people pursue such temporary migration strategies: short-term migration, single long-term migration, or family long-term migration. In China, recent reforms have brought a large 'floating population' of temporary migrants to the major cities (Chapter 3). Large numbers of urban dwellers in these regions are not permanent residents but temporary sojourners who remain deeply involved in a rural community. Latin America presents a distinct contrast: temporary migration is very much the exception, characteristic of some Indian communities.[9]

Temporary migration is predicated on maintaining a rural base. The 'dual-household' migrant leaves wife and children on the farm to grow their own food, perhaps to raise cash crops as well. The strategy is a function of high fertility and limited educational and earning opportunities for women. An added factor in some areas is the lack of compensation for those who leave land that is communally controlled. This is the case in much of Africa South of the Sahara. The 'dual-system' strategy entails social and economic investments that allow the family to return to the village. Either strategy assures the migrant of a measure of security, meagre but more reliable than what the city offers most of its citizenry.[10]

Where women outnumber men in the urban population by a substantial margin, this reflects a pattern of permanent rural–urban migration in which women predominate. The distribution of this pattern—it has been long-established throughout Latin America and the Philippines—invites a cultural explanation for the 'Latin' pattern. However, recent data indicate that this pattern is appearing elsewhere. This suggests a historical transition from a preponderance of men to a preponderance of women in net rural–urban migration: a 'gender transition in rural–urban migration'. It would appear that later marriage, reduced fertility, and greater independence are modifying cultural definitions of women's roles as wives and mothers and set them free to move. Never-married, separated, divorced, and widowed women usually are faced with limited rural opportunities and are attracted by cities that offer the better opportunities in terms of work as well as (re-)marriage (Gugler, 1994). Thus the contrast between India and most Arab and African countries where men predominate in urban populations, and Indonesia and Latin America where they do not, corresponds roughly to their different fertility rates (Table 1.1). The notable exception is China where fertility has been dramatically reduced by drastic government policies.

Major variations in the participation of women in manufacturing may also be related to the regional differences in fertility rates. In India, women hold only 9 per cent of the jobs in manufacturing. And in the few Arab and African countries for which data are available, women usually hold only about 10 per cent of the jobs in manufacturing.[11] In sharp contrast, between one-quarter and close to half of the labour-force in manufacturing is female in China and in the Latin American countries that provide information (International Labour Office, 1994).

Urban Social Organization

The history of urbanization and current patterns of migration have a major impact on urban social organization. In the cities and towns of Africa South of the Sahara that have grown so rapidly over the last three decades, and where there has been substantial return migration to rural areas, the great majority of residents have been born and raised in rural areas. In stark contrast, most urban residents are urban-born in Latin America where urban growth has been considerably slower and where there is little return migration. India had a similarly low rate of urban growth but substantial return migration. In Indonesia return migration continues to be common, while urban growth was faster. The Arab states had similar fast urban growth, and return migration was probably significant, at least in several of the larger countries. China had urban growth as slow as that of India and Latin America, but an assessment of the proportion of migrants in the urban population is wrought with pitfalls because there are at least three specific factors that make the Chinese experience distinct: low population growth implies a larger share of migrants in urban growth; the same implication follows from the campaigns that forced school-leavers to move to rural areas—to the extent that they have not returned since; on the other hand, there was probably very little voluntary return migration because it meant the loss of the privileged status of urban resident, a status exceedingly difficult to obtain under a regime of stringent migration controls.

First generation migrants tend to have significant ties with kin and 'home people'. Most were received by relatives or friends from home when they first arrived in town, some settled amongst them. Ties among first generation migrants of common origin are reinforced where they remain involved in their community of origin. Regional distinctions, in particular the different languages spoken by migrants from various parts of a country such as India—and nearly all African countries are similarly multilingual—foster social networks delineated in terms of region of origin. Such social relationships are reinforced by a common cultural idiom, typically quite distinct from any traditional legacy, to establish an ethnic identity. Such ethnic identities usually cut across divisions of class.

Religion affects urban social organization. Religious practice fashions social relationships and cultural identities. Fundamentalist movements, such as are found in most Arab states, tend to be particularly effective in this respect. A religious identity may coincide or cut across ethnic identities. In either case it becomes more salient if it articulates itself in opposition to other religious identities.

Kinship patterns vary across the Third World and differentially affect urban social organization. The rather independent nuclear family imported by the European colonizers in Latin America and the Caribbean provides only narrowly circumscribed support. When the family comes under stress, severe hardship for some of its members is common. Mothers who have been deserted, divorced, or widowed can expect little support from kin, children may be left to their own devices. In Africa and in India, in contrast, kinship support beyond the nuclear family is more readily available.

The Chinese experience has demonstrated the strong impact state policies can have on urban social organization. Job security and the public allocation of housing, in particular, entailed an unusually high stability in social relations within the neighbourhood and in the work setting. There have been major reforms since the 1970s, but job security has been maintained in state enterprises, and movement amongst the tenants of existing housing remains limited (Chapter 3; Davis, 1996).

The Political Arena

Only a decade ago, most Third World countries were ruled by dictators. While their popularity varied, their power was usually solidly based on the support of the armed forces. Only in exceptional cases did revolutionary movements succeed in challenging them effectively.[12] The two revolutions that were successful in the late 1970s demonstrated the heavy sacrifices such victory required. In Iran, demonstrators took to city streets again and again to confront troops who were shooting to kill. And in Nicaragua, poorly equipped youngsters, led by Sandanista units, battled with the National Guard in one city after another.

Since the 1980s, major political transformations have taken place in a large number of countries. Again, as in Iran and Nicaragua, urban actors, more precisely actors in key cities, particularly capital cities, played the central role. But there was a remarkable break with earlier revolutions: street demonstrations and strikes sufficed time and again to persuade rulers to compose with the opposition, and casualties usually remained quite low. Many strong-men regimes were suddenly found to be quite fragile.

The urban character of these opposition movements, the pivotal importance of the capital city, and the political significance of physical control over symbolic urban space were dramatically demonstrated in China. The dissidents were urban-based, their activities focused on Beijing, and they occupied Tiananmen Square, the capital's most prestigious location, for a month and a half in 1989. However, unlike many other regimes in recent years, the Chinese leadership was not prepared to compose with the opposition, and opted for brutal repression.

There thus remain distinct regional contrasts. Indonesia, like China, is characterized by authoritarian rule, but India has maintained democratic practices since it became independent in 1948. All of the Arab states continue to be ruled by authoritarian regimes—isolated efforts at democratization foundered on the increasingly bitter conflict between secularizing and fundamentalist orientations. In Africa South of the Sahara, the democratization wave brought civilian governments and competitive elections to a number of countries, but some of the most repressive and exploitative regimes are found in this region as well. All of Latin America could pride itself on civilian rule by 1994.

Religious, caste, and ethnic conflicts are articulated in the urban arena. Opposing identities may draw on 'tradition', but they are fashioned in urban confrontations. Religious fundamentalisms are proclaimed by urban intellectuals and attract followings among the urban impoverished. Castes are redefined in changing urban labour markets. And while ethnic identities may refer to rural societies and draw on rural traditions, they are delineated, in some cases invented, in the urban encounter with other groups.

Distinct regional patterns may be discerned. India has known all three types of conflict. The salience of caste, and the conflicts it can engender, is specific to the country. There have been recurrent ethnic conflicts, typically between 'sons of the soil', i.e. people from a city's hinterland, and 'outsiders', i.e. migrants from more distant lands. And recurrent confrontations between Hindus and Moslems have claimed all too many victims since the country was partitioned in 1948.

In China, authoritarian rule has suppressed the expression of communal antagonisms. Liberalization, when it comes, may well set free the articulation of ethnic conflicts. Indonesia has seen murderous confrontations between Javanese and Chinese as well as conflicts between people indigenous to the islands. And there is a distinct prospect of a fundamentalist challenge to the political order.

In the Arab states, much conflict focuses on religion. In some countries secularizing and fundamentalist orientations are locked in bitter opposition. Elsewhere the adherents of different Muslim sects, who may also represent different regions, confront each other.

The great majority of urban residents in Africa are rural-born and most maintain ties with the village. Identities of origin thus are salient. Political conflict and competition over economic opportunities time and again take on ethnic connotations. Today, nearly every African country is deeply divided on ethnic lines. These divisions are exacerbated to the extent that regimes take on ethnic identities and major economic opportunities are seen to be monopolized by members of particular ethnic groups (Chapter 7). In a few countries different ethnic groups are identified with Islam and Christianity respectively, and conflicts are increasingly cast in terms of religious belief and practice.

Latin America appears to be remarkably free of communal conflicts, with the important exception of the few countries where significant Amerindian populations have survived. In the absence of cross-cutting communal identities, stratification is all the more salient in the region (Chapter 8). The substantial urban working class has organized for many decades in trade unions sufficiently strong to secure major benefits and to play at times a major role in national politics (Alves, 1996; Bergquist, 1986; Drake, 1988). The strength of squatter movements which have secured free land for a large proportion of the urban population in a number of Latin American countries is also specific to the region (Castells, 1983). These movements assemble people who are permanently committed to the city and seek both to improve their accommodations and to gain a measure of security for their old age.

In recent years a wide variety of urban social movements organized and demanded civil and political rights throughout much of the South. They derived strength from increasingly sophisticated media and the growing presence of non-governmental organizations. And the authoritarian regimes they challenged could no longer count on the Cold War reflexes of the major powers to support client regimes irrespective of their internal politics. A multitude of social movements appeared, disappeared, and emerged afresh in new forms—and increased the awareness and participation of the citizenry.[13]

The democratization process was particularly successful in Latin America. According to the Freedom House Survey (1994), every country in South America and Central America was 'free' or 'partly free' in 1993, while more than half of the countries in Africa South of the Sahara were 'not free' in 1993. Relatedly, the urban protests sparked by the austerity measures adopted by many debt-ridden countries, typically under pressure from the International Monetary Fund, were most successful in Latin America. Stunned governments frequently rescinded or ameliorated their austerity measures, or provided compensations. In addition, protests sometimes initiated a successful movement to depose a government, or added a push to a teetering regime. Particularly notable, the protests contributed to persuade external actors—foreign governments, the International Monetary Fund, private bankers—to retreat from austerity policies, at least for large-debt countries (Walton and Shefner, 1994).

Quite different urban transitions are taking place in the various regions we here distinguish, and they are different again from the urban transitions the industrialized countries experienced a century ago. But the world keeps shrinking: ideas, people, and goods travel ever faster; the exchange of goods and services keeps increasing; and the political pressures of foreign states and international organizations are quickly felt across the globe. 'Third World' ghettos have emerged in the industrialized countries, and ballot boxes are spreading in poor countries.

Notes

1. Kasarda and Crenshaw (1991) provide a comprehensive review of the burgeoning literature on what is still commonly referred to as Third World urbanization. Stren (1994–5) presents the findings of a large international project surveying past urban research and proposing an agenda for future research in Asia, Africa, and Latin America. For accounts of urbanization in some of the less developed countries not covered here, see Gugler (1980) on Cuba, Levine and Levine (1979) on Papua New Guinea, McGee (1988) on Malaysia, and Portes *et al.* (1994) on Costa Rica, the Dominican Republic, Guatemala, Haiti, and Jamaica.
2. Less developed countries can be usefully categorized in other ways. One obvious difference is in terms of levels of income, e.g. the countries here lumped together as Arab states range all the way from the Sudan with a real GDP of $1620 per capita to Qatar with an estimated $22 380 in 1992. Another approach distinguished less developed countries committed to a socialist paradigm from the rest. Some of the work in this tradition had the merit of including Eastern Europe in its purview. Abu-Lughod, in Chapter 6, uses income criteria as well as political orientation to distinguish five types of Arab states.
3. Bradshaw and Noonan (1996) found Africa South of the Sahara and Latin America to be highly urbanized relative to GNP per capita in a cross-national analysis of the effect of investment dependency and pressures from the International Monetary Fund on the level of urbanization in less-developed countries.
4. Even real GDP per capita fails to reflect welfare adequately. It is an average figure that hides more or less severe inequality in income distribution. Infant mortality, on the other hand, is quite sensitive to inequalities. Thus the infant mortality rate reported for China is remarkably low for such a poor country, even allowing for the effect of low fertility (Table 1.1).
5. Between 1975 and 1990, the contribution of rural–urban migration to urban growth amounted to 50 per cent in nine Asian, 75 per cent in four African, and 49 per cent in eleven Latin American countries (Findley, 1993).
6. Mohan and Hugo, in Chapters 4 and 5, provide detailed discussions of urban growth in India and Indonesia.
7. For a discussion of differences in urban sex ratios among African countries, see Chapter 7; for more comprehensive data and a detailed discussion of the role of gender in rural–urban migration, Gilbert and Gugler (1992: 74–9).
8. I emphasize net migration, because women more commonly migrate at marriage than men and hence frequently predominate in gross migration.
9. For a detailed analysis of contrasting migration strategies among Amerindians in Peru and Mexico, see Moßbrucker (1996).

10. For a more detailed discussion of these migration strategies, see Chapter 7.
11. Women hold about a third of manufacturing jobs in Botswana and Swaziland, countries where men prefer to find work across the border in South Africa.
12. The contrast with national wars of liberation is striking: they succeeded everywhere in throwing off colonial oppression eventually.
13. On the 'new social movements' in Latin America, see Roberts (1996) who observes that new groups found an effective voice: Indian communities, the young and the old, and women.

References

Alves, Maria Helena Moreira (1996), 'The New Labour Movement in Brazil', in Josef Gugler (ed.), *Cities in Asia, Africa, and Latin America: Multiple Perspectives* (Oxford University Press), repr. from Susan Eckstein (1989) (ed.), *Power and Political Protest: Latin American Social Movements* (Berkeley: University of California Press), 278–98.

Bergquist, Charles (1986), *Labor in Latin America: Comparative Essays on Chile, Argentina, Venezuela, and Colombia* (Stanford, Calif.: Stanford University Press).

Bradshaw, York W. and Noonan, Rita (1996), 'Urbanization, Economic Growth, and Women's Labour Force Participation: A Theoretical and Empirical Reassessment', in Gugler (ed.), *Cities in Asia, Africa, and Latin America.*

Castells, Manuel (1983), *The City and the Grassroots: A Cross-Cultural Theory of Urban Social Movements* (Berkeley: University of California Press).

Davis, Deborah (1996), 'Social Transformations of Metropolitan China, 1949–1993', in Gugler (ed.), *Cities in Asia, Africa, and Latin America.*

Drake, Paul W. (1988), 'Urban Labour Movements under Authoritarian Capitalism in the Southern Cone and Brazil, 1964–83', in Josef Gugler (ed.), *The Urbanization of the Third World* (Oxford University Press), 367–98.

Findley, Sally E. (1993), 'The Third World City: Development Policy and Issues', in John D. Kasarda and Allan M. Parnell (eds.), *Third World Cities: Problems, Policies, and Prospects* (Newbury Park, Calif.: Sage), 1–31.

Freedom House Survey Team (1994), *Freedom in the World: The Annual Survey of Political Rights and Civil Liberties 1993–1994* (New York: Freedom House).

Gilbert, Alan and Gugler, Josef (1992 [1982]), *Cities, Poverty and Development: Urbanization in the Third World*, 2nd edn. (Oxford University Press).

Gugler, Josef (1980), ' "A Minimum of Urbanism and a Maximum of Ruralism": The Cuban Experience', *International Journal of Urban and Regional Research*, 4: 516–34.

—— (1994), 'The Gender Transition in Rural–Urban Migration', paper presented at the World Congress of Sociology, Bielefeld, July.
—— (1996), 'Life in a Dual System Revisited: Urban–Rural Ties in Enugu, Nigeria, 1961–1987', in Josef Gugler (ed.), *Cities in Asia, Africa, and Latin America: Multiple Perspectives* (Oxford University Press); repr. rev. from (1991) *World Development*, 19: 399–409.
International Labour Office (1994), *1994 Yearbook of Labour Statistics* (Geneva: International Labour Office).
Kasarda, John D. and Crenshaw, Edward M. (1991), 'Third World Urbanization: Dimensions, Theories, and Determinants', *Annual Review of Sociology*, 17: 467–501.
Levine, Hal B. and Levine, Marlene Wolfzahn (1979), *Urbanization in Papua New Guinea: A Study of Ambivalent Townsmen*, Urbanization in Developing Countries (Cambridge University Press).
McGee, T. G. (1988), 'Industrial Capital, Labour Force Formation and the Urbanization Process in Malaysia', *International Journal of Urban and Regional Research*, 12: 356–74.
Moßbrucker, Harald (1996), 'Amerindian Migration in Peru and Mexico', in Gugler (ed.), *Cities in Asia, Africa, and Latin America.*
Portes, Alejandro, Itzigsohn, José and Dore-Cabral, Carlos (1994), 'Urbanization in the Caribbean Basin: Social Change During the Years of the Crisis', *Latin American Research Review*, 29: 3–37.
Roberts, Bryan (1996), 'The Social Context of Citizenship in Latin America', in Gugler (ed.), *Cities in Asia, Africa, and Latin America.*
Stren, Richard (1994–5) (ed.), *Urban Research in the Developing World*, 4 vols. (Toronto: Centre for Urban and Community Studies, University of Toronto).
United Nations (1994), *1992 Demographic Yearbook* (New York: United Nations).
United Nations Development Programme (1995), *Human Development Report 1995* (New York: Oxford University Press).
Walton, John and Shefner, Jonathan (1994) 'Latin America: Popular Protest and the State', in John Walton and David Seddon (eds.), *Free Markets and Food Riots: The Politics of Global Adjustment* (Oxford: Blackwell), 97–134.
Weisner, Thomas Steven (1972), 'One Family, Two Households: Rural–Urban Ties in Kenya', Ph.D. dissertation (Harvard University).

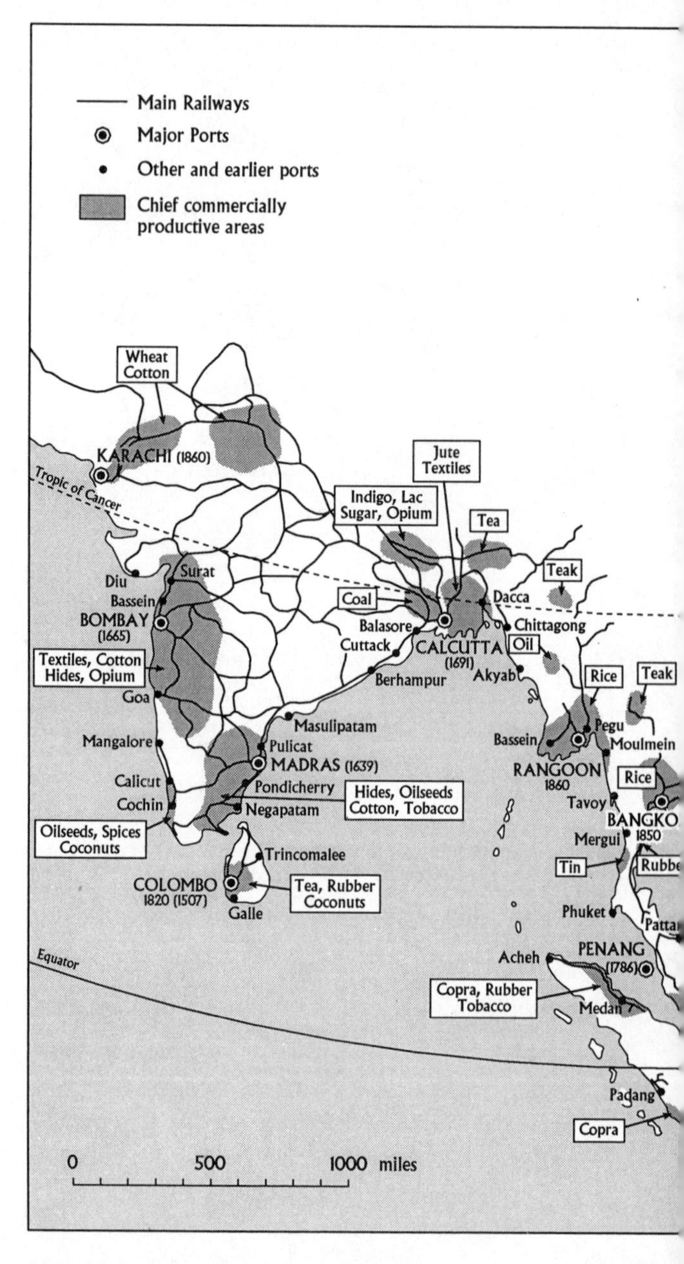

FIG. 2.1 Major Ports and Chief Commercially Productive Areas in Monsoon Asia, 1600–1940

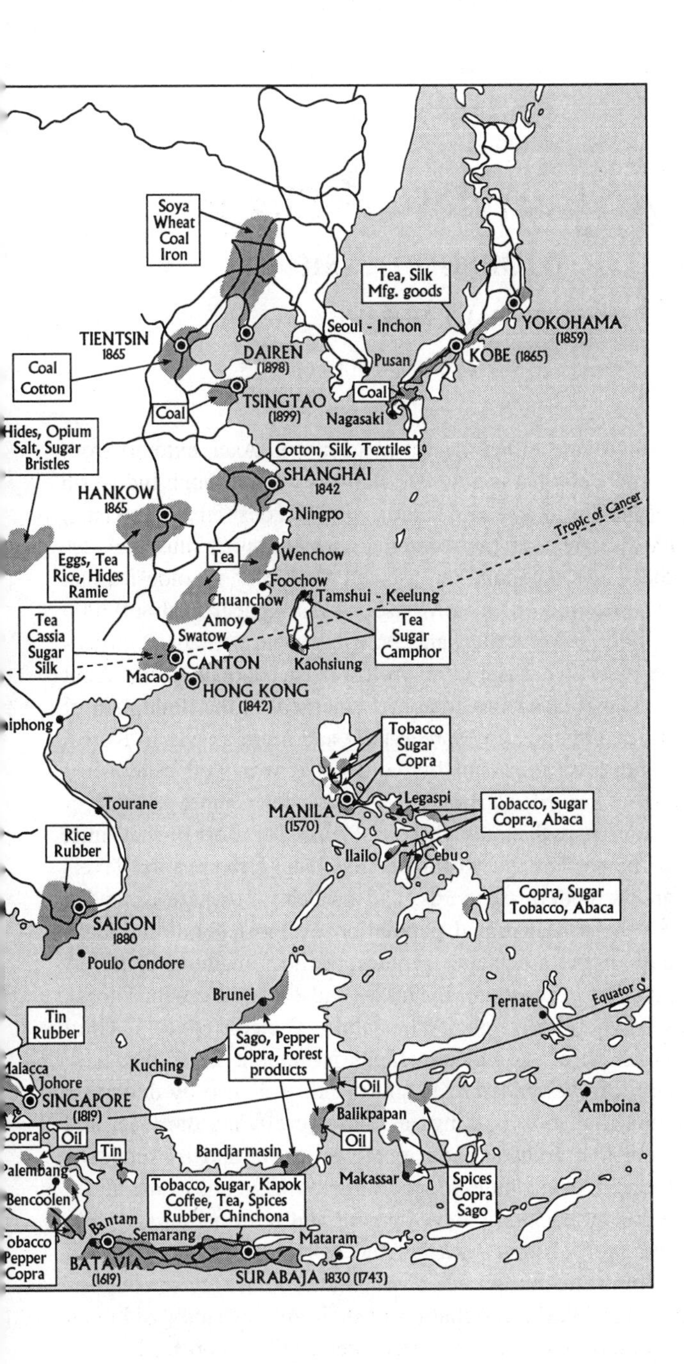

Soya
Wheat
Coal
Iron
Tea, Silk
Mfg. goods
YOKOHAMA
(1859)
Seoul - Inchon
TIENTSIN
1865
DAIREN
(1898)
Pusan
KOBE (1865)
Coal
Cotton
Coal
TSINGTAO
(1899)
Nagasaki
Coal
Hides, Opium
Salt, Sugar
Bristles
Cotton, Silk, Textiles
SHANGHAI
1842
HANKOW
1865
Ningpo
Tropic of Cancer
Wenchow
Eggs, Tea
Rice, Hides
Ramie
Tea
Foochow
Tamshui - Keelung
Chuanchow
Amoy
Tea
Sugar
Camphor
Tea
Cassia
Sugar
Silk
Swatow
CANTON
Kaohsiung
Macao
HONG KONG
(1842)
iphong
Tobacco
Sugar
Copra
Tourane
MANILA
(1570)
Legaspi
Tobacco, Sugar
Copra, Abaca
Rice
Rubber
Ilailo
Cebu
Copra, Sugar
Tobacco, Abaca
SAIGON
1880
Poulo Condore
Brunei
Ternate
Equator
Tin
Rubber
Sago, Pepper
Copra, Forest
products
Kuching
Malacca
Johore
Oil
Amboina
SINGAPORE
(1819)
Balikpapan
Copra
Oil
Oil
Tin
Bandjarmasin
Palembang
Makassar
Spices
Copra
Sago
Bencoolen
Tobacco, Sugar, Kapok
Coffee, Tea, Spices
Rubber, Chinchona
Bantam
Semarang
Mataram
obacco
Pepper
Copra
BATAVIA
(1619)
SURABAJA 1830 (1743)

2

A History of the City in Monsoon Asia

RHOADS MURPHEY

THE area covered here, often known as Monsoon Asia, extends from the western borders of what is now Pakistan eastward through Indonesia and north to Manchuria and Hokkaido, but excludes Tibet, Sinkiang, Mongolia, and the areas of the former Soviet Union. It thus includes nearly half the world's population, as well as one of its oldest urban traditions. The first true cities in Monsoon Asia appeared about 3000 BC along the Indus River valley and its tributaries, perhaps a millennium after the world's earliest cities in lower Mesopotamia. By about 2000 BC the first real cities in China had emerged on the floodplain of the Yellow River. Urbanization spread later and more slowly in Korea, Japan, and South-East Asia, but by AD 1000 it was well established there. Throughout recorded history, there have been more and larger cities in Monsoon Asia than anywhere else, and, for most of that time, more than in the rest of the world put together. Urbanization levels were not high by modern standards, and probably nowhere exceeded 10 per cent of the total national population by most definitions until the second half of the nineteenth century, while remaining considerably lower than that in most areas. This was also the case with Europe until perhaps the end of the eighteenth century. The concept of a 'Third World', to which all of Asia except Japan is thought to belong, is in any case a form of over-generalized shorthand. It is also by definition a relative notion, but reflects things as they were when the idea, and the phrase, were first coined, in the post-war world and its supposed political divisions. Urbanization is, according to many, one of the measures of the degree of economic development, but until perhaps the late eighteenth century Monsoon Asia led the world in such terms, as it had since the third millennium BC. Most Asia scholars are uncomfortable with the Third World designation of their area of study, although if the other common measures are used (per capita incomes, level of industrialization, and so on) most of the area outside Japan currently fits this category.

This chapter provides a brief history of urbanization, by countries, not merely for the historical record itself but also because traditional roots and patterns are still important, and still apparent in the cities of modern and contemporary Monsoon Asia. The first cities of the area, in the Indus Valley and its tributaries, are remarkable not only for their wide areal distribution and numbers, but for their planning. This is most apparent at the major sites yet discovered and partially excavated, Kalibangan and Harappa in the Punjab, and Mohenjo Daro on the lower Indus. Each was laid out on a grid pattern, and in each there is a striking uniformity of most buildings, with a few larger ones, probably for grain storage, possibly for religious purposes, and a great public bath near the centre of the urban complex. These were the world's first planned cities, and they also included a piped water-supply and drains for nearly every house, something not attained elsewhere in cities until the nineteenth century in the West and still not widely available in most cities in the rest of today's world.[1] The cities which arose in China a millennium later were also planned, and from the beginning surrounded by a wall, with gates at each of the four cardinal points and a central complex of palace and temple. These ancient Indian and Chinese cities were surprisingly large; population guesses are difficult, but the walls enclosed a very extensive area; those at Anyang in modern Honan (Henan), the last capital of the Shang dynasty in the twelfth century BC, included within them well over a square mile.[2]

India and Sri Lanka, 320 BC–AD 1948

The continuity of Indian urban history was broken shortly after 2000 BC when the Indus sites were largely abandoned and the centre of settlement migrated eastward to the Ganges valley in the course of the movement of Aryan-speaking peoples into the subcontinent. We have little or no urban evidence from the intervening centuries, but with the rise of the Maurya dynasty in 320 BC (if not before) monumental planned cities reappeared, best known in the case of the Maurya capital of Pataliputra in the central Ganges valley near modern Patna. Little remains since it was built of wood, including its walls, but contemporary Greek accounts describe it as the largest and most splendid city in the world of the time, with broad avenues converging on a magnificent palace, and a well-developed municipal administration and police. The city was some eight by one and a half miles within its walls. Other cities included the port of Tamralipiti near the mouth of the Ganges, and what seem to have been centres of textile, mining, and metal industry (Basham, 1967). Pataliputra was again the imperial capital under the Guptas, from AD 320 to 550, and seems to have recovered most or all of its Mauryan grandeur, although the Guptas, like the Mauryas, ruled only the area north of the Deccan. The south was apparently only briefly and incompletely incorporated even under the Mauryan emperor Ashoka (269–232 BC), but in any case had a different urban history, of which we begin to have some glimpses only in Guptan times and later. The typical southern city was centred on a temple or temple complex, as at the great cities of Madurai and Tanjore, around which the market districts clustered, and the streets dominated by different groups of artisans. The south was chronically divided politically into rival states, but relatively little of their capitals remains except for the temple complexes; there are descriptions also of a port city on the east coast, Puhar, near the site of modern Madras, which seems to have managed a very large trade, mainly with South-East Asia, probably on behalf of south India as a whole (Singer, 1971: 169–70).

Port cities were apparently even more numerous on the Indian west coast, from the time of Mohenjo Daro and what was probably its port of Lothal south of the mouth of the Indus, where the remains of stone docks and warehouses have been found. Later Greek and Roman accounts from the first century AD speak of many others, from the coasts of modern Gujarat southward almost to the tip of the peninsula, including later prominent places such as Goa, Mangalore, Calicut (where da Gama landed in 1498), and Cochin, all centres of a thriving trade across the Indian Ocean from at least Mauryan times and continuing into the modern period. But ever since at least the fourth century BC, the chief urban axis has been the Ganges valley, from Delhi to the Bay of Bengal, where a string of trading and religious cities had emerged by Mauryan times, including the holy city of Benares (now Varanasi). This is still India's most highly urbanized region, as the Ganges valley is its major commercial corridor and contains the largest share of its modern industrial activity.

One of the difficulties of Indian history is the comparative scarcity of surviving written records until relatively recent times, and the concentration of what does survive on the eternal questions of philosophy and religion (plus literature) rather than on more worldly matters, including material bearing on cities. The collapse of the imperial order with the fall of the Guptas left India once more divided into a great number of regional kingdoms of which few records remain except the temples they built in stone or brick, and some stone inscriptions. Somewhat more remains, though thoroughly in ruins, of the city of Vijayanagar, capital of the kingdom of the same name from 1336 to 1565 in central-south India, which was praised by early Portuguese observers as a large and splendid city. The north had meanwhile succumbed in stages to an Islamic conquest by a series of Turco-Afghan dynasties, each of which made Delhi its capital, or rather a series of cities in the Delhi area, from 1206 to 1526, all now in virtually complete ruin, although a few fragments of wall and a few partial buildings still stand. Lahore in the adjacent Punjab also flourished during this period as a large city, but we have few details of other cities of medieval India beyond references to their names as major places or regional capitals.

Firuz Tughluq, Sultan of Delhi from 1351 to 1388, was known for his passion for building and constructed a new capital which is said to have contained forty mosques, thirty colleges, a hundred hospitals, and two hundred new towns in the vicinity of Delhi (Wolpert, 1989: 118). The Delhi Sultanate and its capital were decimated by the conquest and sack of Timur's (Tamerlane's) invasion of 1398. Gujarat on the west coast broke away from Delhi's weakened control and continued to prosper from its role as India's chief port and market for trade westward, through its major port of Surat and at its capital at Ahmadabad. Ahmadabad became a flourishing city, replete with temples and palaces and the houses of rich merchants. Delhi slowly recovered, and was in time largely rebuilt after Timur's devastation, eventually rivalling or exceeding Ahmadabad in size. But its political dominance was increasingly threatened by regional rebellions, and in 1526 it fell to the new conqueror Babur, founder of the Mughal dynasty, which was to create a new and even more splendid capital at Delhi, with a twin capital some 120 miles to the south-east at Agra.

Both cities commanded easy crossing points of the Jumna (Yamuna) River, the major Ganges tributary west of the main stream, and hence controlled the routes of conquest eastward into the rich Ganges valley. They also blocked the easiest routes southward, since the Aravalli Range and the desert of Rajasthan to the west tended to funnel passage from the north-north-west, the source of repeated invasions, into the lowland corridor where Delhi and Agra stood guard. Mughal Delhi and Agra also benefited from the productive alluvial soils of the area between the Jumna and the Ganges, and both sites had seen successive cities, rebuilt after conquests, before the Mughals came, Mughal Delhi being supposedly the seventh capital to be built there. Most of the Delhi built by the Mughal emperor Shah Jahan (reigned 1628–58), and hence often called ShahJahanabad, still stands grouped around the massive red sandstone fort and the huge mosque, Jama Masjid, nearby. Inside the thick walls and gates of the Red Fort were the palaces, audience halls, and pleasure gardens of the Mughal emperors. The walls reached to the foreshores of the Jumna river, while on the other (north-western) side the ground was left open, as it still is, and continues to be used as a fair and parade ground, and site for political rallies or ceremonies, occupied in between times by the temporary stalls of the sellers of a wide variety of goods and services, including snake charmers, jugglers, fortune tellers, scribes, and wandering beggars. Beyond this open area was the confusion of dwellings, shops, and markets, crowded around the Jama Masjid in a tangle of narrow lanes and alleys, most of it surrounded by a wall with gates closed at night but with some houses and commercial functions spilling out beyond the gates. Most of that part of Old Delhi also remains today, relatively little changed since Mughal times, although the modern city has grown widely around it (Saksena, 1962; Hambly, 1968).

Agra had, on a slight rise of ground, its own Mughal Red Fort, much like that at Delhi, and its crowded housing and bazaars near the fort. Somewhat farther out, overlooking the Jumna river, Shah Jahan built the famous Taj Mahal as a memorial to his dead wife. Twenty-three miles to the west, the Emperor Akbar (reigned 1556–1605) built a splendid new capital also in red sandstone at Fatehpur Sikri in gratitude to a local holy man who had promised that he would have a son. Fatehpur Sikri was abandoned after fifteen years when the local water-supply proved inadequate, and still stands much as Akbar left it, a monument to his interest in blending traditional Indian and Islamic-Persian cultures and styles. Lahore, where another Red Fort was built, served as a secondary capital of Punjab, and became an even larger city where, as at Delhi and Agra, the great red sandstone fort and its palaces were largely surrounded by the jumble of a more mundane Indian city (Hambly, 1968, 1987). Ahmadabad and Surat continued to prosper and grow under Mughal rule and benefited from an increasing trade with Europe, especially in fine Indian cottons, many of them made in Gujarat. Amritsar, some forty miles east of Lahore, grew also and became the religious centre of the new religion of Sikhism. Kanpur, Allahabad, and Patna in the Ganges valley flourished as primarily commercial centres, with Benares retaining its religious emphasis as a holy place of Hinduism, and Murshidabad serving as the administrative capital of productive Bengal. In Rajasthan, home of the Rajputs, Jaipur, Ajmer, Jodhpur, and other fortified castle towns functioned mainly as centres of Rajput resistance against the Mughals, especially in the dynasty's last half century under the oppressive reign of Aurangzeb (reigned 1658–1707). The temple towns of the south, for the most part beyond the Mughal domain, and the port cities of the south-west coast, grew as part of a general prosperity and increased trade with the West. Meanwhile, even before the Mughal conquest, da Gama had reached Calicut in 1498 and in 1507 the Portuguese seized Goa and made it their major Asian base. Although they never penetrated far inland and, except for Goa, functioned only as traders at various Indian-controlled ports, they were the forerunners of a new era. In 1639 the British settled and built a fort at Madras, in 1668 they acquired Bombay (then largely uninhabited), and in 1690 founded what was to become Calcutta.

Each of these colonial ports grew slowly at first, but their growth, especially in the nineteenth century, symbolized and reflected the basic changes overtaking India: new commercialization, road and railway building, the beginnings of industrialization, and consequent rapid urbanization. India's cities were no longer dominated by inward-centred administrative or religious foci or local–regional markets but by the burgeoning ports, which faced outward to the world overseas and served as the transmission links for change into the rest of India along the new rail lines. Calcutta became the colonial capital for British India, Bombay the premier port connection with Europe and the early industrial centre, Madras the commercial hub and main port for south India. It was essentially the widening of the market which spurred the growth of each, including the newly improved access to overseas markets, especially with the coming of steamships (which also operated on the Ganges and Indus rivers) and with a major boost from the opening of the Suez Canal in 1869.

By 1800 Calcutta had become the largest of the three, and probably already the largest city in India, as it still is. It owed its growth both to its functions as the administrative capital of an expanding British domain in India, and as the port and dominant service centre and market for the north and east of the subcontinent, including the productive Ganges valley, India's major corridor of internal communication, commercialization, and urbanization. By 1800 Calcutta was already spoken of as 'the second city of the British Empire', second only to London, and with a closely similar mix and balance of functions, like it also peripheral to the main bulk of the country and serving as its sea link to a wider world. Like London, Calcutta grew along one bank of its river, the Hooghly, one of the lesser mouths of the Ganges but comparable in size to the Thames and accessible to most ocean-going ships until their size and draft increased after about 1880, when increasing use was made of outports closer to the river's mouth in the Bay of Bengal, as London used Gravesend at the head of the Thames estuary. At Calcutta itself, some 100 miles up the winding river from the sea, the city and its wharves clustered on the east bank of the Hooghly near the original anchorage and landing place which had led to the choice of this site by the East India Company's agent Job Charnock in 1690. Fort William was built there, commanding the river, by the end of the seventeenth century, and the town, overwhelmingly Indian in population, grew around it, with the larger British houses on the southern edge along the river in an area still called Garden Reach.

Railways began to spread out from the city in the 1850s, with the main station across the Hooghly (not bridged until 1942) on the west bank in the town of Howrah, from where they could better serve Calcutta's principal hinterland, the Ganges valley. Other lines stretched eastward into east Bengal, which became the major source of jute, Calcutta's leading export after processing in the city's jute mills. They marked the beginning of modern power-driven manufacturing in Calcutta, soon joined by cotton textile mills, metal-working industries, and a variety of light consumer goods production. The new manufacturing plants, supplementing the continuing shipbuilding and repair which had begun much earlier, were strung out mainly north of the original city on both banks of the Hooghly and in time made up a ribbon-like urban area forty or fifty miles long. Subsequent urban growth spilled eastward and to some degree westward, away from the river, even though this involved, especially to the east, the use of poorly drained and often swampy land, still plagued by such problems during the summer monsoon (Murphey, 1964; Busteed, 1908; Cotton, 1907). The original settlement had kept largely to the natural levees along the east bank of the Hooghly, which provided somewhat higher ground.

With the growth of the Indian tea industry after 1850, mainly in Assam north-east of Bengal, and tea's rise to prominence as India's premier export by value, Calcutta's further growth was stimulated as the port of tea export and as the service centre for the booming commercial sector of Assam. Finally, India's major coal reserves were found and exploited on a major scale early in the nineteenth century just west of Calcutta in the valley of the Damodar river. Coal moved down the Damodar and railways following it primarily to Calcutta for use and wider distribution, and by the early twentieth century had begun to support a growing iron and steel industry in the Damodar region and adjacent areas, where high grade iron ore was also found, all part of Calcutta's commercially tributary area and helping to swell the city's size. In 1905 Rudyard Kipling published a group of poems celebrating various cities of the world, including Calcutta and Bombay. His lines for Calcutta suited it well:

> Me the sea captain loved, the river built;
> Wealth sought, and kings adventured life to hold.
> Hail Europe! I am Asia, power on silt;
> Death in my hands, but gold![3]

(Jute was often referred to in Calcutta as 'gold on silt' since it thrived in the wet alluvial soils along the muddy foreshores of Bengal's many creeks and rivers; the death rate among the English was very high until the twentieth century, but some of the survivors made fortunes.) Most of East Bengal (now Bangladesh) was economically and culturally tributary to Calcutta, where many East Bengali landowners lived on their rents and helped to support the cultural life of the city. But Dacca (Dhaka) the eastern regional administrative capital and an old trade centre as well, served as a secondary node for the densely populated east, especially for the shipment of jute and the fine handmade cotton textiles for which the city was traditionally famous, Dacca Muslin (derived of course from Muslim—East Bengal became dominantly Muslim after the twelfth century). Dacca stood on the east bank of the main stream of the Ganges, downstream from its junction with the Brahmaputra, and had water transport connections with nearly all of East Bengal, but the main rail connections ran to and from Calcutta. The old port of Chittagong, on the coast below the joint Ganges–Brahmaputra mouth, grew in the colonial period as a regional distribution point by sea and river (Sinha, 1961–2).

Bombay, colonial and modern India's second largest city, grew only slowly until the nineteenth century despite its earlier foundation. Originally a group of some seven islands, later joined by landfill, protecting to landward a large and deep bay, it was hampered in its access to a hinterland by the high and steep range of the Western Ghats less than fifty miles to the east, and by the marauding and still unconquered Marathas, who preyed widely on trade until their final defeat by the British in 1818. At sea, the routes leading to Bombay and along the coast suffered similarly from rampant piracy, finally suppressed only in the nineteenth century. With the completion of the first railway across the Ghats in the 1850s and the coming of steamships, Bombay was at last able to fulfil its great potential. Its harbour was the only one in the whole of India adequate for deep-draught shipping, and its beautifully protected roadstead was large enough to accommodate easily all of the traffic it attracted, rising steeply in amount as the nineteenth century progressed.

The East India Company built a fort there soon after acquiring Bombay, which had been given to the English crown as part of the dowry of the Portuguese princess Catherine of Braganza on her marriage to Charles II of England in 1661, and later passed to the Company in 1668. While it soon became the major British port for contacts westward, its limited access restricted its larger growth beyond its port and harbour functions. The fort, on the peninsula made from linking the islands by landfill, was however soon surrounded to landward by Indian settlement, as at Calcutta, while the wealthier British began to build grander houses northward along the seafront, where the breezes helped moderate the climate, on a ridge above the sea called Malabar Hill, still the city's most fashionable residential district. Apart from the naval and administrative control of the British, Bombay soon became more effectively dominated by Indian entrepreneurs than under the closer colonial control in Calcutta, where British firms tended to be uppermost, with some exceptions. In Bombay, it was primarily Gujaratis, with their long commercial tradition, who rose to the top, together with the Parsee community; 'Parsee' is a variant of 'Persian', and these people migrated from Iran when it was conquered and converted to Islam by the Arab expansion of the eighth century AD, bringing with them their Zoroastrian religion. They settled first in Surat, where they became leading entrepreneurs (and in effect Gujaratis), and then, when the British moved their principal western base to Bombay, followed them there and became early Anglo-Indian collaborators, as many Bengalis did in Calcutta.

It was a Parsee firm, the Wadias, who owned and operated Bombay's major dockyard for shipbuilding and repair and built ships of the line for the British navy out of Indian and Burmese teak, while other Parsees went into trade, banking, and later manufacturing, developing a widespread commercial empire which reached to the China coast and made fortunes in opium and real estate. The Parsee firm of Tata Enterprises, which built India's first large integrated iron and steel mill at Jamshedpur west of Calcutta in 1911, for long the biggest in the British Empire, still dominates the private sector in India today as a vast commercial-industrial conglomerate, and Bombay remains the major Parsee base.

Other Gujaratis were also active entrepreneurs in the boom economy of Bombay, especially with the beginnings of power-driven manufacturing there in the early 1850s, first in cotton textiles, which remained the city's major industrial sector. Bombay was ideally placed to draw raw cotton from India's premier cotton-growing district, in nearby Maharashtra and Gujarat, and could also bring cotton by sea or rail from the Indus valley, the other major producer, while its excellent market access by sea and rail and the presence of Gujarati capital and initiative made a powerful combination. Raw cotton was also exported to feed the mills of Lancashire, a trade which boomed when the American Civil War and the Federal blockade of Southern ports cut off their supply of cotton to Britain. With the opening of the Suez Canal soon after, in 1869, Bombay's growth spurted once more, and by the end of the century the traffic through the port ranked among the largest in the world, as did its textile industry. Bombay's expansion and new industrial employment drew to it a stream of migrants from Maharashtra and Gujarat but also from all over India as well as from abroad (including of course the British but also Chinese, people from the Middle East, and others). Kipling's lines for Bombay were apt:

> Royal and dower-royal, I the Queen,
> Fronting thy richest sea with richer hands;
> A thousand mills roar through me where I glean
> All races from all lands.[4]

('Royal and dower-royal' is of course a reference to the unusual origin of the British claim to Bombay.) Like Calcutta, it rapidly became a cosmopolitan centre, and combined new wealth for a few, both Indians and British, and slum poverty for most. Indian machine-manufactured textiles, cotton mainly in Bombay and Ahmadabad factories and secondarily in Kanpur and other cities, jute in Calcutta, largely drove out higher-priced British and then Japanese competition from the Indian market by the time of the First World War, and invaded low-price markets overseas in Africa and South-East Asia while supplying cheap machine-made yarn to the later developing Chinese cotton textile industry. With the coming of hydroelectric power technology, Bombay derived additional advantage from dams in the rainy Western Ghats, within easy range of the city's power grid.[5]

Madras, though it was the earliest port under *de facto* British control and until the 1760s the major colonial administrative base, was passed in size by both Calcutta and Bombay in the course of the eighteenth century, and left even farther behind by the industrialization of the latter two after 1850. Its harbour was even more inadequate than Calcutta's became as ship sizes increased with steam and steel. Originally a shallow lagoon entered by a small tidal creek protected by Fort St. George, it was by the nineteenth century obliged to have goods transhipped by lighter to and from ships anchored off the beach in an unprotected roadstead, exposed to the monsoon which effectively closed the port for several months each year. Finally in 1881 the first of a series of breakwaters was built, enlarged subsequently to enclose an artificial harbour and docks, although these facilities and sheltered space remain critically inadequate for the port's modern traffic. As at the other colonial Indian establishments, the area around the fort, originally limited to Company officials, with Indians segregated in what was called 'Black Town' adjacent to it, soon became predominantly Indian in addition to containing warehouses, port offices, courts, and other administrative functions in the same pattern as around the forts at Calcutta and Bombay. Wealthier Europeans began to move across the creek southward, along Mount Road and elsewhere, to build larger houses and take advantage of the sea breezes.

Once the minor barrier of the creek, easily bridged, was passed, Madras grew predominantly southward, into what was then relatively open country. Growth rested in part on continued colonial administrative functions for south India as a whole, under the Presidency system which until 1774 divided British India administratively into Calcutta, Bombay, and Madras Presidencies. Even when all of British India was ruled by the Governor-General of Bengal after 1774, each of the other two cities retained considerable regional responsibility. From early in its career, Madras served as the main commercial centre for south

India, a role which was greatly strengthened by the building of railways focused on it after 1860. There was some nineteenth-century development of textile manufacturing in the city, but it grew primarily as the commercial, financial, and cultural central place for the peninsular south, including the role of Madras University, an originally British foundation following the earlier model of Presidency College in Calcutta (the first, later absorbed into Calcutta University) and of Bombay University, all three established by the colonial government in 1857. Madras played the role for the peninsular south which Calcutta played for Bengal and the Ganges valley, Bombay for western India. Like them, it disseminated through its trade hinterland its economic and cultural model, as it also educated growing numbers of the Indian élite in its tributary area.[6]

More so than Calcutta or Bombay, Madras was also a major missionary base; Christianity was attractive to some southerners as an antidote (the conqueror's religion) to what they have continued to see as 'northern domination', and English language an important skill with which to compete in the colonial world. Mission schools, teaching in English, were far more widespread in the south than anywhere else in India, with organizational headquarters in Madras, which also supported a vigorous English-language as well as Tamil press, circulated over much of the south. Madras leadership was not seriously rivalled by the far older indigenous inland cities, such as Hyderabad, still a major conurbation, or Mysore, both cities of the traditional pattern centred around palaces and temples. Bangalore, which grew from a village to a city only in the twentieth century, is a far more modern and open commercial-industrial place, but in most things still tributary to Madras, as is the smaller city of Salem and the old temple complex towns farther south. Tobacco, oil seeds, spices, and fine traditional-style cotton textiles, south India's main commercial products, flowed out via Madras, while cigarettes were manufactured in the city.

In central India, Nagpur became the only major city, largely because it became one of India's largest railway junctions of east–west and north–south routes, and accordingly found some advantage for light industrialization. The latest comer among the colonial port cities was Karachi, a small traditional port for trade mainly with the Persian Gulf, in the Indus delta and hence plagued by silting until this was reduced within an artificial harbour in the 1890s. But the booming commercial production of cotton and wheat in the Indus valley and western Punjab demanded an outlet; some of it moved out via Bombay, but a closer port was clearly desirable, especially after the beginning of steam navigation on the Indus and the further spread of railways. Karachi sprouted warehouses, docks, and commercial facilities, and became the fourth Indian port in volume of shipping and value of exports, while the northern Indus city of Multan grew as a secondary service centre to the newly commercialized agricultural sector.[7]

In nearby Ceylon (since 1975 officially called Sri Lanka), large monumental cities were built as capitals and religious centres beginning in the first or second century BC at Anuradhapura, and later at Polonnaruwa, with a port city of Mannar on the west coast where hoards of Roman coins and pottery have been found. Anuradhapura especially covered a very extensive area and many of its stone buildings still retain some of their columns and paving stones, as at Polonnaruwa, but we have few details about them otherwise. Their size is not a reliable guide to their population, since it is unlikely that the entire built area was fully occupied except on ceremonial occasions; many of the structures seem to have been temples. Nevertheless, some scholars have pointed out that Anuradhapura at its height in the eleventh and twelfth centuries covered an area the size of mid-nineteenth century London. It and Polonnaruwa were largely abandoned with the collapse of the medieval kingdom in the thirteenth century, and the main population centre shifted to the hills of south-central Ceylon, better protected against the earlier chronic invasions from south India (Murphey, 1957).

The Portuguese had established their main Ceylon base in Colombo early in the sixteenth century, attracted in part by the cinnamon gathered in nearby forests and in world demand, and came to control most of the lowland areas, as did the Dutch who succeeded them by 1650, but despite repeated efforts neither ever conquered the Kingdom of Kandy in the highlands, the remnant of indigenous sovereignty. Even after the British absorption of the entire country in 1825, Kandy remained a town rather than a city, as did Jaffna at the island's northern tip, and Trincomalee on the east coast, despite its fine natural harbour, since it was too far from the commercially productive areas.

Colombo leaped ahead of its rival Galle after breakwaters, begun in 1882, progressively enclosed a dredged and protected harbour, replacing the open lagoon which had served as a poor shelter before. Colombo's rapid growth was, however, due also to the boom development of the tea and later rubber and coconut export industries in the island's central hills and fringing lowlands, closer to Colombo than to Galle and hence stimulating investment in port improvements. Roads and railways were built throughout the island but especially in the plantation areas, all focusing on Colombo as the overwhelmingly dominant export point and service centre; it became, as it still is, the only major city, despite the continued limitations of its harbour, but aided by its role as the colonial and, since independence in 1948, the national capital and chief cultural centre.[8]

In general, the colonial period, and especially the railway era beginning in the 1850s, saw a boom in urbanization, led by the British-founded ports but echoed in older cities such as Ahmadabad, Allahabad, Patna, or Kanpur as they too became new commercial–industrial centres. The rail links also stimulated rapid commercialization of the agricultural areas through which they ran, and thus generated a need for new regional service centres. Originally planned mainly for military and administrative purposes, the railways also prompted the growth of garrison towns, especially along the main line from Calcutta up the Ganges valley and on beyond Delhi to Lahore and the north-west frontier, where the largely military towns of Rawalpindi and Peshawar became newly important. Lahore and Delhi also began to grow again, and became once more big cities and the major urban centres for northern India, aided by their new rail connections and becoming involved also in the beginnings of industrialization. In 1911 the decision was made to move the colonial capital from Calcutta to Delhi, a move complete by the early 1930s, beginning Delhi's even more rapid modern growth.

By the Census of 1931, India's population was registered as 11 per cent urban, and the fifteen largest cities contained over seven million people, rising to about ten million by 1941. The war, including the growth of military procurement and the increase in troops and port business, gave a further boost to city growth. By independence in 1947 (and the partition of the subcontinent into the Republic of India and the Islamic State of Pakistan), the total urban population was about sixty million, and approximately 15 per cent of Indians, Pakistanis, and Ceylonese lived in towns or cities. The colonial period had transformed India's urban patterns, provided it with the largest rail network in Asia, almost on a European level, and begun the process of industrialization, all of which were to fuel post-independence growth. Post-1947 urbanization was to be a reflection primarily of major new industrialization, plus administrative functions and new levels of commercialization.

China, 2000 BC–AD 1949

China has almost as long an urban history as India, and given its consistently larger population has supported more cities and more urbanites over most of the past two thousand years than any comparably sized area in the world. We know relatively little about the cities of the Shang and Chou dynasties (*c.*1776–221 BC) beyond their capitals. From the beginnings of the Chinese written language, however, the word for *city* also meant *wall*; walls distinguished it from market towns of a lower order and, at least after the founding of the empire, marked it as the seat of an imperial magistrate. The Shang probably built or controlled other walled cities beyond their migrating capital (apparently shifted with each new ruler) elsewhere on the central part of the north China plain laid down by the silt deposition of the Yellow River. The area of Shang control is shadowy, but probably did not reach eastward to the sea, south into the Yangtze valley, west into the uplands, or north into what is now Shansi province.

Shang culture and urbanism emerged out of the late Neolithic culture of Lung Shan, which by about 2000 BC had built the city of Ao on the site of modern Chengchow near the Yellow River in Honan Province surrounded by large pounded-earth walls and gates enclosing an urban area a mile square; in the course of the Shang period it seems likely that city building must have spread beyond the successive capitals. The archaeological record is less complete for the Yangtze valley and the south, but there is ample evidence there of the emergence of agriculture and bronze technology about as early as in the north, if not earlier, developments likely to have been accompanied by the beginnings of urbanism (K. C. Chang, 1987). With the conquest of Shang by the Chou in or about the eleventh century BC (the date is still disputed), this originally western frontier people, vassals of the Shang, fixed their first capital in the Wei valley near modern Sian (Xian), their home base, and then in 770 BC (the first major definite date in Chinese history) moved it to Loyang in the Yellow River valley (after their Wei valley capital had been sacked by 'barbarians') in a vain effort to maintain control over an ecumene which was growing eastward, northward, and especially southward. The nominal Chou domains were becoming increasingly dominated by rival states, former Chou vassals now the real rulers except around Loyang.

Loyang under the Chou was a larger version of the late Shang capital at Anyang north of the Yellow River, both well placed to guard the agricultural lowlands against incursions from the hills to the west then still inhabited by pre-agricultural peoples, as the Chou had earlier been. Palaces at both capitals were extensive, though built of wood so that little evidence remains, and oriented to face south, like nearly all monumental buildings in China ever since. Royal and aristocratic tombs at each have yielded most of our evidence of Shang and Chou élite culture, but most of the people seem to have lived in crude pit dwellings, strikingly unlike the housing in the cities of the Indus civilization. By later Chou the settled landscape was increasingly dominated by rival contending states, perhaps as many as 200 by the eighth century BC, each rather like a city state centred on its walled capital and controlling the surrounding countryside. The total population of the China ceremonially subject to the Chou was probably about twenty million by 700 BC, already the largest national total in the world. Agriculture became far more productive with the spread of iron tools beginning about 500 BC and the spread of irrigation. Trade became important as the economy generated more surpluses, and the texts of the period speak increasingly of rich merchants, and the appearance of copper coins, all of which suggest the wide growth of cities, though we have little direct evidence. This seems to have been especially true of the middle Yangtze valley, where there was extensive water-borne trade and a prominent role for urban-based merchants. But the state of Ch'u which controlled this area was conquered by the expansionist state of Ch'in based in the Wei valley in the north-west in the wars which marked the final breakup of the Chou order. Ch'in went on to establish the first Chinese empire in 221 BC, uniting the north and conquering the Canton region of the south.

Ch'in imposed its anti-mercantile policies on its new empire and thus choked off what may have been a sharply different urban-based pattern in the Yangtze valley and south. Ch'in Shih Huang Ti, the first emperor, built a splendid new capital and palaces in the Wei valley near modern Sian, to the north of which major burial mounds have recently been excavated, including the terracotta army placed there to guard the first emperor's tomb. Merchants and trade, and thus also cities, were tightly controlled, as potential rivals to the power of the state, which rested on its peasant armies and on agricultural revenues. Primogeniture was abolished as a further move against the growth of hereditary aristocratic power. But the oppressive nature of the Ch'in order led to its collapse in rebellion (in which the capital was destroyed), and to the founding of the Han dynasty in 202 BC after a period of civil war. Han control lasted until AD 220, and it is in that long period that the traditional forms of the Chinese city solidified. They were to remain basically the same until the victory of the Communist revolution of 1949. The Han ruled first from the old centre in the Wei valley at Ch'ang An ('Long Peace') near the site of modern Sian, and after AD 23 from Loyang, closer to the centre of their domains. At both they built large new cities and palace complexes, and left behind in imperial and other élite tombs varied evidence of material life, including clay models of houses and their courtyards, similar to those still to be seen in modern China. Both capitals were probably about the size of imperial Rome. Reconquest of the south added more areas to the Ch'in empire and included the northern part of what is now Vietnam, southern Manchuria, and adjacent northern Korea as well as desert Sinkiang (Xinjiang) in the far north-west. In these areas beyond the sphere of Chinese occupancy, the Han set up garrison towns, as along the Silk Road through Sinkiang, which in some cases marked the beginning of urbanization there and included, especially in Korea and Manchuria, Chinese agricultural colonists. The imperial census of AD 2 counted just short of sixty million people within the empire, almost certainly an under-count but in any case more than the total of the Roman empire at its roughly contemporary height (Fairbank *et al.*, 1989: 21).

To administer this vast area and population, the Han established a system of counties (*hsien* or *xian*), grouped around a walled city where an imperial magistrate and his staff kept order, dispensed justice, and collected taxes. We can trace the spread of Han control, especially southward, through the successive founding of these *hsien* capitals, increasing in numbers also as the population increased in each region (S. D. Chang, 1961). There were of course also trading cities, but they too were presided over by an imperial magistrate, part of whose function was to control trade and merchants and to enforce the state monopolies on commerce in salt and metals (for coinage and weapons). Chinese settlement and subsequent administrative structures spread north into southern Manchuria, but mainly south following the coast and the several tributaries of the Yangtze, avoiding until some centuries later expansion into the mountainous south-west. In northern Vietnam the Han pattern was superimposed on the older Vietnamese city of Hanoi, and in Korea the Han city and garrison of Lo-lang rose on the site of modern Pyongyang. Cities throughout the empire were linked by a system of imperial roads, along which couriers carried dispatches but which also served commerce. Trading cities were concentrated on the extensive waterways of the Yangtze drainage and at its river confluences, as at Wuhan, Chungking (Chongqing), Changsha, Nanking (Nanjing), and along the coast Ningpo, Foochow (Fuzhou), Swatow, and Canton (Guangzhou). Other major cities of the period included the capitals at Loyang and Ch'ang An, Chengdu on a productive irrigated plain in western Szechuan (Sichuan), Taiyuan in Shansi, Tsinan (Chinan) on the lower Yellow River in Shantung, Peking (Beijing) (then called Yenching, or the capital of the old state of Yen), and Shenyang (Mukden), the major Han base in southern Manchuria.

All these, and the smaller *hsien* cities, followed a more or less uniform plan, clearly revealing the imperial imprint on the landscape. This urban model survived the fall of the Han in AD 220 and was reaffirmed by the subsequent dynasties of Sui, T'ang, Sung (Song), Yuan, Ming, and Ch'ing (Qing), the last ending only in 1911. Traditional cities were primarily centres of imperial authority, symbolic monuments of the power and majesty of the Chinese state and of Chinese culture over which it presided. Functionally, they were in most cases predominantly agents of the imperial bureaucracy; directly or indirectly, the largest sector of the urban work-force was employed in that administrative enterprise. These included officials, clerks, scribes, garrison troops, teachers of the classics to aspirant generations of examination candidates, merchants employed by the state in the management of official monopolies of trade or manufacture, artisans whose output went predominantly to the offices and households of all the foregoing, and the vaster army of shopkeepers, coolies, servants, transport workers, butchers, bakers, and candlestick-makers whose service livelihood depended on the city's basic industry, which was most importantly administration. All of these cities included also a commercial and a manufacturing function apart from the official and state-run monopolies in salt, iron, copper, weapons, tax or tribute grain, and foreign trade. In a few, these functions were more important than official and administrative functions, but in none above the level of local market town was the hand of the state not apparent or its business not prominent.

The population of the *hsien* as a whole averaged very roughly about 200 000 (with considerable temporal and regional variations), while that of the *hsien* capital city might average perhaps 10 000–15 000.[9] In most *hsien* its capital would be the only real city; in other more commercialized parts of the country (for example, the lower Yangtze valley) there might be one or more others. But the *hsien* capital was nearly always the largest, and the only base for the official bureaucracy. There the imperial magistrate, through his court and assistants, was officially responsible for everyone and everything in the *hsien*, but in practice it was not possible to carry out close administration of so large a population, the great majority of it rural-agricultural, in an area (again a possibly misleading average) roughly equivalent to an English or eastern American county, some 500 to 800 square miles. Unofficial but often powerful local gentry and peasant village elders or clans managed the bulk of rural and hence of *hsien* affairs.

All *hsien* capitals were walled, and were planned in detail. The walls were as imposing as the rank and size of the city dictated, but in every case were designed to awe and affirm, only secondarily to defend, although of course they might be useful in troubled times. City walls were built in a regular and consistent pattern, with great gates at each of the cardinal compass points, from which broad, straight avenues led to the opposite gate. These intersected in the middle of the city, where there was a ceremonial centre with a plaza, a drum tower, a cluster of official buildings (including the magistrate's offices and court), or a Confucian temple. The major streets, fixing the main axes of the gridiron pattern, divided the city into major quarters, which were sometimes also enclosed by their own walls whose gates, like the city's main gates, were closed at night. Each quarter tended to be functionally specific: transport termini, warehouses, and commercial offices or banks in one sector, retailing (often segregated by street according to commodities) in another, manufacturing in another, and others for academies, universities, booksellers, theatres, the military establishment and its garrison troops, public food markets, and so on. Most people lived in the same structures which housed their work activities, and within each quarter there were regular lanes organized into neighbourhoods. The emphasis, as in the ideology of the Confucian state, was on order and on planned management.

Most cities were founded explicitly by the state, as centres of imperial control, although occasionally (especially in the early imperial period under the Han) a pre-existing urban centre of trade might be designated as an administrative centre and provided with the planned form. Within each *hsien* the capital city was usually located near the middle of the territory, whose areal size varied inversely with the density of population but which was ideally designed as much as possible so as to constitute a coherent geographic region of trade, production, and movement. Where the physical landscape and population density permitted, this might be a stream, watershed or confluence region, or a basin surrounded by hills. The capital city was responsible for the defence as well as administration of the *hsien* as a whole and not merely with the defence of its own walled base. It was truly a *centre*, not an isolated or discrete intrusion.

This in turn was related to the perceived relationship between city and countryside. Traditional China was an overwhelmingly agrarian society; there was no question what the source of wealth was, including the means for the support of the state and its apparatus. Cities were designed to control the countryside, but more importantly to serve it, as the basic reason and sustenance of their existence. Trade and merchants flourished in all Chinese cities, but most commodities were of agricultural or rural origin. The primary responsibility of officials was ensuring the productivity as well as the orderliness of the agricultural countryside since it was this which sustained the empire, its power, its cultural grandeur, and its bureaucratic structures. This was true most immediately for the *hsien* magistrate, the official closest to rural areas, but true also for the emperor himself, at the head of the pyramid; each year to mark the beginning of the new agricultural season he ploughed a ceremonial furrow in the grounds of the Temple of Heaven at the imperial capital and interceded with heaven for good harvests. His first concern was the well-being of the agricultural sector. The close interdependence of city and countryside was far more explicitly recognized, and indeed welcomed, in China than elsewhere.

Change was of course not absent in the long span of the 2000 years of imperial Chinese history. But, especially after about the twelfth century AD, it was not primarily focused in cities, which instead were seen as presiding over the 'Great Harmony', a persistent Chinese ideal in which disruptive change was to be minimized, all groups worked together for the common good, and cities served the countryside as part of a single symbiotic order. What pressures there were for change came far more often from rural areas, as protests against arbitrary city-based power or the exploitative accumulation of urban wealth at the expense of peasants as corruption and self-seeking became more prominent in the last decades of each dynasty. The rural areas were often seen as the base for correcting the excesses of the city, including the overthrow of urban-based power which marked the collapse of a dynasty, to be succeeded by a new dynasty which often had significant peasant or rural origins. Urban merchants, the spearhead of change in the modern West, were in traditional China too closely involved with and nurtured by the official system to fight against it. They had little to gain, unlike their Western counterparts, by upsetting the apple cart, either by changing the rules or by putting a new group in power. Most of the big merchants were from urban gentry families; many were in effect officials involved in one or another of the many state trade monopolies, but all depended on some form of official connection or patronage. If that were lost, as happened of course from time to time to individuals, and to all when the dynasty or government fell, the result was total ruin.

Cities were centres of action in great variety, but in general not of institutional change, which indeed their major efforts were directed to preventing. They were splendid places, probably the most splendid in the world as well as the largest and richest, as Marco Polo tried to persuade an unbelieving Europe, who nicknamed him 'millione', a sort of Venetian Baron von Münchhausen, although his book is corroborated by others. Well before his time, while we know too little of Han Ch'ang An or Loyang, the Tang capital, again at Ch'ang An from the early seventh to the tenth centuries, may have had about a million people inside its great walls and nearly another million outside them. The city was dominated by its palace complex, facing south, and was divided into quarters by its great avenues. It had two or three large market areas, in different parts of the city, and was by far the largest city in the world up to that time and for many centuries thereafter, a cosmopolitan centre which attracted traders, philosophers, and travellers from western Asia as well as Japan, Korea, and India (Wright, 1965).[10] The Northern Sung capital at Kaifeng (960–1126), east of the great bend of the lower Yellow River, was almost equally large, and was also a major industrial centre for the smelting of iron and steel as well as a major market centre, in addition to its function as the imperial capital. It was, however, the Southern Sung capital at Hangchow (Hangzhou, 1127–1279), at the terminus of the Grand Canal south of the Yangtze, which won the most fulsome comments from the small group of early European travellers to China, beginning with Marco Polo in the thirteenth century. They described it accurately as by far the largest and 'noblest' city in the world of its time, in the words of the fourteenth century Italian traveller John of Marignolli 'the first, the biggest, the richest, the most populous, and altogether the most marvellous city that exists on the face of the earth'. His contemporary Ibn Battuta says that the city was composed of six towns each larger than anything in the West and three days' journey across (Yule, 1886).[11]

Yet all of these European observers saw Hangchow only after its greatest days had passed and it was under Mongol control after they had overrun the southern Sung domains. Even then Hangchow remained a vibrant centre of trade, and of urban culture, as well as of imperial administration. But the collapse of the Mongol order brought a new Chinese dynasty to power, the Ming, who first built a huge new capital at Nanking and then, in 1421, moved their capital to Peking, where the Mongols had ruled. There they built a splendid new walled city, as Nanking was, on the site of the remains of Mongol Tai Tu (Polo's 'Cambaluc', derived from 'Khanbaligh' or 'city of the Khan') but far larger. The walls were forty feet high and over fourteen miles around (Nanking's had been sixty feet high and over twenty miles around, the longest city wall in the world), and formed a square with nine gates, with the Imperial City in the centre surrounded by its own walls, containing the inner walls of the imperial palace and the Forbidden City, some five miles around. The last Chinese dynasty, the Ch'ing (Qing, 1644–1911), restored most of Ming Peking's grandeur, and added a southern extension, also surrounded by a wall with seven gates, while rebuilding the imperial palaces in the form in which we see them today. Ming and Ch'ing Peking contained about a million people within the walls, and a large population outside them.

Throughout the long span of Chinese history and with only brief breaks, the imperial capital has remained in the north, partly out of tradition, partly to guard what was consistently seen as the frontier most exposed to invasion from the northern and north-western steppes. There was thus a progression from west to east, as the centre of nomadic threat migrated, from Chou and Han capitals in the Wei valley, to Loyang, back to Wei valley Ch'ang An under the Tang, eastward to Kaifeng under the Sung, and finally to Peking under the Ming and Ch'ing, to guard the empire against first the Mongols and then the ultimately victorious Manchus from Manchuria, last in a long succession of originally nomadic tribes which harried the northern borders. Southern capitals, Hangchow and Nanking, were discredited because of their association with defeat (as again under the Kuomintang (Nationalist) government of Chiang Kai-shek from 1927 to 1937 and from 1945 to 1949). Chinese tradition came to equate the north with military power and the south with a softer emphasis, on commerce and merchants—despised in theory. Rice-eaters (southerners) could not make good soldiers, it was said, and no capital in that area could succeed. Successive northern capitals had nevertheless to be fed to a large degree on rice imported from the south, via the Grand Canal from the Yangtze valley or via its predecessors, first begun under the Ch'in. Under Mongol rule the Grand Canal was extended from Kaifeng to Peking, carrying the tribute grain which was a major part of imperial tax revenues until the collapse of the Ch'ing.

The north as a whole was a periodic grain-deficiency area, worsening as population increased and land deteriorated after millennia of use and drastic erosion. The far more productive south became, at least from late Chou times, the major urbanized area, aided also by the availability of water transport within the vast Yangtze system, along the coast, and in the lesser but still extensive waterways of the Hsi (West) River with its mouth near Canton. The chief urbanized region was the lower Yangtze and the delta, from Nanking to Hangchow and Suchow, where by the early nineteenth century probably about 10 per cent of the population lived in towns or cities. The Canton delta, on a slightly smaller scale, was similarly urbanized, and there were other smaller clusters in the central Yangtze valley around Wuhan, along the southeast coast, and in Szechuan.

The modern West began to impinge marginally on China during the Ming, when Portuguese, Dutch, and English traders appeared at Canton and nearby Macao, but the real Western impact was delayed by Chinese intransigence until their defeat by the British in 1840–2 and the consequent opening of first five and then many more ports to foreign trade and residence, under what came to be called the Unequal Treaties. In these treaty ports, as they were called, Westerners began to develop the kind of city they had built at home, dominated by trade and merchants and largely outside imperial (state) control, where private enterprise and profit-seeking flourished as they had never been able to do under the traditional Chinese system, and where gains were secure or could be invested in trade and the beginnings of manufacturing without fear of state taxation or confiscation. Chinese merchants as well as labourers flocked to them, and also refugees and political dissidents. The Ch'ing were in their long decline, plagued by repeated major rebellions (which drove many thousands from their homes to seek refuge in the foreign-controlled ports) and by the rise of openly revolutionary groups, many of whom also found refuge in foreign Shanghai and other treaty ports. By the 1930s there were over a hundred treaty ports, with Shanghai, which had become the chief commercial centre for the entire Yangtze valley and China's leading port for foreign trade, as the largest, followed by Tientsin (Tianjin), Hankou (Hangkow—now part of the tri-city urban area called Wuhan in the central Yangtze valley), Nanking, Canton, and Dairen (Dalian), the chief port of Manchuria. Other treaty ports which became large cities included Tsingtao (Qingdao) in Shantung, Foochow, Amoy, Chungking, and Changsha.

Hong Kong, by 1937 with over a million people, was granted outright to Britain in 1841 and served as an entrepôt for the trade of the entire China coast as well as being in effect the outport for Canton. The only major cities which remained outside the treaty port system were Peking (no longer the capital after 1927) and Chengdu in western Szechuan. But except for Tsingtao (and to a lesser extent Dairen) none of the treaty ports were new cities but merely areas of foreign settlement and administration (though not formal sovereignty) which had grown on the edges of a pre-existing Chinese city which was already a major trade centre, which is why the foreigners wanted to settle there. And the foreign settlements in every case were from the beginning overwhelmingly Chinese in population, foreigners remaining a relative handful, as in India. By about 1900, Shanghai reached the million mark in population and about the same time passed Peking as China's largest city, a position it still retains; by 1937 it totalled about four million.[12]

After 1895, the rapid growth of the treaty ports was further stimulated by the beginnings of power-driven manufacturing there, most of all in Shanghai (which contained about half of China's modern industry from 1900 to 1950) but also in Tientsin, Hankou, Dairen, and the other major treaty ports. This, and the rise of modern Western-style banking and insurance, river and coastal steam navigation, the beginnings of railways, and the improved access to overseas markets, created in the treaty ports not only China's fastest growing and largest cities but urbanism of a new kind, a model of the sorts of development which China had lacked and which it now needed in order to catch up with the West. The Westerners indeed saw the treaty ports as desirable models for China's development and expected that their example would transform the rest of the country. In fact they remained tiny and isolated islands of Western-style urbanism, while the rest of both urban and rural China remained largely unchanged. The treaty ports were resented as alien, and as symbols of China's humiliation and exploitation by foreigners. But they did contain the seeds of fundamental change, economic, institutional, and ideological, since they were the chief bases also for China's growing group of intellectuals seeking radical change as the cure for China's poverty, political weakness, and technological backwardness, and finding the example of the modern West as demonstrating in the treaty ports a source of new strength which China needed. The Nationalist Party of Sun Yat-sen was based primarily to begin with in Canton and Hong Kong, and the treaty ports remained its chief base of support and ideas. The Chinese Communist Party was founded in Shanghai in 1921, the major centre of political ferment and the leading centre of refuge for dissidents and revolutionaries.

Nearly all the Chinese who lived and worked in the treaty ports were profoundly affected by their exposure to Western ways and ideas, but these semi-colonial ports did not remake the rest of China in their image, as did happen to a greater extent in India as Calcutta, Bombay, Madras, and Karachi served as centres of change for much of the rest of their hinterlands. After the anti-Japanese war from 1937 to 1945 and the invaders' occupation of nearly all the major treaty ports, an enfeebled Nationalist government, returned to Nanking, was in 1949 overthrown by the Communists. In the course of their rise to power, primarily from rural bases, the Communists had become anti-urban and specifically anti-treaty port, as foreign excrescences as well as Nationalist party bases. Anti-urbanism had a long populist history in China, but it was newly galling to have all of China's biggest cities except for Peking dominated by foreigners and developing in a clearly Western mode, a very different reaction than that of most Indians to the rise of the colonial ports there, even though they had been actually founded by foreigners who became colonial masters (Murphey, 1977; Murphey, 1980: 25–33).

Korea and Japan, AD 313–1945

Korean urban-based civilization developed earlier than in Japan and spread eastward from there as the main transmission route from still earlier Chinese urban culture. After the fall of the Han dynasty, much of the Chinese colonization there survived, but in AD 313 the nascent Korean kingdom of Koguryo absorbed the Han base of Lo-lang and spread its control over the northern two-thirds of the peninsula. In 427 it moved its capital from the Yalu River valley near the Chinese border to the site of Lo-lang, and built a city there called Pyongyang, still the capital of modern North Korea, where magnificent stone royal and noble tombs decorated with Chinese-style paintings survive to this day. The two rival Korean kingdoms in the south, Paekche and Silla, had their own capital cities (including the ancestor of modern Kwangju), but we have little information about them. By the late seventh century Silla conquered all its rivals, unified Korea for the first time, and ruled it from its old capital in the extreme south-east near modern Kyongju, where Silla promoted contacts with T'ang China and the wholesale adoption of Chinese culture, including an astronomical observatory and a national university. Five subsidiary capitals were built to administer other regions of the country, but Kyongju seems to have been the only large city, very much in the Chinese style, although we have few details. Silla was conquered by a new state called Koryo (based on Koguryo—the name Korea derives from it) in 935, which built its capital at Kaesong (north of Seoul), also in the planned Chinese style and dominated by a great palace, with secondary capitals at Pyongyang, Seoul, and Kyongju.

Korean wealth and sophistication were heavily concentrated in the capitals, while the people and countryside elsewhere remained relatively poor and primitive by comparison with T'ang China. Koryo power was destroyed by the Mongol invasion of the thirteenth century, and, in the next century a new dynasty, the Yi, emerged which was to rule Korea until the Japanese take-over in 1910. The Yi capital was fixed at Seoul, whose roughly central location gave it an advantage, and the other old regional capitals were also rebuilt, but other towns and commercial centres were slow to develop, as Korean economic growth continued to lag. Nevertheless, by the end of the seventeenth century Seoul had a population of over 200 000, and a money economy was spreading, as at the same time more rapidly in Japan, but most of Korea remained rural and uncommercialized. By the eighteenth century there was however new urban growth at Pusan on the south-east coast, the chief port for trade with Japan across the Tsushima Strait. With the gradual spread of more commercialization and the beginnings of railways built by the Japanese, Seoul's population was over one million by 1940, after the Japanese take-over of the country as a formal colony in 1910, Pyongyong and Pusan about 400 000, Taegu in the south about 260 000, Inchon (the port of Seoul) over 200 000, and Kwangju about 100 000.[13]

Urbanization came relatively late to Japan, as part of the spread of Chinese culture via Korea. This had begun by at least the first century BC, but we have no evidence of anything resembling a city until the seventh century AD, when the Yamato state built a Chinese-style capital at Naniwa (now included in Osaka). A century later, stimulated by repeated Japanese missions to China to seek the sources of T'ang culture, a new and larger capital on the T'ang pattern was built at Nara, some thirty miles to the east, much influenced by Ch'ang An but without its walls and a fraction of its overall size. In 794 a new capital was built at Heian (modern Kyoto), also based on Ch'ang An but unwalled, slightly larger than Nara, and which was to serve as the official Japanese capital until 1868. Court culture at Heian became highly refined, concerned with aesthetic nuances, but most of the rest of the country remained not only rural but poor and relatively primitive. By the twelfth century, after regular contact with China had largely ceased, Chinese patterns in all things had been adapted to Japanese needs and preferences, and economic and cultural growth had spread more widely. We have little evidence of city-building beyond Heian (Kyoto) until a new political order in the late twelfth century headed by the Minamoto family established a new capital, or perhaps better, regional base, at Kamakura (now a Tokyo suburb), from which they imposed their military rule until 1339. They were in turn overthrown by the Ashikaga, which again used Kyoto as its base, until its power dissolved into civil war, ending with the first really effective unification of Japan under the Tokugawa Shogunate, from 1600 to 1867 (Storry, 1982; Sansom, 1946).

The Tokugawa Shoguns (officially the emperor's military lieutenants) retained the formal imperial capital at Kyoto but ruled effectively from their fortified base at Edo (modern Tokyo), keeping order elsewhere through a feudal-style structure of dependencies presided over by local lords, the *daimyo*. Tokugawa order stimulated a renewed growth of trade, and with it a merchant class and flourishing towns. These trends had begun under the Ashikaga in the fifteenth and sixteenth centuries when Japanese had become active in the trade with China and Korea (and also as pirates) as well as developing domestic commercial networks. By the fourteenth century money had largely replaced rice as the chief medium of exchange, and Japan soon began to export manufactured goods, such as swords and works of art. The decline of Ashikaga power tended to make many areas autonomous; those with a good commercial location, such as harbours, increasingly used them for trade. Osaka became a national centre for the rice trade, and the third largest city in the country, grouped around its fortified castle. Sakai, now part of Osaka, became a *de facto* free city, like an Italian city state or the Hanse towns of medieval Europe, walled and governed by its rich merchants and with its own army. Kyoto itself was part of this commercial growth, and by the end of the Ashikaga probably had a population close to 300 000, by far the largest city in the country, although court and administrative functionaries, servants, labourers, and artisans probably far outnumbered merchants.

The Tokugawa imposed their own central rule over the *daimyo* domain which had become increasingly autonomous under the declining Ashikaga, and conquered their castle-protected bases which had arisen in the course of the sixteenth century. Most were rebuilt as points of Tokugawa control in the centre of each domain, and the commercial towns which clustered around most of them brought under central supervision. These castle towns were the origin of nearly all of modern Japan's large- and middle-sized cities (Hall, 1955). Sakai and its few parallel merchant cities were overrun and brought under the control of Edo, the real Tokugawa capital, where the Shogun enlarged his own walled castle and palace on a grander scale. By or before 1800, Edo had reached a million in population and may have been the largest city in the world, rivalled only by Manchu Peking, and aided by its twin function as a major port, which also brought in food and luxury goods to feed the populace and to amuse the palace, the court, and the élite. Edo's size was further swollen by the Tokugawa rule that the regional *daimyo* must spend alternate years there and must leave their wives and heirs as hostages during their absence, which involved them in maintaining large establishments in the city, and in supporting great numbers of artisans to provide them as well as the Shogun's court with luxuries. Troops were another major part of the city's population. Osaka, the major commercial centre, and Kyoto, still the official imperial capital, approached half a million each by 1800. The larger castle towns of the *daimyo*, such as Nagoya or Kanazawa, were probably about 100 000, while the port city of Nagasaki, founded in the 1560s as the chief base for trade with China and with the early Westerners, may have been nearly as large.[14]

The Tokugawa closed Japan to foreign trade in 1635, fearing its disruptive influences, but a few Chinese merchants and two Dutch ships per year were permitted to trade at Nagasaki so that the city there continued to grow slowly as Japan's only 'window on the world'. Meanwhile in Edo, rich merchant families became increasingly prosperous, both in trade and in the supply of luxury goods to the aristocratic and *samurai* élite, and as their daughters married impoverished or indebted noblemen their unofficial status rose too. Edo supported a rich urban culture patronized by merchants and aristocrats alike, including the famous 'floating world' of the pleasure quarters outside the walls of the castle. There were closely similar developments in Osaka and Kyoto, and on a smaller scale at Nagoya, Kanazawa, and other lesser towns. Perhaps as many as half of male urbanites were literate to some degree by late Tokugawa, and urban culture was sophisticated, but despite the relatively rapid urbanization since 1600 townspeople were probably considerably below 10 per cent of the population even by the end of the Tokugawa in 1867, although Japan was soon to begin its fast modern growth.

The advent to power of a radically reformist government in the Meiji Restoration of 1868 opened an era of change even more rapid than experienced by Western countries since the Industrial Revolution. Change was sought in order to save Japan from a Western take-over, and it followed the Western path to modern strength: forced-draught industrialization, railways, full-scale commercialization, renewed and expanded foreign trade, all of which fed an accelerating urbanization. By 1895 12 per cent of the population of forty-two million lived in cities of 10 000 or more, and by the 1930s 45 per cent of a population then totalling nearly seventy million. By the eve of the Pacific war in 1940, Tokyo (its name since 1868, meaning 'eastern capital', the emperor having moved there from Kyoto with the Meiji Restoration) was fast overtaking London and New York as again the world's largest city, with a population of 6 779 000.[15] Yokohama rose to big city status as the new outport for Tokyo, from its origins as a treaty port opened to foreign trade and residence (as in China) in 1859, on the site of a small fishing village; the same was true for Kobe, Osaka's new outport, a treaty port as of 1863 and soon a major city in its own right. Osaka, Kyoto, and Nagoya also grew very rapidly, Osaka especially so as Japan's second city and commercial hub, in some ways a Japanese parallel to Chicago though with far older origins. Tokyo remained the chief centre of Westernization and its new styles, its size swollen by its augmented national government functions and by its role as the headquarters for virtually all of the large new Japanese corporations and industrial combines. Tokyo also led the nation in educational institutions. The Pacific War devastated nearly all Japan's cities, especially Tokyo, but their recovery after about 1947 was spectacular. What has been called the Japanese economic miracle lifted Japan especially after the 1950s into the forefront of modern world success, while at the same time bringing the urbanization level to one of the world's highest, over 80 per cent, with Tokyo a giant conurbation of some twenty million by the late 1980s. Japan clearly does not belong in a 'Third World' category by any measure.

South-East Asia, AD 200–1945

South-East Asia is a somewhat loosely integrated and fragmented area lying between India, China, and the Pacific cultures of Melanesia and Micronesia. It includes Burma, Thailand, Malaya, Laos, Cambodia, and Vietnam on the mainland of the Indo-Chinese peninsula, plus the island archipelagos of what are now Indonesia and the Philippines. It has been profoundly impacted throughout its recorded history by influences from India and China, while many of its people appear to have migrated originally from south-west China and the Sino-Tibetan borderlands. The origins of the Malays, their culture and language, which in a broad sense include the inhabitants of both Indonesia and the Philippines, are obscured by time, but they almost certainly came from the Asian mainland. Agriculture and associated village settlements are as old in South-East Asia as anywhere in the world, perhaps older, but we have no evidence for genuine cities before the spread of Indian civilization, including Hinduism, Buddhism, and Indian political and art forms, into the area after the first century AD. Northern Vietnam and the city of Hanoi are an exception, since Hanoi was the capital of the Vietnamese–Cantonese kingdom of Nan Yueh, or Nam Viet, before the Ch'in conquest, and the region's later absorption into the Han empire, during which Hanoi is best regarded as a Chinese city in the traditional mould.

By the second century AD Chinese records speak of central and southern Vietnam as being controlled by the Indianized states of Champa, with its capital near Tourane, and Funan, based in the Mekong delta and with its capital there near the sea. The rulers claimed descent from Indian Brahmin princes, but the Chinese accounts, all we have for this period, say little more about these capital cities except that they were walled. Funan depended importantly on trade, and by the fifth century may have controlled lesser states around the Gulf of Siam to the westward. The Khmers had been moving into the Indo-Chinese peninsula, and by the sixth century had built a capital city in the middle Mekong valley, gradually absorbing Funan. About the same time, another group of migrants from the north, the Mon, established several small kingdoms to the westward, including a city at Nakom Patom near modern Bangkok, and another at Prome in the Irrawaddy delta of Burma. Mon rule was displaced by the later southward movement of the Thais into modern Thailand, and of the Burmans into the Irrawaddy valley, but the Mon and Khmer states were thoroughly Indianized and passed this heritage on to their successors, including the religion of Buddhism (Coedes, 1968; Chandler, 1983).

By the fifth century AD, several small Indianized states had also arisen on the coasts of Sumatra, Java, and Borneo in what is now Indonesia, dependent largely on sea trade, with urban control bases at strategic points. The Sumatran kingdom of Palembang rose to prominence about the seventh century, where it could control trade through the Straits of Malacca, and built near modern Palembang what is described as 'a city on rafts'. From this base there developed the great maritime empire of Sri Vijaya, which from the eighth century until the eleventh century also controlled most of Malaya and West Java. In central and east Java, the Hindu-Buddhist kingdom of Sailendra presided over a great cultural renaissance and built monumental temples and Buddhist stupas, including the immense stupa of Borobodur associated with a city there, begun about 772 and still standing, though the city remains have largely vanished.

On the mainland, the Khmer kingdom began in the ninth century to build the first city of Angkor, in the lower Mekong valley near the lake of Tonle Sap, which may have been larger than ancient Rome, at least in terms of the area covered (Coedes, 1968; Fisher, 1971; Hall, 1968), but it was more a religious and symbolic centre of temples than a Western-style city, a walled model or replica of the cosmos and the sacred Mount Meru. Khmer engineers provided it with a water supply from the river and lake through an intricate system of canals and storage reservoirs. The city was restored and rebuilt after an invasion from the east in 1177 and renamed Angkor Thom, now served by a network of roads radiating from it and with a subsidiary cluster of temples nearby at Angkor Wat, the ruins of which, invaded by jungle, still stand, the largest religious building in the world. Angkor (Angkor Thom) remained the capital of the Khmer empire, which came to include peninsular Burma and southern Thailand as well as the whole Mekong valley, from the Chinese borders to the border of Sri Vijaya domains in Malaya, until 1432. Meanwhile, the Vietnamese were moving south from their northern base at Hanoi, and eventually fixed their capital at Hue, roughly midway down the coast of modern Vietnam, where the city was centred around a walled citadel, with the market and merchants' houses on the outskirts. The Burmans continued their migrations southward at the expense of the Mon, and in the early ninth century built an impressive capital at Pagan (south of Mandalay). A Shan dynasty replaced it in 1287, with its capital at Ava (between Pagan and Mandalay) until 1531, while the Mon at Pegu and the Burmans at Toungoo (both in lower Burma) maintained their separate capitals. In the sixteenth century, the Toungoo kingdom conquered both Ava and Pegu, and used both as its new capitals. By 1350, the southward-moving Thais, originally from south-west China, had founded their capital at Ayuthia, just to the north of modern Bangkok, and from there spread their control over much of the former Khmer domains, capturing Angkor itself in 1431.

In the archipelago, Sri Vijaya's power slowly ebbed, it lost control over several of its strategic ports, and yielded by the fourteenth century to the new Java-based but also maritime empire of Madjapahit, which included the former Sri Vijaya holdings in Malaya. There the walled city of Malacca rose to prominence in the fifteenth century as the best placed control point for east–west trade where the Straits narrow. Although it probably never reached or exceeded 10 000 before 1500, Malacca became the chief entrepôt for insular and peninsular South-East Asia and also controlled much of the trade by sea between India and China, with the Sultan's palace inside the walls and the houses of rich Chinese and other merchants outside across a small river with a bridge of boats. It was an obvious magnet for the early European adventurers of the Age of Exploration, and was seized by the Portuguese in 1511, to cement their control of the sea lanes. Ayuthia, the Thai capital, was sacked by Burmese forces in 1767, but the Thais founded a new capital at Bangkok in 1782, to become one of South-East Asia's big cities and major ports in the course of the nineteenth century. Ayuthia, and early Bangkok, were built in concentric circles, the innermost with the royal palaces, outer ones containing aristocrats' houses and their servants, and on the edges the Chinese mercantile quarter (Coedes, 1968; Fisher, 1971; Steinberg, 1971).

Portuguese, Dutch, and English traders sought footholds first in existing coastal trade bases, including the lower Burma ports of Syriam and Moulmein, the southern Thai port of Pattani, the port city of Acheh at the northern tip of Sumatra, the southern Vietnamese port of Saigon, and a widespread group of ports in the spice trade scattered through the Indonesian archipelago as far east as Amboina and Ternate in the Moluccas, or Spice Islands. There was Chinese trade with the Philippines probably from Han times, but the islands as a whole were culturally and economically undeveloped before the Spanish conquest of the sixteenth century, with its administrative and commercial capital at Manila, the site of a small Chinese merchant settlement for several previous centuries, where the Spanish built a walled city, Intramuros. It served as the centre for increased trade with China and with Spanish America as well as within the islands. A few smaller ports on the other major islands also grew under Spanish control, but the main commercial and urban development of the Philippines came under American control only after 1899.

The Portuguese commercial empire in Asia was based on control of the sea lanes rather than on territorial bases, but they were progressively replaced by the British in India, Burma, and Malaya, by the French in Vietnam, Cambodia, and Laos, and by the Dutch in what is now Indonesia (Boxer, 1961, 1965; Parkinson, 1937). In the course of this full colonialism, many of the old trading ports were eclipsed, especially in the age of the steamship with its greater cargo capacity and the concentration of trade in a few particularly advantageous locations. Thailand alone retained its independence, but Bangkok was transformed in the nineteenth century to become one of South-East Asia's major port cities as the export trade to wider markets prospered under Western management, which imposed on the Thais the same set of 'unequal treaties' for foreign privileges as in China and Japan. Rangoon, its parallel, similarly became a major port and city (from a small group of temples) only under British colonial control from 1852, serving the whole of the productive Irrawaddy valley as its outlet and commercial as well as administrative centre, with a secondary base at Mandalay, accessible by steam navigation on the Irrawaddy. In 1872 at the first census effort, Rangoon fell just short of 100 000 and Mandalay was about double that; by 1901, Rangoon was nearly 250 000, while Mandalay had actually shrunk. In Malaya, the British East India Company founded a new base at Penang on the north-west coast in 1786, and when this proved inadequate to tap the trade with South-East Asia and China (Malaya itself then being still undeveloped commercially), moved their chief Malay base to a new foundation at Singapore in 1819. Singapore soon supplanted Malacca as the key to the Straits passage (the Dutch had taken it from the Portuguese in 1641, and the British from the Dutch in 1795, dates which make convenient indicators of the rise and decline of different Western powers in South-East Asia since the Portuguese capture of Malacca in 1511), and it soon boomed as the major entrepôt for South-East Asian trade as a whole, aided by its large and excellent harbour but with a dominantly Chinese population.

The French took southern Vietnam in 1862, and soon built their colonial capital at Saigon into another new port city, eclipsing older Vietnamese cities and ports but with a secondary colonial and commercial centre at Hanoi and its French-built port of Haiphong. All of Burma and Malaya, Vietnam, Cambodia, and Laos came respectively under British and French control by the 1880s. The Dutch colonial capital for their Indonesian territories at Batavia (now Jakarta) in West Java, founded in 1619, became, as their control spread, the chief city and trade centre for insular South-East Asia in the course of the nineteenth century, with another newly founded Dutch base and port at Surabaya for east Javanese trade (Murphey, 1969).

By the twentieth century, urbanism in South-East Asia was dominated by these foreign-founded or foreign-dominated coastal ports, by then augmented by the British colonial capital of Kuala Lumpur, a new largely Chinese boom town in the tin and rubber belt of the west coast and served by a new partly artificial harbour at Port Swettenham, and by the beginnings of modern city growth at Iloilo, Cebu, and Davao in the Philippines. The colonial ports had become by far the largest cities in each country, in some cases the only real city, while their traditional inland capitals decayed or fell into ruins. The colonial ports were located to best serve the new plantation and cash crop sector which now dominated the commercial scene in South-East Asia: rice for export in lower Burma and teak and oil farther up the Irrawaddy valley, export rice and rubber from lower Thailand, tin and rubber from west coastal Malaya, rice and rubber from the Mekong delta of southern Vietnam, and a great variety of tropical crops from Indonesia and the Philippines, plus oil and tin (see Figure 2.1) but with plantation crops concentrated in east coastal Sumatra, Java, Luzon (where Manila lies), and the coastal areas of other main Philippine islands. Trade goods were thus easily assembled by sea or river at the major port, for export to the West, a 'drainage economy', aided by some railway building, which has parallels with late colonial India. By the end of colonialism in the ashes of the Second World War, Jakarta, Manila, Saigon, Bangkok, Singapore, Saigon, and Surabaya had all reached or passed the million mark in population, and nearly all the other large cities such as Hanoi-Haiphong, Rangoon, Semarang (in east Java), or Medan (in the north of Sumatra) were ports of dominantly or entirely foreign origin and character, though as in the rest of monsoon Asia's cities with only small numbers of foreign residents. The new coastal ports have also become the major national cultural centres in each country, and each inevitably became the political capital at independence.

Only Japan could claim that the foreign hand had never been dominant in Tokyo, or in any of its then three biggest cities (Osaka and Nagoya after Tokyo). Seoul was Korean in origin, and a major place under indigenous control, but from 1910 Japanese property. Peking escaped treaty port status, but lost its capital functions to Nanking from 1927 to 1949 and was outdistanced by treaty port Shanghai, as it still is, while nearly all of China's other big cities rose to modern prominence as foreign-dominated treaty ports. In India, Delhi could legitimately claim ancient status as a capital but was overshadowed by totally foreign-founded Calcutta, Bombay, and Madras, the colonial ports, which with Karachi handled over 90 per cent of India's foreign trade at independence. A similar proportion of China's foreign trade by then passed through the major treaty ports of Shanghai, Tientsin, Dairen, Hankou, and Canton–Hong Kong, which like their Indian parallels also dominated the country's cultural and intellectual life.

All of these colonial or semi-colonial cities which together virtually monopolized the urban scene were both hybrid Asian-Western and cosmopolitan centres, attracting immigrants from their own and other countries. All lines of internal communication focused on them, as well as the sea routes, and it was from them that railways were built. These cities were the breeding grounds of a new Asia which in every country, including Japan, showed strong Western influences. Their cosmopolitan character was, however, also the result of their mixture of peoples in which actual Westerners remained in a small minority: Indians (including Parsee merchants) in Rangoon, Singapore, and the China coast ports, Gujaratis in Calcutta, Cantonese in Shanghai, and even larger number of Chinese, primarily from the south-east coast, in all the port cities of commercially booming South-East Asia, where they became the major entrepreneurial group, usually in their own quarters of each city—and the numerical majority of the populations of Singapore and Bangkok.

To begin with, Western traders had competed with Asian and Arab merchants at existing ports, but over the three centuries after da Gama, the superiority of their ships and their naval guns won them first control over the sea lanes, and then the beginnings of territorial holdings, starting with coastal footholds in the ports they founded and fortified, often on previously unoccupied sites ignored by the indigenous powers.

Nineteenth-century industrialization at home gave the Westerners enormously greater power, over Asian states which were in any case then in decline, and at the same time enormous new appetites for Asian goods, including industrial raw materials. It was access, to commercial hinterlands internally, and by sea to markets abroad, which determined the rise of all the colonial or semi-colonial ports. Where their harbours were inadequate for larger shipping after about 1870, as at Calcutta, Madras, Karachi, Colombo, Bangkok, Saigon, Batavia, Shanghai, Tientsin, and to a lesser degree all the others except for Bombay, Singapore, Manila, Hong Kong, Dairen, Tsingtao, Yokohama, and Kobe, they were dredged and deepened or provided with artificial harbours. A great many of them were deltaic or grew around a shallow lagoon, and presented formidable problems for shipping, including offshore bars which still require constant dredging. But river valleys, deltas, and coastal plains are commonly productive places, and the valleys themselves provide easy access inland. Most of the pre-European ports had better harbours, at least for the ships of the time, but a more widely commercialized age bypassed them in favour of places which could accommodate and fill larger ships and at the same time could serve a wider and more productive hinterland made tributary to them by river valleys and/or railways (Murphey, 1969).

One final set of problems common to nearly all of the colonial port cities was associated with the disadvantages of their local sites. Extollers of the benefits of the Western presence in Asia and of the model which Western action presented to the 'backward' East were fond of pointing to the cities which arose under Western management along the swampy foreshores of tidal rivers or on jungly or rocky coasts as the very embodiment of 'progress' Western style. But despite the filling of swampy ground and the bunding (embankment) of streams, flooding has remained a chronic problem at all but a few, ground water levels are high, drainage difficult, and building an often awkward and expensive matter. Most of these ports acquired a dreadful reputation for morbidity, especially in the early days when the death rate among Europeans was indeed shockingly high, attributed to 'fevers' (probably including lethal forms of malaria) unknown to outsiders. Parkinson (1937: 71–4) estimates, on a variety of evidence, that less than a quarter of those who went out from Britain to the East India Company's bases in India, Ceylon, South-East Asia, and China between 1765 and 1820 survived to return—and it was consistently their aim to return as quickly as possible after 'shaking the pagoda tree'—i.e. making a fast fortune. The French and Dutch records were little better. No doubt it was in part this situation which explains the attitude which Adam Smith (1776: 605) found hard to accept on the part of East India Company servants:

> It is a very singular government in which every member of the administration wishes to get out of the country . . . as soon as he can, and to whose interest, the day after he has left it and carried his whole fortune with him, it is perfectly indifferent, though the whole country was swallowed up by an earthquake.

The same attitudes were to remain dominant among the foreigners in the treaty ports of China, Japan, and Thailand, and among the British, Dutch, French, and Americans in South-East Asia.

It may seem surprising that Westerners accomplished so much change in Asia under these circumstances. But the commercial motives which primarily drove them led to activities with powerful catalytic effects, far beyond their port settlements and despite the brief Asian careers of most individuals. The planting of tiny islands of modern commercial development in the new ports on the fringes of each country eventually set in motion changes which spread over much of Monsoon Asia as ripples cover a pond from the casting of a single stone. In the phrase of Martin (1951: 37):

> It has been observed that of all the European nations who have planted distant settlements, the English have invariably shown the least regard for the proper selection of localities for the sites of their colonial cities; and I think this must in general be ascribed to the commercial spirit taking the lead . . . the embouchures of great rivers were the first object of desire.

This puts the point neatly, but is hardly fair in suggesting that French Saigon or Haiphong, Dutch Batavia or Medan, or Spanish-American Manila were any better. Low-lying Batavia, for example, with its creeks and artificial canals clogged with refuse and breeding disease and its vectors, was as notorious as Calcutta, and the others were much the same. Hong Kong, though with a bad reputation for health in its early years, was in general a fortuitous exception, on a hilly island off the coast.

The water which surrounded these port cities, often supplemented by dug canals, was in fact a symbol of their origin and role as foreign beach-heads, links with the wider world by sea which sustained them and which also defended them. It was naval guns which gave these settlements their early security. As the first British consul at the water-surrounded foreign settlement at Shanghai put it in 1843: 'By our ships our power can be seen, and if necessary felt.'[16] It was indeed by their ships that the power of the Westerners could be seen, economic as well as military, symbols of the industrial, commercial, and technological revolution which had made the West the arbiter of so much of modern Asia's development. Right up to the final success of the Communist revolution in China in 1949, foreign warships were anchored prominently off the Bund of Shanghai's waterfront and in similar centre-stage positions in the other treaty ports, landing sailors or marines when necessary to 'keep order' and supplemented by gunboats on the inland rivers. Such a display of course helped to heighten the Chinese feeling that these cities, now the largest in the country and its leaders in most respects, were alien intrusions, to be rejected and wiped away on the clean slate of revolution.

The influence of the colonial and semi-colonial port cities on Asian welfare has long been debated, and it is understandable that many Asians have seen them as a drain of Asian wealth to Western profit. The cities were clearly enough funnels through which primary production and treasure were exported to the West and manufactured goods brought in, to the frequent detriment of domestic producers. But, in part through this very function, these cities also acted as the spearheads and catalysts of Asia's modernization. The demonstration and backwash effects from even their crassest military, commercial, and industrial operations, the 'progress' they were seen to represent, and their investments in transport, education, publication, and municipal facilities generated the changes which are still remaking the whole of Monsoon Asia. The colonial ports have rightly been reclaimed by a renascent Asia, but only after they had first brought Asia into the wider modern world.

Notes

1. For a fuller account of the cities of the Indus civilization, see Possehl (1979).
2. For more detail, see K. C. Chang (1987).
3. Rudyard Kipling, 'Song of the Cities' (1915: 53). On colonial cities in general, see Fox (1970).
4. Rudyard Kipling, 'Song of the Cities' (1915: 54).

5. See Edwardes (1902); Mehta (1954); Morris (1965); Dandekar (1986); and Wadia (1957).
6. On the growth of Madras, see Barlow (1921) and Lewandowski (1980).
7. On Karachi, see Feldman (1960), on Ahmadabad, Gillion (1968).
8. On the growth of Colombo, see Murphey (forthcoming).
9. Averages are as usual misleading, especially for a country the size of China over a 2000-year period, quite apart from the grossly approximate nature of the population data. The average *hsien* in Han times was probably closer to 50 000 and by late Ch'ing, with a national population three or four times greater, about 300 000.
10. For more general studies of the traditional Chinese city, see Balazs (1954) and Eberhard (1956).
11. Ibn Battuta was presumably guilty of gross exaggeration, at least in travel time. On Sung Hangchow, see Gernet (1965), on Nanking, Mote (1977).
12. On Shanghai's population and its more general growth, see Murphey (1953), on the treaty ports as a whole, Murphey (1977).
13. The most convenient brief account of Korean history is in Fairbank *et al.* (1989: 277–323, 609–18, and 907–24), but see also Henthorn (1971) and Lee (1985). The city population figures given here are from McCune (1950: 327).
14. The figures are from Fairbank *et al.* (1989: 414 ff.).
15. The figures are from Fairbank *et al.* (1989: 664).
16. Quoted in Lanning (1923: 134).

References

Balazs, E. (1954), 'Les villes Chinoises', *Recueils de la Société Jean Bodin*, 6: 225–40.

Barlow, G. (1921), *The Story of Madras* (Madras: G. Natesan).

Basham, A. L. (1967), *The Wonder that was India*, 3rd edn. (New York: Macmillan).

Boxer, C. R. (1961), *Four Centuries of Portuguese Expansion* (Johannesburg: Witwatersrand University Press).

—— (1965), *The Dutch Seaborne Empire* (London: Knopf).

Busteed, H. E. (1908), *Echoes from Old Calcutta* (London: W. Thacker).

Chandler, D. P. (1983), *A History of Cambodia* (Boulder, Colo.: Westview Press).

Chang, K. C. (1987), *The Archeology of Ancient China*, 4th edn. (New Haven: Yale University Press).

Chang, S. D. (1961), 'Some Aspects of the Urban Geography of the Chinese Hsien Capitals', *Annals of the Association of American Geographers*, 23–45.

Coedes, G. (1968), *The Indianized States of Southeast Asia*, ed. W. Vella (Honolulu: East–West Center Press).

Cotton, H. E. (1907), *Calcutta Old and New* (Calcutta: Spink).

Dandekar, H. C. (1986), *Men to Bombay, Women at Home: Urban Influence on Sugao Village, Deccan Maharashtra, India, 1942–1982*, Michigan Papers on South and Southeast Asia (Ann Arbor: Center for South and Southeast Asian Studies, University of Michigan).

Eberhard, W. (1956), 'Data on the Structure of the Chinese City', *Economic Development and Cultural Change*, 4: 253–68.

Edwardes, S. M. (1902), *The Rise of Bombay* (Bombay: The Times of India Press).

Fairbank, J. K., Reischauer, E. O., and Craig, A. (1989), *East Asia: Tradition and Transformation* (Boston: Houghton Mifflin).

Feldman, H. (1960), *Karachi Through a Hundred Years* (Karachi: Oxford University Press).

Fisher, C. A. (1971), *Southeast Asia* (London: Methuen).

Fox, R. G. (1970) (ed.), *Urban India* (Durham, NC: Duke University Press).

Gernet, J. (1965), *Daily Life in China on the Eve of the Mongol Invasion*, trans. H. M. Wright (London: Macmillan).

Gillion, K. L. (1968), *Ahmedabad* (Berkeley: University of California Press).

Hall, D. G. E. (1968), *A History of Southeast Asia* (London: Macmillan).

Hall, J. W. (1955), 'The Castle Town and Japan's Modern Urbanization', *Far Eastern Quarterly*, 15: 37–56.

Hambly, G. (1968), *Cities of Mughal India* (London: Elek Books).

—— (1987), *The Cambridge Economic History of India* (Cambridge University Press).

Henthorn, W. E. (1971), *A History of Korea* (Glencoe, Ill.: Free Press).

Kipling, R. (1915), *The Five Nations and the Seven Seas* (New York: Doubleday).

Lanning, G. (1923), *A History of Shanghai* (Shanghai: Kelly and Walsh).

Lee, K. B. (1985), *A New History of Korea* (Cambridge, Mass.: Harvard University Press).

Lewandowski, S. (1980), *Migration and Ethnicity in Urban India: Kerala Migrants in Madras, 1870–1970* (New Delhi: Manohar).

McCune, G. M. (1950), *Korea Today* (Cambridge, Mass.: Harvard University Press).

Martin, J. R. (1951), 'Notes on the medical topography of Calcutta', in *Census of India*, app. XVI, vol. i, part ii. 37–53.

Mehta, S. D. (1954), *The Cotton Mills of India, 1854–1954* (Bombay: Textile Association).

Morris, M. D. (1965), *The Emergence of an Industrial Labor-force* (Berkeley: University of California Press).

Mote, F. W. (1977), 'The Transformation of Nanking', in G. W. Skinner (ed.), *The City in Late Imperial China* (Stanford, Calif.: Stanford University Press).

Murphey, R. (1953), *Shanghai: Key to Modern China* (Cambridge, Mass.: Harvard University Press).

—— (1957), 'The City in the Swamp: Aspects of the Site and early Growth of Calcutta', *Geographical Journal*, 130: 241–56.

—— (1964), 'The Ruin of Ancient Ceylon', *Journal of Asian Studies*, 16: 181–200.

—— (1969), 'Traditionalism and Colonialism: Changing Urban Roles in East Asia', *Journal of Asian Studies*, 29: 67–84.

—— (1977), *The Outsiders: The Western Experience in India and China* (Ann Arbor: University of Michigan Press).

—— (1980), *The Fading of the Maoist Vision: City and Country in China's Development* (New York: Methuen).

—— (forthcoming), 'Port Cities and the Transformation of Asia: Colombo as Prototype', in D. Basu (ed.), *Colonial Port Cities in Asia.*

Parkinson, C. N. (1937), *Trade in the Eastern Seas* (Cambridge University Press).

Possehl, G. L. (1979) (ed.), *Ancient Cities of the Indus* (Durham, NC: Carolina Academic Press).

Saksena, B. P. (1962), *History of Shah Jahan of Delhi* (Allahabad: The Indian Press).

Sansom, G. (1946), *Japan: A Short Cultural History* (London: Appleton-Century).

Singer, M. (1971), 'Beyond tradition,' *Comparative Studies*, 13: 160–95.

Sinha, K. N. (1961–2), *The Economic History of Bengal*, 2 vols. (Calcutta: Verry).

Smith, A. (1776), *An Enquiry into the Nature and Causes of the Wealth of Nations* (London: Strahan and Cadell).

Steinberg, D. J. (1971) (ed.), *In Search of Southeast Asia* (New York: Praeger).

Storry, R. (1982), *A History of Modern Japan* (London: Penguin).

Wadia, A. R. (1957), *The Bombay Dockyard* (Bombay: Privately printed).

Wolpert, S. (1989), *A New History of India*, 3rd edn. (New York: Oxford University Press).

Wright, A. (1965), 'Ch'ang An', *Journal of Asian Studies*, 24: 167–79.

Yule, H. (1886) (ed.), *Cathay and the Way Thither* (London: Hakluyt Society).

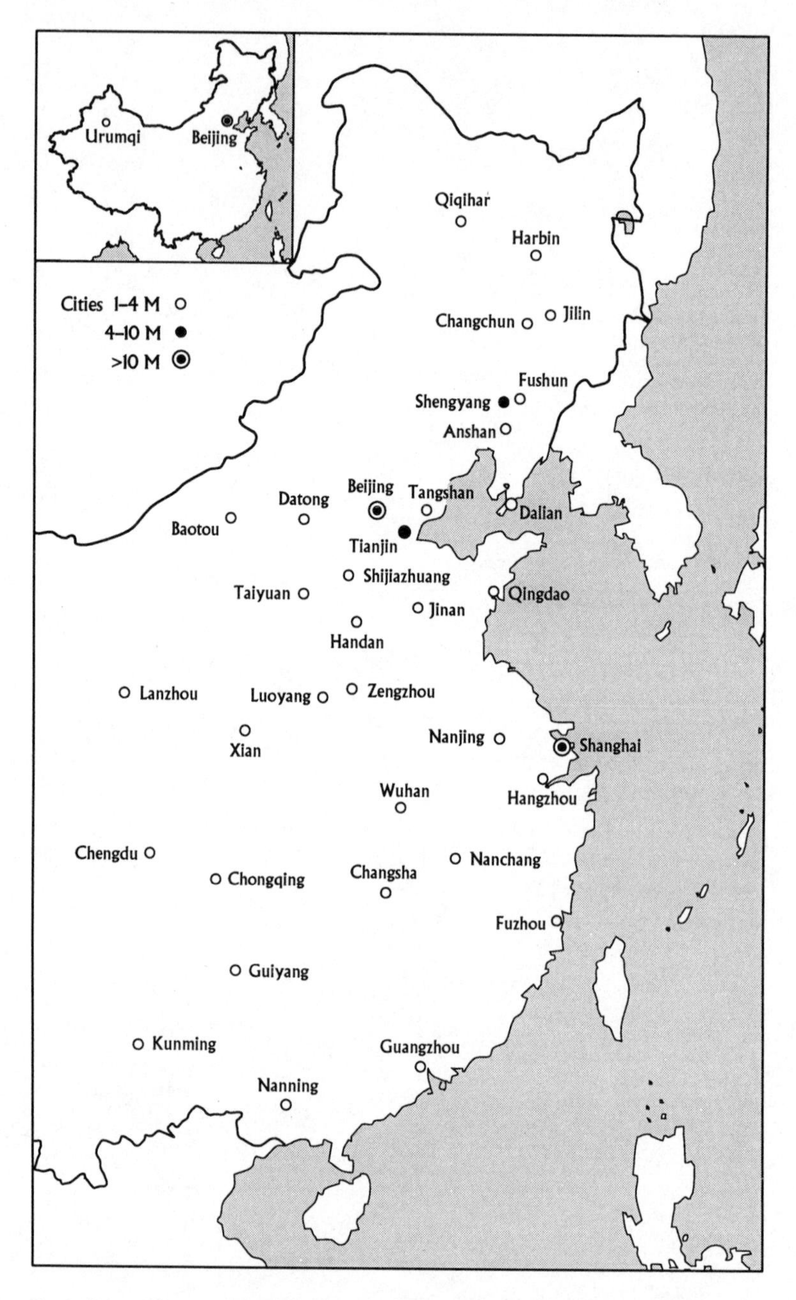

FIG. 3.1. Urban Agglomerations in China with more than 1 Million Inhabitants in 1990

3

Urbanization in China: Reassessing an Evolving Model

XIANGMING CHEN AND WILLIAM L. PARISH

A CENTRAL question about urbanization processes in Third World countries is whether they will converge upon a general path travelled by Western industrialized countries or become increasingly divergent due to different historical and national conditions. If the answer is the latter, as suggested by recent cross-national evidence (Findley, 1993; Gilbert, 1992; Smith and London, 1990), two additional questions need to be addressed. First, how is the distinctive pattern of urbanization in each developing country shaped by a combination of historical, political, economic, and geographical factors unique to that country? Second, what does the change in the distinctive urbanization experience in a developing country reveal about the dynamics underlying Third World urbanization in general? We address these questions by focusing on Chinese urban development. China is currently in a period of transition that questions many older patterns. What was formerly distinctive about China may disappear as China adds more market elements to its economy. In this chapter, we reassess the Chinese model by focusing on the changing pattern of urbanization in China during the 1980s and early 1990s.

Urban Population and Labour-Force

Though problems with varying definitions of urban are common around the world, these problems are particularly acute in China, where frequently changing administrative definitions make it difficult to attach precise meanings to urban trends. Nevertheless, based on several careful

Chen's research and writing for this chapter was supported in part by small grants from the American Sociological Association, the Pacific Cultural Foundation on Taiwan, and a Chinese studies postdoctoral fellowship from the American Council of Learned Societies and the Social Science Research Council during 1993–4.

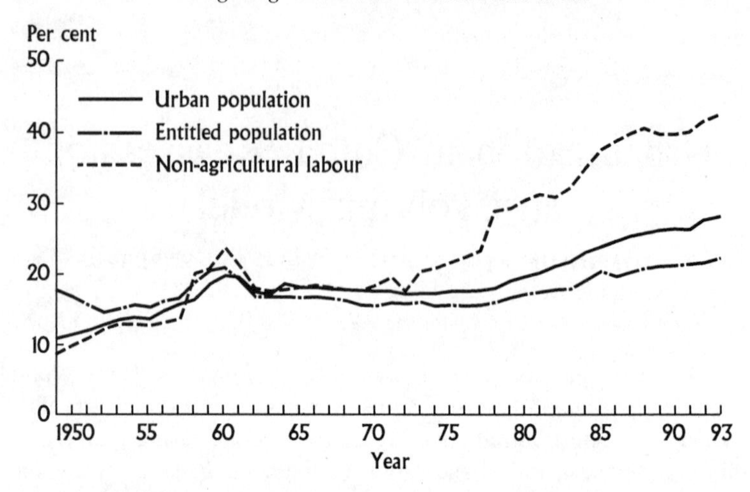

FIG. 3.2 China's Urban and Labour-Force Trends, 1949–1993
Sources: CSSB (1990*a*: 570–1; 1992*a*: 77, 101; 1993: 81, 677; 1994: 296); Parish (1990: 7).

commentaries on changing Chinese definitions and on new adjustments within China's own statistical series, it is possible to provide a fairly clear account of urbanization (Chan and Xu, 1985; Chen, 1991; Martin, 1992; Orleans, 1982; Parish, 1990).

Three indices of urbanization and movement away from agriculture reflect past trends (Figure 3.2). Our data for 'urban population' is, by and large, that population living in contiguously populated urban neighbourhoods. Because Chinese cities and towns are overbounded, often including much of the surrounding rural hinterland and even administratively subordinate rural counties, one must restrict the data to urban neighbourhoods. This restriction is particularly important in the 1980s, when including the broad administrative definition of urban would lead to a figure twice that in Figure 3.2. The 28.1 per cent urban for 1993 is closer to reality, though it may be understated by 1 to 2 percentage points because of a large informal, floating population in many major cities that does not appear in official statistics. Including that informal urban population raises the 1993 per cent urban to slightly over 30 per cent of the total population. And it is the 30 per cent urban that we will assume is the best approximation for our purposes.

Our second indicator of urbanization trends is the percentage of the entitled population. Starting in the early 1950s, everyone got a household registration which labelled them and their children as either entitled (non-agricultural) or non-entitled (agricultural). The entitled were first in line for highly subsidized housing, food, and state-sector jobs with many fringe benefits, including health care, disability, and pension payments. The non-entitled enjoyed almost none of these benefits. Thus formal entry into the entitled, urban sector was extremely attractive, leading on average to consumption levels 2.5 to 3 times higher than in the countryside. While attractive to individuals, urban entitlements were very unattractive to state planners, who by the middle 1950s found highly subsidized urban living standards increasingly difficult to support. As a result, by the early 1960s, central planners imposed a bamboo curtain between city and countryside, effectively walling off urban entitlement from the great majority of the rural population and their children. Many who had slipped illegally into cities in the late 1950s were driven back into the countryside in the early 1960s. With a relaxation of the rationing of food and other items in the 1980s, the attempt to exclude people from towns and cities weakened. But even as the numbers in towns and cities began to grow once again, full entitlement through urban household registration was rarely given. This increasingly promoted a two-class system. While perhaps as much as 30 per cent of the population was in towns and cities by 1993, including the 'floating' population, only about two-thirds of this number were entitled to full urban benefits.[1]

A third index of trends related to urbanization is the percentage of the labour-force outside agriculture. For many years, urbanization and the non-agricultural labour-force moved in close parallel (Figure 3.2). However, by the late 1970s and early 1980s the two trends began to diverge. One reason for this was the government policy that tried to limit urban subsidies by promoting rural industry in place of urban industry. When new agricultural reforms freed farmers from field agriculture in the late 1970s and early 1980s, many of them flocked not to cities but to new enterprises in their home villages and nearby market towns to which they could commute daily. Thus, the percentage working outside agriculture rose to 43 per cent by 1993, even while the percentage in cities and towns remained significantly lower.[2]

The overall conclusion that one draws from both urbanization and non-agricultural labour-force trends is that China has approximated the urbanization pattern of several other large, low-income, Asian nations. It is closest to Indonesia, which also has somewhat over 40 per cent of its labour-force outside agriculture and about 30 per cent of its population resident in towns and cities. Other large countries with similar levels of urbanization include India at 27 per cent and Pakistan at 32 per cent urban (World Bank, 1991). During 1985–90, China's urban population grew by 3.8 per cent per year, joining the rank of twenty-one Third World countries that averaged over 3 per cent, including the countries in sub-Saharan Africa, Bangladesh, and Indonesia (Findley, 1993: 6).

The trend data also shed light on the Chinese model compared to urbanization in state socialist societies and Third World countries. Murray and Szelényi (1984) note three urbanization patterns in existing socialist states, ranked according to the intensity of their anti-urban bias. De-urbanization, referring to a sharp decrease in the urban population through forced administrative mechanisms, fits the situation in South Vietnam and Kampuchea in the wake of their socialist revolutions. Under-urbanization features a much faster expansion of industrial employment than the growth of the urban population, such as the Stalinist period in the former Soviet Union. Zero urban growth, which emphasizes rural–urban balance, characterizes the 1960–79 period in China and Cuba up to 1965. China's slow pace of urbanization in the 1960s and 1970s resulted from 'anti-urban' policies, including strict control on rural–urban migration and suppression of general urban consumption (Chan, 1992), except the privileged consumption of high-ranking party and state officials. Anti-urbanism in China had its origins, one suspects, not only in the high costs of providing highly subsidized urban living standards but also traditional ideas about urban-based merchants having lower social status than peasants and the rural origin of the Chinese Communist Party (Kirkby, 1985: 6–7). These are the policies that are often identified with the 'uniqueness' of China's earlier model of urban development, especially in contrast to the seeming over-urbanization of many Third World countries, where urban population growth outstrips that of urban industrial employment even though some of the surplus urban labour is absorbed by the service sector.

Does the Chinese experience offer any lesson about the possibility of industrialization without urbanization? China's emphasis in the 1960s and 1970s was not just in restraining urban population growth but also in turning cities from being centres of consumption to centres of production. The emphasis was on maximum output growth through high rates of saving, on heavy industry as opposed to light industry, and on discriminatory policies toward agriculture. Reviewing this period, Chan (1992) concludes that China was successful in maximizing the industrial output/labour and industrial output/urban population ratios by minimizing the socioeconomic consequences of urbanization, such as the growth of service and non-agricultural employment, and of the non-working population. Initial increases in the industrial output/labour ratio were achieved from a very low beginning level of productivity and a high level of industrial investment. In this extensive growth pattern, copied from the Stalinist period in the Soviet Union, China both sharply increased capital intensity in industrial production and increased the proportion of urbanites at work—as a result of lower birth rates, fewer dependent children, and more women in the urban labour-force. With these patterns, China's industrialization outpaced urbanization for a time. In the 1960s and 1970s, while industrial output continued to surge at almost 10 per cent per year and accounted for more and more of total output, the urban population grew only very slowly and as a percentage of total population remained almost constant.

From a comparative-historical perspective, we conclude that China has experienced three distinct periods of urbanization. In the initial 1950s period there was rapid urbanization, particularly in the late 1950s when the failure of the massive commune experiments and the sharp fall in agricultural production drove many people from the countryside into towns and cities. China's rapid urbanization in the 1950s was not dissimilar to Third World countries at the initial stage of industrialization. The second period of the 1960s through 1970s was one of urban containment, with constant or even falling levels of urbanization. Given the 1950s, the second period falls somewhere between under-urbanization and zero urban growth (see Cell, 1980) with regard to Murray and Szelényi's three models of socialist urbanization. Then, finally, with the relaxation of administrative controls, China's urban population began to grow again in the 1980s, especially in the second half of the decade and the early 1990s. The third period resembles the pattern and pace of Third World urbanization, although some features of the earlier Chinese socialist model have been kept, such as restricting the growth of the entitled population and encouraging new non-agricultural employment in rural enterprises. The recent elimination of grain and edible oil coupons (crucial components of the urban household registration system) in Beijing and other large cities has further weakened whatever practical constraints that the registration system may have on urban growth.

Moreover, several past trends will be difficult to sustain. Currently, rural industry tends to concentrate around major cities and in economically rich and transport-dense regions, while rarely spreading into the hinterlands of China (Perkins, 1990). Without massive investments in transport, cheap electricity, and other aspects of infrastructure, rural industry will fail to spread and perhaps even stagnate (Blank and Parish, 1990). Indeed, rural industry employment stagnated briefly after 1988. Given these trends, some scholars called for an emphasis not on rural industrialization but small cities (Sung, 1990). The large, informal, 'floating' population in Shanghai, Beijing, and other major metropolitan areas reflects some of these pressures. Thus, following the fading of the older, control-oriented under-urbanization model that emulated patterns in the Soviet Union during Stalin's time, it remains uncertain what will be the new Chinese model.

City Hierarchy and Rural–Urban Balance

Around the world, city–countryside gaps and city-size distributions help index whether government policies are biased in favour of cities over the countryside and toward a few metropolitan cities over against the full range of cities and towns. Resource allocations between the city and the village in many contemporary Third World countries reflect urban priorities or 'urban bias' (Bates, 1981; Hoselitz, 1953; Lipton, 1977). Within the city system, urban primacy (the top one or two cities being disproportionately large) or lack of 'balance' in the city-size distribution is thought to be both a cause and effect of underdevelopment (Kasarda and Crenshaw, 1991). While many Third World countries have tried to redirect city growth in a 'balanced' direction through such administrative measures as 'growth poles', the evidence on the success of these policies is mixed (Richardson, 1987).

In its earlier Stalinist-Type emphases, China was unable to escape the problems of urban bias. Despite periodic rhetorics in favour of peasants, nominal urban consumption standards remained about 2.5 times higher in urban compared to rural areas, reaching a peak of 2.9 : 1 in 1978 (CSSB, 1993: 280). With the addition of hidden urban subsidies, urbanites enjoyed at least a 3 : 1 consumption advantage over villagers (Johnson, 1991). In the early 1980s, this gap narrowed somewhat, reaching a nominal low of 2.2 : 1 in 1985 (Parish, 1987: 74) and producing complaints from urbanites that suburban farmers were making more than they. However, with new salary increases and subsidies, the nominal consumption gap had grown to an all-time high of 3.1 : 1 by 1990 (CSSB, 1993: 280). While this gap is smaller than in some

African societies, where the gap can be 8 : 1, it is significantly higher than in some Asian societies such as India, where the gap is only 2 : 1 (Lipton, 1977). As already suggested, what may narrow this gap in the future is not so much rising rural living standards as the lowering of urban standards. With more and more urban residents denied full urban entitlement and with housing and other benefits and amenities increasingly distributed at market prices, the hidden subsidies that so elevate current urban living standards could be diminished.

Another issue is the gap in income and growth among cities. Historically, China never had a primate city structure. In ways similar to India, China was a large, complex society with an even spread of cities of different sizes throughout the pre-1911 empire (Skinner, 1977). Because of foreign contact, a few foreign-trade cities began to outpace all other cities, but still as late as 1949 the largest city, Shanghai, was not disproportionately large compared to all other cities in China (Ginsburg, 1961: 36). Nevertheless, post-1949 policy was biased against these foreign-trade cities. To promote production rather than consumption, to move strategically vulnerable facilities from coastal areas to the militarily safer central and western regions, and to promote regional equality, resources were drained from large coastal cities such as Shanghai and reinvested in the interior. As a result of these kinds of policies, large cities such as Shanghai grew hardly at all during the first three decades of Communist rule (see Chen, 1988: 232). The capital, Beijing, did continue to grow, as one would expect in a centrally administered society. But other large cities did not, and in the late 1970s China's distribution of cities was more evenly balanced toward small and medium cities than it had been in 1949 (Parish, 1990).

Whether this 'balanced' size and regional pattern can continue is uncertain. Post-1978 policies have consequences that favour both large, coastal cities and some smaller cities in the interior. The increasing devolution of taxing authority and the control of foreign investment funds to many cities favours large, coastal cities and their immediate hinterlands. Foreign investment is concentrated around Shanghai, Guangzhou, and other coastal areas (Chen, 1994). It is in these regions that growth is fastest, urban unemployment lowest, and the attraction for official and unofficial in-migrants greatest. These tendencies all suggest much more rapid urban growth in the coastal regions, particularly in and around major foreign-trade centres. For example, Shanghai and its hinterland are once again growing rapidly, especially since 1990, much as before 1949.

Administratively, there are some countervailing tendencies that will do little to halt the movement toward coastal areas but may still favour smaller urban locales. After 1982, the regulations on granting places 'town' and 'city' status were greatly relaxed. Town and city status gave a locale far greater autonomy in collecting taxes, dispersing revenue, and handling foreign investment (Yeh and Xu, 1990). The result was that from the mid-1980s to the early 1990s, many places rushed to have themselves reclassified, often expanding boundaries to include more rural population within supposed urban boundaries, and searching for loopholes that gave certain coastal trade areas and frontier regions distinct advantages (for the new regulations, see CASS, 1986).[3] The consequences are clearest for 'cities', which enjoy greater financial autonomy than towns. All told, the number of 'cities' grew from 239 in 1982 to 570 in 1993, a net increase of 331 'cities', which was more than the number of cities that existed in 1982. As one would expect, the overwhelming majority of the added 'cities' were in the two smallest size categories, accounting for 95 per cent of all new cities. Proportionally (not absolutely), China's city hierarchy shrank at the top and expanded at the bottom. In 1982 the groups of cities with over one million and less than 200 000 people made up respectively 8.4 per cent and 50.6 per cent of all cities, whereas the two figures were 5.8 per cent and 60.2 per cent in 1993 (Figure 3.3). This change resulted both from restarting the growth of large cities and from liberalizing the rules for creating new cities.

There were similar tendencies with the lower level urban reclassification of 'towns'. Previously, in 1963, the state had tightened the criteria for towns, causing many towns to be reclassified as rural places. As a result, China's towns dropped from 5404 in 1953 to 2664 in 1982. But in 1984 the criteria for establishing towns were loosened to speed up the development of small towns (*xiaochengzhen*). By the end of 1987, the number of towns almost tripled to a total of 7280, and exceeded 15 000 by 1993 (CSSB, 1994: 17). Thus, through 1993 China added more than 300 new 'cities' and over 10 000 new towns, thereby redefining millions of rural population as urban by administrative fiat.

What do China's recent policy shifts mean both domestically and in comparison to other developing countries? Though the wholesale redesignation of places as cities departs from past policy, the limits on who is entitled to full urban benefits continues much as in the past. For example, from 1982 to 1993 as many formerly rural places were

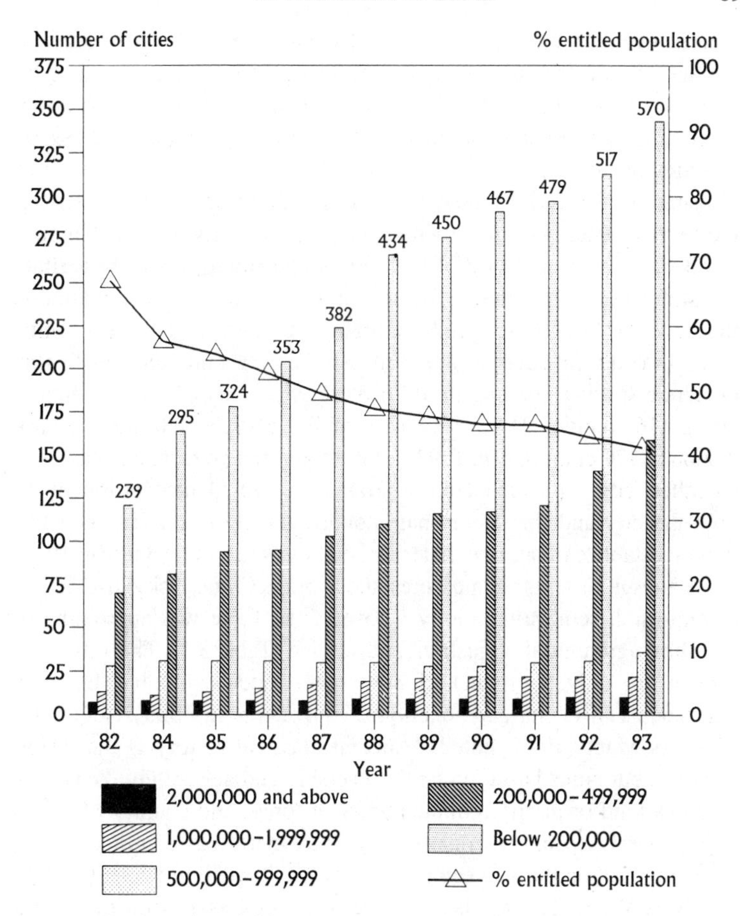

FIG. 3.3. Chinese Cities by Size Categories and Entitled Population, 1982–1993

Sources: CSSB (1983, 1985, 1986, 1987, 1988, 1989, 1990*b*, 1991*b*, 1992*b*, 1993, 1994).

labelled as cities, the percentage of people in those cities entitled to full benefits declined from 67 per cent to 41 per cent (Figure 3.3).

Similarly, as local administrators inflated their boundaries so as to get enough population to qualify for city status, and the additional taxing and budgeting authority that status brings, the occupational composition of 'cities' was diluted. As one would expect, the tendency to inflate one's boundaries was particularly severe in the smaller cities, where (even excluding subordinate rural counties) the percentage of the agricultural labour-force rose (see Table 3.1). In contrast, the cities with over 500 000 population (excluding subordinate counties) saw their share of the agricultural labour-force dropping slightly during 1989–93. In cities of all size categories (with or without subordinate counties), the share of industrial labour-force declined during 1989–93, whereas the service sector's employment grew proportionally. Overall, in cities of all sizes, over one-third of the 1993 labour-force was in agriculture. To stimulate rural industrialization and city–countryside integration, the state in 1983 introduced an urban administrative reform, which incorporated rural counties adjacent to major cities into the latter's span of administrative control. In 1976 there were only 39 cities administering 116 counties. By 1989, a total of 702 counties fell into the jurisdiction of 170 cities (Chen, 1993). This administrative reorganization—the so-called city-controlled county strategy—created new problems for both the city and the subordinate counties, such as conflict over limited investable capital, raw materials, and markets (Zhou and Hu, 1992).

As a result of all these measures, the average Chinese city in 1993 was a semi-rural, semi-urban entity whose labour-force was approximately one-third agricultural, industrial, and service (Table 3.1). This city structure is not only a product of new state policies, but has historical antecedents in China and functional equivalents in other Asian countries. As Marx (1973) noted: 'Ancient classical history is the history of cities, but cities based on land ownership and agriculture; Asian history is a kind of undifferentiated unity of towns and country (the large city, properly speaking, must be regarded merely as a princely camp superimposed on the real economic structure)' (cited in Szelényi, 1981: 12). Max Weber (1958) also saw the Asian city as a 'semi-rural' unit. Historically, Chinese cities lacked separation from the surrounding countryside, as official administrative lines tended to cut across the urban–rural divide (Whyte and Parish, 1984: 10).

More recently, throughout Asia, there are new metropolitan regions that integrate large cities and their surrounding rural areas and smaller cities. These regions include Sunan (southern Jiangsu) with Shanghai as the core, southern Liaoning with Shenyang the centre, the Jakarta region in Indonesia, the greater Delhi region in India, the corridor between Bangkok and its airport in Thailand, and southern Okayama in Japan. Characterizing them as 'dispersed' or 'extended' metropolises, Ginsburg (1991*a*: 42) argues that they reflect a massive extension of urban influences over the countryside and a significant blurring of the

TABLE 3.1 Labour-Force in Chinese Cities: Size Category by Sector, 1989 and 1993

Size category[a]	Year	% Excluding subordinate counties			% Including subordinate counties		
		Agriculture	Industry	Services	Agriculture	Industry	Services
2 Million and above	1989	8	54	38	27	44	29
	1993	6	51	43	26	42	32
1 000 000–1 999 999	1989	13	55	32	43	35	22
	1993	13	52	35	41	34	25
500 000–999 999	1989	12	57	31	36	36	28
	1993	10	58	32	43	35	22
200 000–499 999	1989	34	42	24	59	24	17
	1993	39	36	25	57	23	20
Below 200 000	1989	57	25	18	62	20	18
	1993	56	24	20	59	20	21
All cities	1989	35	40	25	51	28	21
	1993	37	37	26	49	28	23

[a] Size categories by nonagricultural (entitled) population alone.

Sources: CSSB, 1990*b*: 8–17; 1994: 296–305.

traditional distinctions between 'urban' and 'rural'. Two factors are involved. In China, economic decentralization and rural industrialization has created subcontracting relationships between industrial enterprises in large cities and small village or township enterprises in their surrounding areas. In China and some other Asian countries, this new metropolitan complex is facilitated by the development of low-tech rather than high-tech transport facilities, such as improved all-weather roads, canals in the lower Yangtze River delta, and commuter trains near Calcutta (Ginsburg, 1991*b*: p. xiv).

This section notes the interaction between historical, economic, and ecological conditions and recent state policy. In China, as elsewhere, urban bias is very difficult to erase, because it is maintained by the entrenched interests of urban political élites. The entitled few hold on jealously to their privileges, providing interest groups who work through informal channels to maintain position. Balanced growth of cities relies heavily on an earlier spread of cities of various sizes in a historically complex society. The extended metropolis, while reflecting problems of bottlenecks in urban industry that turns to more flexible rural industry for relief and conscious policy intervention to restrain urban costs, also reflects structural tendencies that are common among several Asian societies.

Rural–Urban Migration

The increasing linkages between cities and their hinterland involves both rural industrialization and rural–urban migration. Despite significant labour absorption by rural enterprises, this was insufficient to absorb all newly released rural labour. As a result, many farmers moved from rural to urban areas. During 1978–93, an average of one-fifth of the entrants in the urban labour-force (those in non-agricultural employment in cities and towns) came from rural areas (Table 3.2). There was an ebb and flow in the rate that new labour came from the countryside. Low in the early 1980s, the absorption of rural labour accelerated during the economic liberalization of 1985–9. Then following the sharp entrenchment in the money supply undertaken to control the inflation of 1988–9, new employment opportunities decreased. This retrenchment had several consequences. The private sector could not get loans, and new employment opportunities shrank from 1989 through

TABLE 3.2 New Entrants to China's Urban Labour-Force, 1978–1993 (selected years)

New entrants	Year										
	1978	1980	1984	1985	1987	1988	1989	1990	1991	1992	1993
A. *Source* (%)											
Local pool	50.6	69.2	62.3	61.7	51.6	50.1	44.7	43.3	38.6	40.3	41.4
Rural pool	27.2	14.1	17.0	18.4	20.9	19.0	19.4	15.0	18.3	21.7	26.2
College graduates[a]	7.0	8.9	11.4	10.9	14.6	15.5	23.4	21.4	22.6	25.4	27.4
Other sources	15.3	7.8	9.3	9.0	13.0	15.5	12.6	20.3	20.5	12.5	5.0
B. *Destination* (%)											
State sector[b]	72.1	63.6	57.6	61.3	62.5	58.3	59.2	60.5	47.5	49.9	44.0
Collective sector	27.9	30.9	27.3	25.1	26.8	31.2	31.0	30.0	35.6	29.6	28.7
Private sector	—	5.6	15.1	13.6	10.8	10.5	6.0	5.1	7.8	10.0	13.5
Other sectors	—	—	—	—	—	—	3.8	4.4	9.2	10.5	13.9
TOTAL (000s)	5444	9000	7215	8136	7991	8443	6198	7850	7650	7360	7050
C. *Unemployment*											
Total unemployment rate (%)	5.3	4.9	1.9	1.8	2.0	2.0	2.6	2.5	2.3	2.3	2.6
Youth unemployment rate (%)	47.0	70.6	83.1	82.6	85.0	82.8	81.8	81.6	81.9	83.2	84.0

[a] Includes vocational school graduates.

[b] Data prior to 1989 include new employees in other sectors, i.e. state–collective joint ventures, collective–private joint ventures, Chinese–foreign joint ventures.

Sources: CSSB, 1991*a*: 116; 1992*a*: 118; 1993: 119; 1994: 106.

1991 (panel B, Table 3.2). Similarly, total unemployment increased (panel C). And, as a secondary consequence, and partially as a result of government efforts to keep rural migrants out of cities, rural labour absorption declined by 1990 (panel A).

This proved but a short interlude, however, and after the economic recovery starting in 1992, rural labour absorption once again accelerated, along with the growth in private and other kinds of employment. A 1987 survey found that rural migrants constituted 45.2 per cent of all immigrants in cities and towns (Shi and Liu, 1990: 209). During 1980–9, the number of in-migrants in Shanghai, Beijing, and Tianjin—China's three largest cities, reached 492 235, 542 248, and 265 059, which accounted for 38, 26, and 40 per cent of the cumulative net increase in each of these cities' populations (CSSB, 1990*a*: 624–9). These proportions approximate the migrant contribution (40 per cent) to urban growth in Third World countries (Findley, 1993: 16; Kasarda and Crenshaw, 1991: 474).

Permanent *versus* Temporary Migration

The above statistics, however, fail to differentiate *permanent in-migrants* and *temporary migrants* in China. Permanent in-migrants were defined according to the narrow Chinese criterion, i.e. those persons who have legally obtained household registration, while temporary migrants were defined as those who had moved to their current residence without an official change of registration and had been living there for less than a year (see Goldstein, 1993). Temporary migrants, also known as the 'floating' population, are a major new demographic and economic phenomenon. In Shanghai, the number of temporary migrants rose from an estimated number of 560 000 in 1984 to 2.19 million in 1988. In Beijing, there were only about 300 000 temporary migrants in 1978; the number grew to 1.31 million in 1988 (Li and Hu, 1991: 7). A late 1980s survey estimated there were 701 million temporary migrants in nine cities of over one million people, approximately 23 per cent of their official resident populations (Li and Hu, 1991: 67). In these large metropolises, the proportion of 'floaters' vary widely, from 10 per cent in the chilly northeastern city of Harbin to about 30 per cent in the bustling Guangzhou near Hong Kong (Solinger, 1995: 128).

The length of stay for temporary migrants has also increased. Most of them stayed for a few days in the early 1980s. By 1988, those who stayed for over one month made up 62 per cent of the total, 29 per cent of which lived in places of destination for over one year (Li and Hu, 1991: 9). In other words, approximately one-third of the temporary migrants stayed long enough to be counted as permanent migrants according to the one-year residence definition. The stay of many 'floaters' is dictated by the seasonality of rural production, whereas others stay on limited short-term contracts, which may range from a few months for a housemaid to 2–3 years for young girls in textile mills. These migrants generally view their lives in the city as temporary in the light of the urban household registration system (Solinger, 1995: 129).

The temporary migrants are characterized primarily by their orientation to finding job opportunities in cities. Before 1980, over two-thirds of the temporary migrants were in places of destination to visit relatives and friends, to live away from home under entrustment, or to seek medical treatment. A 1988 survey showed that 48 per cent of all temporary migrants were job-seekers or job-holders. The figure for Shanghai was 61 per cent. Second, that 59 per cent of the temporary migrants in cities of one million plus population came from rural areas. Third, that most of the temporary migrants were young, male, and with low education. Of the 313 000 temporary migrants in eleven large cities, 72 per cent were men, and 55 per cent of both men and women fell into the 18–35 age range. Of the temporary migrants of working age, 9 per cent had functional illiteracy (including semi-illiteracy), 24 per cent primary, and 48 per cent junior middle school education, compared with 7, 19 and 41 per cent for the permanent residents (Li and Hu, 1991: 9–14). This profile has begun to resemble that of most migrants in large cities of Third World countries.

The impact of large temporary migration on cities is mixed. On the one hand, temporary mobility has helped to ameliorate rural labour shortages and to redistribute the population to meet labour-force needs in urban areas, especially in the booming construction and service sectors. The flourishing of urban free markets, staffed by temporary rural vendors of all types, has enriched the daily supply of consumer goods for urban residents (see next section). Since temporary migration does not involve official state intervention, it can respond to labour-market needs more quickly than controlled permanent migration (Goldstein and Goldstein, 1991: 47). Hired in the place of ordinary urban labour at lower wages and benefits, 'floaters' ironically help keep the financially troubled and burdened state factories in business (Solinger, 1995). On the other hand, the rapid growth of temporary migration places added strains on the already stretched urban infrastructure, crowding public transportation, raising the need for food and water supply, and contributing to crime. In 1987, the inclusion of the so-called 'floating population' in twenty-five cities of over one million people reduced the water supply for daily use from 175 litres per person to 146 litres per person, a drop of 17 per cent. In Shanghai, 'floaters' accounted for 7 per cent of the reported crimes in 1983, and 31 per cent in 1989 (Li and Hu, 1991).

To respond to the serious socioeconomic consequences of temporary migration, the central and local governments have adopted several policy measures. In 1984, the central government stipulated that peasants could settle in small cities and towns so long as they took care of their own housing (through renting or self-construction) and food needs. Several places have begun to welcome temporary in-migration, creating special agencies to supervise the migrants and informing them about housing and jobs. These types of migrants are particularly common in the Pearl River Delta region near Hong Kong. In the Delta, foreign investment, particularly from Hong Kong, and rapid local industrialization have created labour shortages that attract labourers from other regions (Xu and Li, 1990: 55–8). Movement from the central and western (inland) regions to the eastern (coastal) region accounted for 90 per cent of the total temporary labour migration (Li and Hu, 1991).

This section has revealed another evolving aspect of the Chinese model of urban development, namely, readjusting a rigid migration policy in response to the shifting labour demands unleashed by market-oriented economic reform. In an attempt to curb the tidal wave of temporary rural–urban migration, China has kept the basic elements of the household registration system in place. The large 'floating population,' however, has already begun to weaken the 'urban public goods regime' of the entitled population based on household registration, as urban bureaucrats try to balance their loyalty to the old system and established habits against the promotion of markets by allowing rural migrants into cities and reducing entitled urbanites' welfare benefits (Solinger, 1995).

Quality of Life in Chinese Cities

Chinese cities were in many ways barren in the middle 1970s. During two decades of forced Stalinist industrialization, housing, consumer goods, and other consumer services had been sacrificed on behalf of heavy industry. Few stores were open, restaurants were scarce, a long list of goods was rationed, and housing provided only a paltry 3.6 square metres of floor space per person in 1978 (Chen and Gao, 1993). Much has changed since.

Services

In an attempt to solve the urban unemployment problem and to alleviate discontent from so few goods and services, investment in light industry was increased, rules for neighbourhood collective and private enterprises were liberalized, and farmers were allowed to market their goods in cities once again. Personal incomes in cities rose considerably. The appearance of cities changed radically in the decade of 1980s. Stores reopened and new vegetable markets, bicycle repair, tailor shops, and other kinds of service units blossomed. Shoppers reappeared, and Chinese cities began to take on some of the hustle and bustle one expects from other cities in Asia. Visually, clothing became much more colourful. More goods appeared on shelves. And, as a by-product, per capita garbage production increased about one-fourth (CSSB, 1991*a*: 669).

Systematic data (see Table 3.3) show that in cities of all sizes (except the smallest cities), there were more service establishments, telephones, public buses, living-space, roads, and green areas on a population basis in 1991 than in 1984. With regard to higher education and medical services, larger cities improved more than smaller cities, largely because the latter included a growing number of formerly rural counties.

Despite all these changes, Chinese cities are not replicas of other Asian cities. Old heavy industries continue to get more investment funds than light industry. As a result the labour-force remains heavily tilted toward blue-collar, manual occupations, with 43 per cent of the non-agricultural labour-force remaining in industry—down some from the 48 per cent of 1978, but still much higher than the developing country average of 27 per cent. Though the percentage of the non-agricultural labour-force in commerce increased from 12 per cent in

TABLE 3.3 Quality of Life Indicators for Chinese Cities, 1984 and 1991

Indicator	Year	Size Category					
		2 million and above	1 000 000–1 999 999	500 000–999 999	200 000–499 999	Below 200 000	All cities
Services:							
Retail, catering and service units per 10 000 persons	1984	94	81	103	98	95	96
	1991	124	158	157	145	127	137
Telephones per 100 persons	1984	3.0	2.2	1.8	1.2	0.7	1.6
	1991	8.2	5.0	4.9	2.9	1.2	3.1
Public buses per 10 000 persons	1984	4.9	3.7	2.4	1.4	0.7	2.2
	1991	5.2	3.9	2.6	1.7	0.5	1.9
Cinemas and theatres per 10 000 persons	1984	0.1	0.1	0.2	0.2	0.2	0.2
	1991	0.1	0.1	0.1	0.1	0.1	0.1
Public library books per capita	1984	1.6	1.7	1.1	0.6	0.4	0.9
	1991	1.8	1.4	0.9	0.4	0.2	0.6
Education:							
College students per 10 000 persons	1984	126	147	96	38	22	69
	1991	166	160	80	40	12	59

Medical Care:							
Hospital beds per	1984	49.3	56.2	57.3	46.0	33.5	45.7
10 000 persons	1991	62.6	69.8	66.5	42.3	24.8	41.9
Doctors per	1984	49	46	41	30	22	34
10 000 persons	1991	52	49	44	28	16	29
Housing:							
Living area	1984	4.8	5.0	4.9	4.9	5.0	4.9
per capita (sq. metre)	1991	6.6	6.7	6.3	6.8	7.6	6.9
Transportation:							
Paved road per 10 000	1984	2.1	2.7	2.1	1.6	1.1	1.7
persons (kilometre)	1991	3.5	3.6	3.6	2.7	1.7	2.5
Environment:							
Green area per 10 000	1984	9.3	21.4	9.3	6.4	5.9	8.7
persons (sq. metre)	1991	13.4	16.9	16.9	9.4	4.5	9.1

Note: Data for cities proper, excluding subordinate counties.

Sources: CSSB, 1985: 195–9, 456–9; 1992*b*: 30–7.

1978 to 19 per cent in 1991, this remains below the developing country average of 21 per cent (CSSB, 1992*a*, 101; Whyte and Parish, 1984, 36).

Social Security

More problematic than consumer services was the fate of social security provisions. Historically, some of the benefits enjoyed by the entitled portion of the urban population were automatic—e.g. subsidized food grains, better education, almost free subsidized housing. However, many other benefits—health care, pensions, disability insurance—were available only to the 68 per cent (80 per cent in 1978) of the urban labour-force employed in the state sector in 1993 (CSSB, 1994: 84). This meant that over half of all urban women, never working in state enterprises, lived out their old age bereft of pension and health benefits, relying instead on their sons much as in more traditional times (Tian, 1991: 169).

For those who were covered, of course, the benefits produced major advantages in everyday life. Though spartan, adequate basic food rations, extensive public health work, and other measures to provide a floor under living standards led to extremely high life expectancies and a basic security in living conditions that are rare in most cities of the developing world (see, e.g. Jamison *et al.*, 1984; Whyte and Parish, 1984; ch. 4). Today, for those in the state-subsidized programmes or able to afford their own care, doctors and hospitals are even more readily available, providing much better access to these facilities than almost any other country at this level of development.

The overall situation with respect to health care, old age support, and other types of floors under living standards is becoming increasingly complex, however. Although some measures will continue, e.g. public health provisions, employees in some state units have now to co-pay for some portions of medication, which had been covered completely by the state. And with most people enjoying much higher incomes with which to buy ever more available food and consumer goods, many of the special rationing and subsidized food programmes of the past are unnecessary. Nevertheless, for some groups, the basis of urban living standards is increasingly problematic, particularly among those groups who never worked in state industries. As already indicated, perhaps as much as one-third of urban residents are no longer entitled to subsidized housing, subsidized food, and other similar benefits.

Moreover, more than half of all new urban employees are outside the state sector (see Table 3.2), and thus largely bereft of pensions, health insurance, and other benefits. Even in the state sector, the attempts to reform the economy mean that more inefficient state enterprises will go bankrupt, throwing employees onto the street with few benefits. The greater mobility of employees makes the old system of benefits based on the work unit extremely cumbersome. The old system sufficed when virtually all employees spent their whole career in a single work unit. It will suffice no more.

All of these issues are creating great ferment in Chinese academic and bureaucratic circles (see, e.g. Xie, 1991, ch. 3). Increasingly, China must move toward a social benefits system that more closely approximates that in market societies. One can assume full employment no more—unemployment insurance must be invented, which would be a historical first for a country of China's size and level of development. Old age benefits must be tied not to specific work-places but to a universal old age insurance system. Much of what one used to think of as uniquely Chinese will pass away. The passage will be slower than in the Soviet Union or Eastern Europe, for in China the state sector remains somewhat more vibrant than in formerly socialist Europe. But the passage will come, nevertheless. In the midst of the passage, there will be many anomalies. In one sector will be those without a secure living standard but huge current incomes. In another sector will be those with a guaranteed decent living standard—but only modest current incomes. These sorts of anomalies should provide ample sources of discontent and the mobilization of possible interest groups that can be used for political purposes.

Housing

Another area of needed reform is housing. In many respects housing has vastly improved in the last decade. Average floor space in 1991 was almost double its 1978 low of 3.6 square metres per person (Table 3.3). The percentage of residences with their own kitchen, their own toilet, running water, and natural gas has risen dramatically (CSSB, 1991*a*). Most cities remain alive with new housing construction. Despite all this activity, several issues remain, including maldistribution and poor maintenance. As Chen and Gao (1993) showed, the percentage of residences with less than two square metres of space per person, with three generations living in the same room, and other indicators of crowding did not decrease that much. In 1987, one-fourth of urban households had no proper housing or lived in crowded conditions (Xie, 1991: 294). As late as 1992, there were still 4.4 million households in China's 517 cities lacking sufficient living space; 385 355 (8.7 per cent) of those households averaged less than two square metres per person (CMOC, 1993: 161).

These problems persist in part because of the continuing absence of a market in housing. Over half of all urban housing continues to be owned not by the city or individuals but by work units. This means that rich work units, which already had reasonable housing, can afford to build even better housing for their employees. Poor work units can afford to build little or none. Rents, at 1 or 2 per cent of take-home pay, are too low to pay even for maintenance, much less new building. Moreover, low rents provide no incentive for the over-housed to leave their ample facilities. Parents who once got claim to an extra bedroom as their children were growing up hang on to the extra bedroom, and sometimes even an extra apartment, once the children have moved away. Only with rising rents will the over-housed have an incentive to free up space. This has led to a host of proposals to marketize rents, starting with a very gradual increase in rents and ending perhaps with more selling of homes on the open market (see Tolley, 1991; Xie, 1991). Rents have been raised ever so slightly, and in some cities there are more vigorous experiments in marketizing housing. Will such experiments relieve the tremendous pressure from employees on enterprises to divert investment funds from new production facilities into housing? With increased rents, adequate funds might finally become available to better maintain the existing stock which can appear shoddy soon after having been built.

Land

A closely related problem is that of urban land markets. The outside observer might be excused for thinking that in a centrally planned economy the allocation of land was tightly controlled from above. It is not. Once work units acquire rights to land they tend to assume legal rights. When there is no market price to give them an incentive to move when they become less efficient, even the least efficient enterprises can continue to occupy valuable space. Taxation as well is often poorly tied to implicit land values, thus providing little incentive for more efficient use. The result is a system of land ownership which some Chinese scholars term not state ownership or societal ownership but instead 'feudal ownership', with each work unit a fiefdom unto itself. Land is often traded in makeshift barter arrangements that depend as much on personal relationships as on the true value of the land. The result on a broad scale is inefficiency in land use. The most efficient producers with the greatest need for a central location in a large city have difficulty getting access to those locations. An indicator of these difficulties is a very weak pattern of agglomeration economies. In other societies, productivity tends to be higher in larger cities. In China those patterns are largely absent (Lim and Wood, 1985, ch. 5; Perkins, 1990). Here, too, there is active discussion of how to bring market elements into ownership rights, rent, and taxation of land, a discussion given urgency and form by the increasing frequency of more market-like agreements with foreign investors. Land reform took a major step in 1988, when the Chinese parliament modified the constitution to permit the transfer of land-use right. In 1990, the Chinese government began to allow foreign real estate investors to develop large tracts of land. By the end of 1990, the state sold use right for 732 parcels of land covering 4888 acres (CSLA, 1991: 70).

Satisfaction

The pattern of investment in urban services has had distinct consequences for how urban residents perceive the quality of their lives. Since the early 1980s, the city government of China's third largest city, Tianjin, has sponsored an annual survey on the quality of urban service delivery. To no one's surprise, the number one complaint by urban residents in 1983 was difficulty with commercial services—being forced to stand in line, too few goods available, etc. In time, with more farmers coming into cities to sell foods in open private markets and with the greater production of food and other consumer goods in general, market concerns receded to be replaced with concerns about transportation, and the quality of child care (e.g. oral communication). There are also a few American-style quality-of-life surveys in Shanghai and Tianjin by American and Chinese researchers. In these surveys, the number one complaint is about housing and the second most common complaint is about low pay and the inability to advance on the job (Lin, Leung, and Lu, 1991). Thus, citizen response is very consistent with what one knows about overall investment patterns. Areas of neglect lead to complaints that filter back to government leaders and provide fodder for interest group manœuvring and eventual policy changes.

Conclusion

To return to the question of whether China has a distinct urban pattern, the answer is that it did have a distinct pattern. For a time in the 1960s and 1970s, China was able to an extremely unusual degree not only to constrain urban growth but also to change the whole structure of cities. By emphasizing high rates of saving and the channelling of this forced saving into heavy industry, and by mobilizing virtually all working-age adults into the labour-force, cities changed in emphasis from consumption to production, and rates of quantitative industrial growth accelerated. Unlike cities in many other developing countries, the informal, service sector was small, and most people were thoroughly enmeshed in tight neighbourhood and work-unit organizations.

All of this reshaping was at tremendous cost, however. With the emphasis on heavy industry, employment expansion slowed, leading by the early 1970s to increased unemployment, particularly among youth. The attempt to solve the problem by sending 16 million urban youth to the countryside for varying lengths of time only heightened parental anxiety, and created the subsequent problem of finding jobs for the returned youth (Gold, 1980). The siphoning of current income away from salaries and housing and consumer goods, meant that salaries were frozen for long periods of time, housing stock per person shrunk and maintenance deteriorated, consumer goods disappeared or were available only through cumbersome rationing procedures, and fewer people were available to provide meals, laundry, and other services.

It was the backlash against these socioeconomic problems that helped fuel popular support for post-1978 reform policies. With these reforms, many of the formerly distinctive features of Chinese cities have been weakened or shown to be inappropriate for a part-market, part-planned economy. Major interest groups will fight dramatic changes in the old system. Workers, once accustomed to nearly free housing, subsidized food, lifetime job security, and other entitlements, are extremely discontent when there is any attempt to remove these privileges. The increasing subsidy levels for the entitled portion of the urban population in the 1980s hints at the ability of some of the workers to mobilize through informal networks and secure privileges that mimic the urban bias tendencies of other developing societies (see, e.g. Tidrick and Chen, 1987).

Market reform has created new but uneven opportunities for the different groups of Chinese urbanites. Those in the private sector have got rich quick, while an increasing number of state-sector employees, including the recently retired, have seen their standard of living stagnate. The old path of urban social mobility, characterized by going from high school and college to lifetime employment in state enterprises, has become inaccessible to some, unattractive to others. Some of the 'new urban rich' have used the college degree as a credential to start a profitable business, while others have simply bypassed college education. Despite diverse economic opportunities, the reform has not generated channels for the newly aligned economic interest groups to develop corresponding political power and voice. The Tiananmen tragedy in 1989 illustrates that it was easy for the government to suppress an organized urban political movement under the condition of fragmented interest groups. In this increasingly differentiated urban environment, the economically wealthy may be content with not gaining political clout for the foreseeable future. The urban destitute, on the other hand, are denied a political voice, even if they attempt to have one. Although the less-advantaged groups may count on the state to maintain a decent living standard, they will face increasingly less available state resources in the future as urban privatization deepens.

However, much as in other developing societies, China's analogue to the informal sector—new neighbourhood-run collective enterprises, urban private enterprises, and rural subcontracting and other enterprises—are the most dynamic part of the economy. With few of the entitlements enjoyed formerly by state workers, sometimes higher salaries, and far greater flexibility and work enthusiasm, these enterprises may in time come to rival the old state sector. If this reformer's dream is realized, China will come to appear much like other developing societies, where a highly heterogenous informal sector makes up for many of the deficits in the formal sector of employment (see Castells and Portes, 1989; Kasarda and Crenshaw, 1991; de Soto, 1989). Barring drastic changes in unemployment insurance, old age assistance, and other service delivery programmes, Chinese cities and towns will become highly fragmented, with different workers and citizens subject to very different life chances. Already, families think carefully in which sector of the labour market they want to pursue their fortune. In popular mythology, the best solution for a family is to diversify their risks, with some members working in the staid state sector, offering low but secure wages and many fringe benefits and other members working in the much more dynamic non-state sector with few fringe benefits and greater risk of failure but also much greater chances of high incomes and rapid upward mobility when market conditions are ripe.

In this chapter, we have shown that while urbanization in China in the 1980s and early 1990s accelerated in response to rapid industrialization, the pattern of city growth reflects how state policy anticipates and reacts to the consequences of new patterns of economic development. The Chinese city has evolved into an increasingly semi-rural and semi-urban entity, which shows initial signs of sprawling into the adjacent hinterland. There is growing variation in the structure of urban life, and in the perceptions that urban residents have about the quality of that life. We conclude that as urban economic reform proceeds it is creating a totally new set of dynamics that makes much of the old structure inappropriate for current circumstances.

Notes

1. Some of the 'entitled population' live in the countryside, e.g. rural administrators and school-teachers. This explains why the percentage entitled exceeded the percentage urban in the 1950s, and it suggests that two-thirds entitled among all urban may be a generous estimate for 1993.
2. The 1978–9 uptick in percentage non-agricultural labour includes a redefinition of non-agricultural to include villagers working outside agriculture. The pre-1979 figures are thus understated by a few percentage points. Note, also, that some villages, particularly on the outskirts of cities and towns, have become industrialized that they might be more properly counted as urban. If so, the percentage urban figure would move up more towards 40 per cent.
3. Since 1986 two major criteria have guided city certification. First, any town, whose non-agricultural population exceeds 60 000 and whose GNP is larger than two million yuan, can be certified as a city. Towns that do not meet the population and GNP standards but are important towns in minority and remote regions, important industrial and mining bases, famous tourist areas, transport hubs, or border entries can be certified as cities, if it is necessary. Second, for counties with a population of 500 000 or less, the county-seat towns whose non-agricultural population exceeds 100 000 and whose GNP is over three million yuan, can be granted city status to replace county status. For counties with over 500 000, the county-seat towns can be certified cities from counties if their non-agricultural population exceeds 120 000

and GNP surpasses four million yuan (Zhang, 1990: 63). Qufu in Shandong province was certified as a city for being the hometown of Confucius and attractive to overseas and domestic tourists even though its non-agricultural population was only 53 500 in 1987 (CSSB, 1988: 27).

References

Bates, Robert H. (1981), *Markets and States in Tropical Africa* (Berkeley: University of California Press).

Blank, Grant and Parish, William L. (1990), 'Rural Industry and Nonfarm Employment: Comparative Perspectives', in Reginald Kwok, William L. Parish, and Anthony Gar-on Yeh, with Xu Xueqiang (eds.), *Chinese Urban Reform: What Model Now?* (Armonk, NY: M. E. Sharpe), 109–39.

CASS (Chinese Academy of Social Science, Population Research Center) (1986), *Zhongguo Renkou Nianjian* [China Population Statistical Yearbook] (Beijing: Chinese Social Sciences Press).

Castell, Manuel and Portes, Alejandro (1989), 'World Underneath: The Origins, Dynamics, and Effects of the Informal Economy', in Alejandro Portes, Manuel Castells, and Lauren A. Benton (eds.), *The Informal Economy: Studies in Advanced and Less Developed Economies* (Baltimore: Johns Hopkins University Press), 11–40.

Cell, Charles P. (1980), 'The Urban–Rural Contradiction in the Maoist Era: The Pattern of Deurbanization in China', *Comparative Urban Research*, 7: 48–62.

Chan, Kam Wing (1992), 'Economic Growth Strategy and Urbanization Policies in China, 1949–1982', *International Journal of Urban and Regional Research*, 16: 275–305.

—— and Xu, Xueqiang (1985), 'Urban Population Growth and Urbanization in China Since 1984', *China Quarterly*, 104: 583–613.

Chen, Xiangming (1988), 'Giant Cities and the Urban Hierarchy in China', in Mattei Dogan and John D. Kasarda (eds.), *The Metropolis Era: A World of Giant Cities* (Newbury Park, Calif.: Sage Publications), 225–51.

—— (1991), 'China's City Hierarchy, Urban Policy and Spatial Development in the 1980s', *Urban Studies*, 28: 346–71.

—— (1993), 'China's Urbanization and City Growth: A Retrospective and Prospective View' (in Chinese) in Yue-man Yeung (ed.), *China's Cities and Regional Development: Prospect for the 21st Century* (Hong Kong: Institute of Asia-Pacific Studies), 182–202.

—— (1994), 'The New Spatial Division of Labor and Commodity Chains in the Greater South China Economic Region', in Gary Gereffi and Miguel Korzeniewicz (eds.), *Commodity Chains and Global Capitalism* (Westport, Conn.: Greenwood Press), 165–86.

—— and Gao, Xiaoyuan (1993), 'China's Urban Housing Development in the Shift from Redistribution to Decentralization', *Social Problems*, 40: 266–83.

CMOC (China's Ministry of Construction) (1993), *Chengshi Jianshe Tongji Nianbao* [Urban Construction Statistical Manual] (Beijing: Ministry of Construction).

CSLA (China's State Land Administration) (1991), *Zhongguo Tudi Guanli Gaikuang* [Survey of Land Management in China] (Beijing: People's University of China Press).

CSSB (China's State Statistical Bureau) (1983, 1991*a*, 1992*a*, 1993, 1994), *Zhongguo Tongji Nianjian* [China Statistical Yearbook] (Beijing: China Statistics Press).

—— (1990*a*), *Zhongguo Renkou Tongji Nianjian* [China Population Statistical Yearbook] (Beijing: Science and Technology Press).

—— (1985, 1986, 1987, 1988, 1989, 1990*b*, 1991*b*, 1992*b*), *Zhongguo Chengshi Tongji Nianjian* [China Urban Statistical Yearbook] (Beijing: China Statistics Press).

Findley, Sally E. (1993), 'The Third World City: Development Policy and Issues', in John D. Kasarda and Allan M. Parnell (eds.), *Third World Cities: Problems, Policies, and Prospects* (Newbury Park, Calif.: Sage Publications), 1–31.

Gilbert, Alan (1992), 'Third World Cities: Housing, Infrastructure and Servicing', *Urban Studies*, 29: 435–60.

Ginsburg, Norton (1961), *Atlas of Economic Development* (University of Chicago Press).

—— (1991*a*), 'Extended Metropolitan Regions in Asia: A New Spatial Paradigm', in Norton Ginsburg, Bruce Koppel, and T. G. McGee (eds.), *The Extended Metropolis: Settlement Transition in Asia* (Honolulu: University of Hawaii Press), 27–46.

—— (1991*b*), 'Preface', in Ginsburg, Koppel, and McGee (eds.), *The Extended Metropolis*, pp. xiii–xviii.

Gold, Thomas B. (1980), 'Back to the City: The Return of Shanghai's Educated Youth', *China Quarterly*, 84: 755–80.

Goldstein, Sidney (1993), 'The Impact of Temporary Migration on Urban Places: Thailand and China as Case Studies', in Kasarda and Parnell (eds.), *Third World Cities*, 199–219.

—— and Goldstein, Alice (1991), *Permanent and Temporary Migration Differentials in China*, Papers of the East–West Population Institute No. 117 (Honolulu: East–West Center).

Hoselitz, Bert F. (1953), 'The Role of Cities in the Economic Growth of Underdeveloped Countries', *Journal of Political Economy*, 61: 195–203.

Jamison, Dean T., Evans, John R., King, Timothy, Porter, Ian, Prescott, Nicholas, and Prost, Andre (1984), *China: The Health Sector*. (Washington, DC: World Bank).

Johnson, D. Gale (1991), *The People's Republic of China*, Country Studies No. 8. (San Francisco: International Center for Economic Growth).
Kasarda, John D. and Crenshaw, Edward M. (1991), 'Third World Urbanization: Dimensions, Theories, and Determinants', in W. Richard Scott and Judith Blake (eds.), *Annual Review of Sociology* (Palo Alto, Calif.: Annual Reviews Inc.), 467–501.
Kirkby, Richard J. R. (1985), *Urbanization in China: Town and Country in a Developing Economy 1949–2000* AD (London: Croom Helm).
Li, Mengbai and Hu, Xin (1991) (eds.), *Liudong Renkou Dui Dachengshi Fazhan De Yingxiang Ji Duice* [The Impact of Floating Population on the Development of Large Cities and Their Policy Responses] (Beijing: Economics Daily Press).
Lim, Edwin and Wood, Adrian (1985), *China: Long-Term Development Issues and Options*, World Bank Report (Baltimore: Johns Hopkins University Press).
Lin, Nan, Leung, Shu-yin and Lu, Han-long (1991), 'Quality of Life in Urban China: A Model and Data from Shanghai', presented at the Conference on China and the Chinese, The American Enterprise Institute, Washington, DC, January.
Lipton, Michael (1977), *Why Poor People Stay Poor: Urban Bias in World Development* (Cambridge, Mass.: Harvard University Press).
Martin, Michael F. (1992), 'Defining China's Rural Population', *China Quarterly*, 130: 392–401.
Marx, Karl (1973), *Grundrisse: The Foundations of the Critique of Political Economy*, trans. Martin Nicolans (New York: Vintage).
Murray, Pearce and Szelényi, Ivan (1984), 'The City in the Transition to Socialism', *International Journal of Urban and Regional Research*, 8: 90–108.
Orleans, Leo A. (1982), 'China's Urban Population: Concepts, Conglomerations, and Concerns', in *China Under the Four Modernizations*, part I (Joint Economic Committee, US Congress, Washington, DC: US Government Printing Office).
Parish, William L. (1987), 'Urban Policy in Centralized Economies: China', in George S. Tolley and Vinod Thomas (eds.), *The Economics of Urbanization and Urban Policies in Developing Countries* (Washington, DC: World Bank), 73–84.
—— (1990), 'What Model Now?', in Reginald Kwok *et al.* (eds.), *Chinese Urban Reform*, 3–16.
Perkins, Dwight H. (1990), 'The Influence of Economic Reform on China's Urbanization', in Reginald Kwok *et al.* (eds.), *Chinese Urban Reform*, 78–106.
Richardson, Harry W. (1987), 'Whither National Urban Policy in Developing Countries?', *Urban Studies*, 23: 227–44.
Shi, Ruohua and Liu, Xuefeng (1990), 'Research on the Surplus Rural Labour and Urbanization in China' (in Chinese) in Ruohua Shi (ed.), *Zhongguo*

Nongcun Shengyu Laodongli Zhuanyi Wenti Yanjiu [The Shift of China's Surplus Rural Labour] (Beijing: China Prospect Press), 3–35.

Skinner, G. Williams (1977) (ed.), *The City in Late Imperial China* (Stanford, Calif.: Stanford University Press).

Smith, David A. and London, Bruce (1990), 'Convergence in World Urbanization? A Quantitative Assessment', *Urban Affairs Quarterly*, 25: 574–90.

Solinger, Dorothy J. (1995), 'China's Urban Transients in the Transition from Socialism and the Collapse of the Communist "Urban Public Goods Regime"', *Comparative Politics*, 27: 127–46.

Soto, Hernando de (1989), *The Other Path* (New York: Harper and Row).

Sung, Shuwei (1990), 'On New Style Middle Level Urban Centres' (in Chinese), *Chengshi Guihua*, 1: 3–9, 17.

Szelényi, Ivan (1981), 'Structural Changes of and Alternatives to Capitalist Development in the Contemporary Urban and Regional Systems', *International Journal of Urban and Regional Research*, 5: 1–14.

Tian, Xueyuan (1991) (ed.), *Zhongguo Liaonian Renkou* [China's Aged Population] (Beijing: Chinese Economic Press).

Tidrick, Gene and Chen, Jiyuan (1987) (eds.), *China's Industrial Reform* (New York: Oxford University Press).

Tolley, George S. (1991), *Urban Housing Reform in China*, World Bank Discussion Paper No. 123 (Washington, DC: World Bank).

Weber, Max (1958), *The City*, trans. and ed. Don Martindale and Gertrud Neuwirth (New York: Free Press).

Whyte, Martin King and Parish, William L. (1984), *Urban Life in Contemporary China* (University of Chicago Press).

World Bank (1991), *World Development Report 1991* (Oxford University Press).

Xie, Bai-san (1991), *China's Economic Policies, Theories, and Reform since 1949* (Shanghai: Fudan University Press).

Xu, Xueqiang and Li, Si-ming (1990), 'China's Open Door Policy and Urbanization in the Pearl River Delta Region', *International Journal of Urban and Regional Research*, 4: 49–69.

Yeh, Anthony Gar-On and Xu, Xueqiang (1990), 'Changes in City Size and Regional Distribution, 1953–1986', in Reginald Kwok *et al.* (eds.), *Chinese Urban Reform*, 45–61.

Zhang, Yidi (1990) (ed.), *Chengshi Tongji Gongzuo Shiyong Shouce* [A Handbook of Urban Statistical Work] (Beijing: China Statistics Press).

Zhou, Yixing and Hu, Dapeng (1992), 'A Survey Analysis of the Economic Impact of City-Controlled-County System on the Administratively Subordinate Counties' (in Chinese), *Chengshi Jingji*, 3: 51–7.

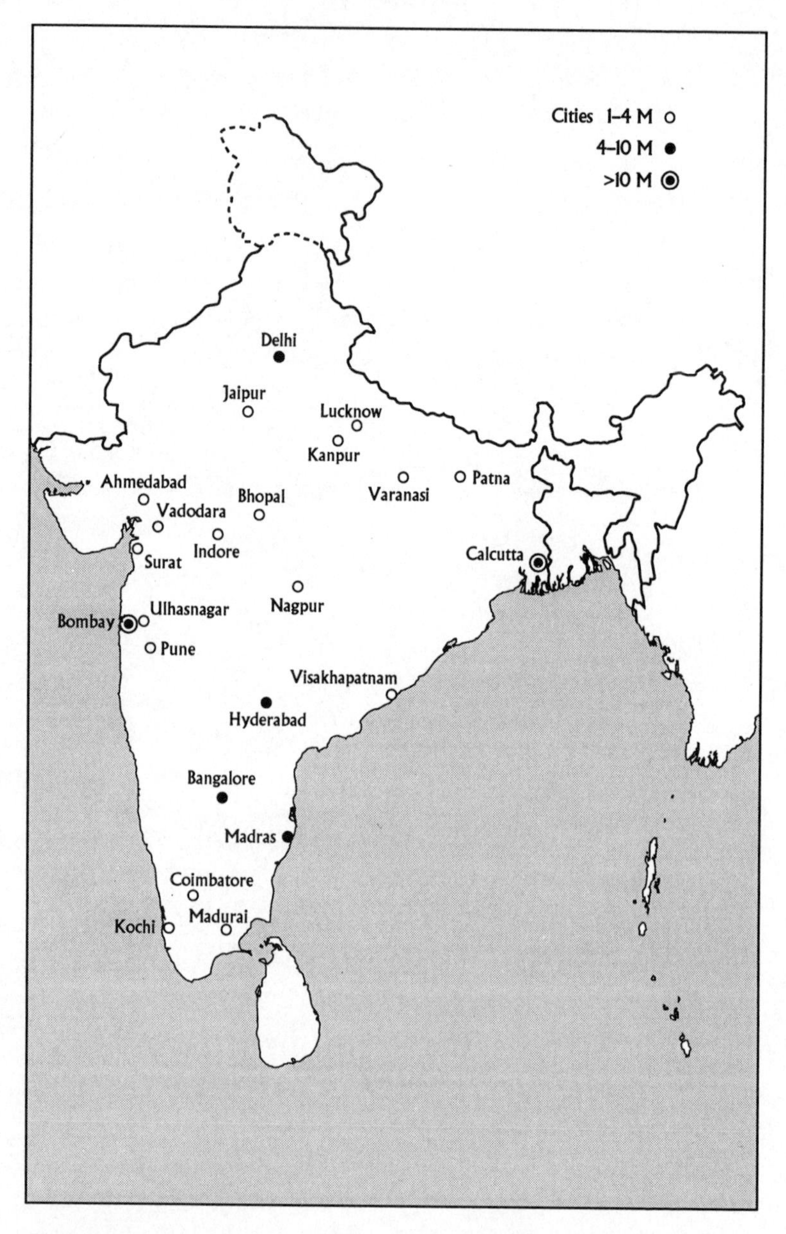

FIG. 4.1 Urban Agglomerations in India with More than 1 Million Inhabitants in 1990

4

Urbanization in India: Patterns and Emerging Policy Issues

RAKESH MOHAN

Urbanization in India: The Record

URBANIZATION in India has been relatively slow over the past forty or fifty years as compared with many other developing countries. Yet, as shown by the 1991 population census, its urban population of 217 million places India along with China as the countries with the largest urban systems in the world. Although, even in 1991, the urban population was just over a quarter of total population, the absolute magnitude is becoming so large that the country will have to give much greater attention than it has done in the past to devising policies which enable the country's urban system to cope with such growth.

Table 4.1 presents the urbanization experience of India since 1901 (for a more detailed treatment, see Mohan and Pant, 1982; Mohan, 1985; Mills and Becker, 1986). While total urban population increased eightfold between 1901 and 1991, the number of urban settlements only doubled. Thus most of the growth arose from the enlargement of existing towns at every level, rather than from the addition of new towns. This implies that the majority of settlements now classified as towns, and especially the larger cities, have exhibited urban characteristics for a very long time. A large number of villages are at the borderline, but only a small number of them have graduated to town status. The majority of regions in India have had settled cultivation for long. The spatial distribution and number of settlements reflect this long history. The principal function of most small towns is that of serving the rural surroundings as market and service centres. Thus, their number and spatial distribution reflect the magnitude of demand

The views reflected in this paper are those of the author and should not be attributed to the Government of India. The author is deeply indebted to Josef Gugler for comments on the first draft.

TABLE 4.1 Growth of Urban Population in India, 1901–1991[a]

Census	No. of towns[b]	Total urban population (millions)	Population in towns above 20 000 (millions)	Level of urbanization[c] (per cent)	Annual growth-rate of urban population (% per year)	Annual growth-rate of rural population (% per year)	URGD[d] (col. 6–col. 7)	Annual growth-rate of population in towns above 20 000 (% per year)
1901	1811	25.6	13.5	11.0	—	—	—	—
1911	1754	25.6	13.8	10.4	0.0	0.61	−0.61	0.22
1921	1894	27.7	15.5	11.3	0.79	−0.18	0.97	1.16
1931	2017	33.0	19.6	12.2	1.77	0.94	0.03	2.37
1941	2190	43.6	28.7	14.1	2.82	1.11	1.71	3.90
1951	2795	61.6	43.2	17.6	3.52	0.82	2.70	4.17
1961	2270	77.6	61.4	18.3	2.34	1.88	0.46	3.58
1971	2476	107.0	89.6	20.2	3.26	1.97	1.29	3.85
1981	3245	156.2	134.9	23.7	3.86	1.75	2.11	4.18
1991	3609	212.9	190.2	26.1	3.15	1.80	1.35	3.50

[a] Excluding Assam and Jammu and Kashmir.
[b] Constituent towns of urban agglomerations are not counted as separate units.
[c] Proportion of urban to total population.
[d] Urban–Rural Growth Differential.

Source: Census of India, 1991.

for their services from the hinterland. There is then a hierarchy of settlements in each region and sub-region and it appears that this hierarchy has remained stable for a very long time. However, in areas where the distribution of existing towns is sparse, a large number of new towns can be expected to appear as and when economic and population growth takes place, particularly when enhancement of agricultural prosperity occurs.

The data over ninety years do not reveal an obvious pattern of urban growth over the decades. Observing the annual rate of urban population growth (Table 4.1, col. 6), there had been a steady acceleration of growth from 1911 to 1951, a slow-down during 1951 to 1961, acceleration again from 1971 to 1981, and again a deceleration to 1991. Some of these observed changes have been plagued with difficulties related to the definition of urban settlements. According to the Indian census, a settlement is defined as urban when its population is over 5000, its population density is over 400 per hectare, and 75 per cent of its male labour-force is engaged in non-agricultural pursuits. In addition, however, a settlement can also be defined as urban by government notification; and the census authorities also have discretion to classify as urban 'some places having distinct urban characteristics even if such places do not strictly satisfy all the criteria mentioned above'. Since the definitional problems of urban classification affect the inclusion or exclusion of the smallest towns between the censuses, it is helpful to use a cut-off point of 20 000 population to better observe urbanization trends. The trends are then purged of the noise generated by classification problems at the lower end of the urban settlement spectrum. Few towns which achieve 20 000 population lose urban characteristics to the extent that they have to be then declassified as urban. Observation of this pattern also reveals acceleration from 1951 to 1981, though the 1971–81 acceleration is less marked (col. 9). The deceleration in growth of the 1980s is quite evident. Thus classification problems are not significant in seeking an explanation for the deceleration in urbanization observed during the last decade.

Another issue of interest illuminated by Table 4.1 is the rate of rural population growth. It may be expected over time that, as the share of agriculture in the economy falls, the rate of rural population growth would progressively slow down. This was observed during 1971–81 but the rate of growth increased again during 1981–91. The rate of rural population growth had accelerated significantly during the 1960s: years which had been particularly characterized by low economic growth resulting from prolonged drought, and other associated economic dislocations. During the 1980s, however, the Indian economy exhibited its highest decadal growth record, and particularly in the industrial sector. Thus the increase of rural population growth during the 1980s is a cause for concern: the higher industrial growth of the 1980s has simply not brought benefits on a large scale.

The biggest surprise thrown up by the 1991 census is the significant decline recorded in urban population growth relative to the previous 1971–81 decade. The total recorded urban population of 217 million is approximately 15–20 million less than projected (Mohan, 1985; Census of India, 1989). The expectation had been that the level of urbanization in 1991 would be of the order of 27 to 27.5 per cent. In fact, the level is about 25.7 per cent. This slow-down is particularly surprising since there has been no significant fall in total population growth and the total population in 1991 is quite comparable to what had been expected. It is also difficult to ascribe the fall in the rate of urban population growth to data errors or to interpretation problems arising from the data. Whereas it is probably the case that the 1981 urban population was somewhat overstated due to the wholesale administrative notification of towns in some states, an analysis of the data suggests that the recorded slow-down is a genuine one and cannot be attributed to data problems. Given the magnitude of the slow-down, it is also unlikely that any undercount of the urban population is responsible for the recorded deceleration in urban growth. The deceleration of urban growth is apparent at all levels of the urban structure.

The deceleration in urban growth may be seen as a very welcome development by many people. It could be attributed by some to the success of the many rural development programmes instituted in India during the 1980s, such as the Rural Labour Employment Guarantee Programme (RLEGP), the Integrated Rural Development Programme (IRDP), the Jawahar Rozgar Yojana (JRY), and the like. However, I view it as a rather disturbing signal suggesting that we have failed to cope with the demands of industrialization and urbanization in adequate fashion. The economy has generated too few jobs in the non-agricultural economy, particularly in the manufacturing sector. Contributing to this may have been inadequate infrastructure investment in urban areas. The lack of jobs compounded by a worsening quality of life in urban areas may have discouraged would-be rural migrants searching for better livelihoods in urban areas. The deceleration of urban growth could as well be contributing to greater rural immiserisation rather than being caused by greater rural prosperity. The growth of value-added in industry and in the tertiary sector was in the range of 6 to 8 per cent per annum in the 1980s, whereas that of agriculture was in the range of 2.5 to 3 per cent per year during the same period. Prima facie, unless there has been large-scale rural industrialization, it would seem that the benefits of higher industrial and overall economic growth have not been as widespread as they should have been and are not being shared with would-be urban immigrants.

Urbanization, Industrialization, and Development

Urbanization has long been associated with industrialization, indeed, they have been considered synonymous. But manufacturing activities in cities are of a relatively recent vintage. Cities evolved originally as market or trading centres and have traditionally been known more for their service sector activities than for industry. The situation is now turning full circle. In contrast to the early days of the industrial revolution, when the proportion of employment in the urban service sector usually declined with rapid industrialization and urbanization, in less developed countries today services have tended to expand as fast as industry so that the process of urbanization is a movement of people to both industry and service from agricultural activities. What is common to both experiences—early Western urbanization and current LDC urbanization—is a consistent increase in agricultural productivity. What is somewhat different for developing countries now is that as latecomers to the development process, the labour productivity in industry is also very high. It is only the service sector that lags in productivity changes and therefore employs much of the incremental urban labour. Another difference is that overall population growth is higher in developing countries today than in the now-developed countries in their earlier stages of development. The growth-rate of cities is therefore high even without migration.

Technical changes in agriculture which raise farm productivity have to provide for the growing rural population in addition to the burgeoning urban population. In times when overall population growth-rates were less than 0.5 per cent, a 3 per cent growth in agricultural production was tremendous. Another major difference from the European experience is the coexistence of high and low technology in LDC cities. High population growth and technological development make it possible for cities to have multimillion populations at low levels of income. This was just not possible in the last century. Advances in public health and technology related to waste disposal and sewerage, for example, have made it possible for cities to exist with large populations and densities. The third major difference may be the decline in relative prices of transportation and communication.

Why do we expect increasing urbanization with industrialization? What is the effect of population growth in this process? Why is a slowdown in urbanization in India in the 1980s regarded as a surprise? Although we still do not have a very good idea of the structural causes of urbanization, various economic models suggest the following framework for thinking.

Urbanization is a natural consequence of economic change that takes place as a country develops. Certain activities are better performed in, indeed require, agglomerations of people while others do not. The location of activities has therefore to be seen in the total context of activities existing in a country and their development in the future. Agglomeration economies are very important for reducing the costs of new firms as they enter the manufacturing world. There are also many economies of scale in the provision of urban infrastructure. Service activities such as banking and insurance also exhibit economies of scale. Economic activities thrive in the presence of many other economic activities. Hence, agglomeration of economic activities and people, that is urbanization, should be seen as positive for overall development. Hence, it should be supported by policy actions. International and historical patterns suggest that urbanization accelerates as an economy industrializes, slowing down once again only after a 60 to 70 per cent level of urbanization has been achieved. The puzzle in India is that, during the 1980s, when industry and overall gross national product grew at unprecedented rates, the rate of growth of urbanization slowed down.

In such a framework, deceleration of urban growth in a developing economy is indeed a cause for concern. This could be caused by a deceleration in rural productivity growth. It is possible that the slow growth of agricultural productivity, except in certain pockets of the country, is not releasing agricultural labour from rural areas. The evidence of increasing disparity between states and regions in agricultural productivity is accompanied by a reduction in disparities in industrial productivity throughout the 1970s and 1980s. Inappropriate technology choice or product composition in the country's industrialization could also lead to lower absorption of labour in urban areas. It is possible that this may have been caused by a faulty tariff structure providing greater protection for capital-using industries (Kelkar and Kumar, 1990). The change in labour laws during the late 1970s and early 1980s making labour adjustment more difficult in the organized sector could also have led to inappropriate technology choice. The proportion of labour-force employed in manufacturing fell for the first time in the 1980s since 1931. Lastly, and most important for urban policy, inadequate investment in urban infrastructure would inhibit both industrial growth as well as tertiary sector growth.

What is clear for the 1990s and beyond is that the urban population is going to grow indefinitely into the future in India. Means must be found to enable urban growth that effectively absorbs the many rural migrants looking for ways to improve their livelihoods. Such movement also contributes to the increased possibility of improvements in labour productivity and earnings in the rural areas. But the resource needs of urbanization are high; the management of cities is becoming increasingly complex; and the urbanized populace is much more demanding.

The administrative and fiscal problems that arise from such demands will not be solved by wishing that cities would grow less quickly, or that population growth would slow down. The growth of the fastest growing cities such as Delhi, Bangalore, Bhubaneshwar, Bhopal, Durgapur and Bhilai, Faridabad (near Delhi), Bhiwandi (near Bombay), Haldia (near Calcutta), and Surat (in Gujarat), have been due mainly to fast industrial growth, or special infrastructure investment related to capital cities. India has always had an ambivalent attitude toward industrialization and urbanization. Decongestion of cities and industrial dispersal have been important components of our industrial policy. We seem to want industrialization but not the urban population growth that goes with it.

In forty years, India's level of urbanization increased from about 17 per cent in 1951 to 25.7 per cent in 1991. This slow growth made it feasible for the government to ignore the demands of urbanization. The structural transformation of the economy that has taken place since independence is expected to intensify over the next two or three decades. The share of agriculture has already fallen to less than a third in national income whereas labour engaged in agriculture continues to account for almost two-thirds of the labour-force. This shrinkage will continue as industry grows at rates of over 8 per cent a year while agriculture, at best, grows at 3 per cent a year. The massive transformation in labour-force structure which must match this transformation in output structure has yet to occur: it will happen in the next thirty to fifty years. The spatial implications are likely to be qualitatively different during this period than in the past. The much overdue shift of labour from agricultural and rural pursuits to industrial and urban pursuits will imply a massive spatial shift of population from rural to urban areas. What is of importance in this process is that even if overall population growth slows down, the level of urbanization will continue to grow apace.

Lest the deceleration in urbanization in India during the 1980s lull observers of the Indian development process into complacency regarding India's urban future and its requirements, it is necessary to investigate the possible reasons for this unusual deceleration.

The Structure of Employment

The main clue to the reasons behind the deceleration in urbanization during the 1980s is found by looking at the changes in the structure of employment. One of the most important characteristics of the acceleration of urbanization in the previous decade was the marked acceleration in employment in manufacturing in urban areas. It may be expected that with continuing industrialization and development, the structure of the labour-force would change from being predominantly agricultural towards manufacturing and services. This transformation has been taking place at a very slow pace in India (see Table 4.2). It was for the first time since the turn of the century that, in 1981, the proportion of people employed in agricultural and allied activities fell substantially below 70 per cent. That fall continued during the 1980s, despite a significant step up in the growth-rate in the rural labour-force, but the change in employment structure has been lagging far behind the transformation in the share of output where the share of agriculture in GNP is now below a third.

The fall in the share of primary sector employment may not seem significant to outside observers. This fall, first observed over the 1970s and then sustained in the 1980s, is the beginning of the long overdue structural change in the employment structure of the Indian economy. A drop in share of 7 per cent in primary sector employment is then associated with substantial changes in secondary and tertiary sector employment and consequently with accelerated urbanization.

TABLE 4.2 Employment Structure in India, 1961–1991

A. Total Male main workers[a]

Category	Per cent				Decadal rate of growth (per cent)		
	1961	1971	1981	1991	1961–71	1971–1981	1981–91
Primary[b]	68.0	69.9	65.6	62.9	18.8	11.8	19.7
Secondary[c]	12.7	11.9	14.5	13.8	8.2	45.8	18.8
Tertiary[d]	19.3	18.3	19.9	23.3	9.4	29.3	46.5
TOTAL	100.0	100.0	100.0	100.0	15.6	19.0	24.9
(000s)	129 015	149 147	177 545	221 676			

B. Male main workers (urban areas)

Category	Per cent				Decadal rate of growth (per cent)		
	1961	1971	1981	1991	1961–71	1971–1981	1981–91
Primary[b]	10.2	11.5	11.7	12.1	44.1	44.0	40.6
Secondary[c]	32.6	33.0	35.4	32.1	29.6	52.3	23.3
Tertiary[d]	57.2	55.5	52.9	55.8	24.3	35.3	43.6
TOTAL URBAN	100.0	100.0	100.0	100.0	28.1	42.0	36.0
(000s)	22 394	28 674	40 714	55 384			

[a] Only male main workers have been considered. Main workers are defined as members of the labour-force who have worked in the reference year for more than half the year. Other 'marginal workers' are not regular workers. Changes in definitions between censuses impede comparisons for female workers.
[b] Cultivators, agricultural labour, livestock, and fishery and forests.
[c] Mining, manufacturing, and construction.
[d] Trade and commerce, transport, and other services.

Source: Census of India (various); Mohan and Thottan, 1992.

What was noteworthy in the 1981 census was the distinct increase in the share of manufacturing and allied activities. In sharp contrast, their share was down in the 1991 census, despite a major acceleration in industrial output growth. This feature of the change in employment composition in the 1980s would seem to be the key explanatory factor behind the deceleration of urbanization during this decade. In fact, the growth of employment in manufacturing was significantly below the growth of total employment. This has been compensated by substantial acceleration of employment in the tertiary sector, which increased at almost double the rate of total employment growth. In this sector, interestingly, the rate of growth was higher in the rural areas than in the urban areas. Still, the increase in the growth of rural employment was essentially in the category of agricultural labour (as distinguished from 'cultivators'). The indications therefore are that the 1980s witnessed great stress in the availability of productive jobs. It would appear that the growth in labour-force was then absorbed by the residual sectors, with large increases in workers classified as agricultural labour or in the tertiary sector, both in rural and urban areas.

Organized sector data on industrial employment shows remarkable stagnation in the 1980s. The census data have confirmed this slowdown. It seems that growth in unorganized sector employment in industry has not been high either. As mentioned, this result is particularly surprising in view of the very significant increase in industrial sector output growth in the 1980s. The commonly observed pattern of economies undergoing industrialization and urbanization is that secondary and tertiary sector employment growth accelerates during similar periods of income growth. The slow-down in the secondary sector or manufacturing employment usually occurs at a much later stage of high income levels. Is this feature of the 1991 census in India an aberration or a peculiar change in urbanization patterns resulting from changes in manufacturing technologies? Or was there overstocking in manufacturing employment in the 1970s leading to correction and consequent stagnation in the 1980s? Or was it a result of inappropriate macroeconomic, tariff, and trade policies resulting in inappropriate technology choice and lower employment intensities?

Both the acceleration of urbanization along with urban industrial employment in the 1970s and the deceleration of urbanization and urban industrial employment in the 1980s point to a strong correlation between urbanization and industrialization. But, as might be expected, this correlation is between urbanization and industrial employment growth, and not as much with industrial output growth. Thus the record of slow urbanization in India is really because of slow growth in industrial employment.

City Size Distribution

The discussion on urbanization in India has long been preoccupied with what is often termed the lopsided and dysfunctional growth of large cities. There is a widespread belief that large cities have grown faster than and at the expense of small and medium-sized towns. There is also the complementary belief that the stream of rural migrants has been and continues to be disproportionately to the metropolitan cities. In fact, towns and cities of all sizes have been growing at roughly comparable rates in India for at least the last four decades. It is simply not true that large cities have grown faster than small towns. It is a purely statistical artefact that the proportion of the urban population living in Class I cities (over 100 000 population) continues to form a larger and larger proportion of total urban population.

The belief about faster growth of larger cities has persisted because, until the 1991 census, tabulations have usually been based on size classes. Table 4.3 is an example of the kind of table that is usually used to show that larger cities are growing faster than smaller towns. It may be observed that the number of cities in each size class changes between each census. Naturally, in the highest size class (Class I cities) no cities devolve out of it, while many graduate into it. Indeed, what is true is that the proportion of total urban population which lives in cities and towns above any population cut-off point continues to increase. Hence the illusion is created that cities in the highest size class are growing very fast. However, what is notable in this table is deceleration of urban growth at every level during the 1980s.

In comparing growth-rates of any size class of cities across decades we are in effect comparing non-comparable entities. For example, the 1981–91 growth-rate computed for Class I cities is between the population of 216 cities in 1981 and the population of 296 cities in 1991. Not surprisingly, it is high. Tables 4.4 and 4.5 give the tabulation that should be used in comparing the growth experience of different-sized cities and towns. These tables take towns according to their classification in 1971 (1981) and compute growth-rates by comparing the total population of towns in each class in 1971 (1981) with the total population

TABLE 4.3 Distribution and Growth of Urban Population by Size Classes in India, 1961–1991[a]

Town classification	Population in size class				Growth-rate (% p.a.)		
	(%)[b,c] 1961	(%)[b,c] 1971	(%)[b,c] 1981	(%)[b,c] 1991	1961–71	1971–81	1981–91
Class I (100 000 +)	51.4 (102)	57.2 (148)	60.4 (216)	65.2 (296)	4.32	4.60	3.92
Class II (50 000–100 000)	11.2 (129)	10.9 (173)	11.6 (270)	11.0 (341)	3.49	4.22	2.51
Class III (20 000–50 000)	16.9 (437)	18.0 (558)	14.4 (738)	13.2 (927)	2.60	2.53	2.28
Class IV (10 000–20 000)	12.8 (719)	10.9 (827)	9.5 (1053)	7.8 (1135)	1.74	2.10	1.02
Class V (5000–10 000)	6.9 (711)	4.5 (623)	3.6 (739)	2.6 (725)	–1.09	1.45	–0.13
Class VI (<5000)	0.8 (172)	0.4 (147)	0.5 (229)	0.3 (185)	–2.18	4.86	–2.42
TOTAL	100.0 (2270)	100.0 (2476)	100.0 (3245)	100.0 (3609)	3.27	3.86	3.13
TOTAL URBAN POPULATION (m)	77.6	107.0	156.2	212.9			

[a] Excluding Assam and Jammu and Kashmir.
[b] Constituent towns of urban agglomerations are not counted as separate units.
[c] Figures in brackets are the number of towns in each size class.

Source: Census of India, 1991.

TABLE 4.4 Annual Growth-Rate of Urban Population by Size of Town in India, 1971–1981[a]

Size class	No. of towns 1971	Total population 1971 (000s)	Total population 1981 (000s)	Growth-rate[b] % per year	Growth-rate[b] % over decade
Class I (100 000 +)	145	60 122	85 801	3.62	42.7
Class II (50 000–100 000)	178	12 030	16 874	3.44	40.3
Class III (20 000–50 000)	560	17 170	23 712	3.28	38.1
Class IV (10 000–20 000)	818	11 656	16 107	3.29	38.2
Class V (5000–10 000)	594	4300	6264	3.83	45.6
TOTAL	2295	105 278	148 758	3.52	41.3

[a] Excluding Assam and Jammu and Kashmir.

[b] The growth-rates are calculated by comparing the total population of towns in each size class according to their classification in the 1971 Census as compared with the total population of the same towns in the 1981 Census, e.g. the growth-rate of 3.62 per cent per year for Class I towns in 1971 refers to the growth between 1971 and 1981 of the 145 towns classified as Class I in 1971.

Sources: Census of India, 1975, 1981.

of the same towns in 1981 (1991) irrespective of their classification in the latter census. The picture emerging is quite different from that in Table 4.3. While the Class I cities have grown somewhat faster than smaller towns the differences are not very large in general. Hence it should be clear that no general statement can be made on the growth trends of different-sized towns and cities. Similar results were found in an earlier study by M. K. Jain (1977) which showed that there was no appreciable difference between the growth-rates of different-sized cities between 1951 and 1961 and between 1961 and 1971. There had, however, been an acceleration in the overall rate of growth of population in each size class between each census since 1951 until the 1991 census.

How do these observations affect urbanization policy? They show that a fixation of thinking with problems of large cities and consequently wishful thinking about the role of small and medium towns as alternatives to the growth of large cities is misplaced. In any case, the

TABLE 4.5 Annual Growth-Rate of Urban Population by Size of Town in India, 1981–1991[a]

Size class	No. of towns 1981	Total population 1981 (000s)	Total population 1991 (000s)	Growth-rate[b]	
				Per cent per year	Per cent over decade
Class I (100 000 +)	218	96 058	129 190	3.01	34.5
Class II (50 000–100 000)	269	18 103	23 825	2.78	31.6
Class III 20 000–50 000)	706	21 496	27 853	2.62	29.6
Class IV (10 000–20 000)	1007	14 303	18 367	2.53	28.4
Class V (5000–10 000)	684	5162	6711	2.66	30.0
TOTAL	2884	155 125	205 948	2.88	32.8

[a] Excluding Assam and Jammu and Kashmir.

[b] The growth-rates are calculated by comparing the total population of towns in each size class according to their classification in the 1981 census as compared with the total population of the same towns in the 1991 census.

Source: Census of India, 1991.

arithmetic of the distribution of the urban population is such that the idea of small and medium towns hiving off the migration to large cities is not practicable. The arithmetic is simple. If we want to limit the growth of the four largest cities in the country to the natural growth-rate of 2.0 per cent a year, as many as 400 to 500 towns[1] with an average population of 50 000 would have to grow at 5 per cent a year. This illustration is not given to argue that small and medium towns should not be given attention. It is given to show that attention to small and medium towns will not have an appreciable effect on the growth of large cities. There may be good reasons to make investments in these towns, but there should be no illusion that they will solve the problems of large cities. The problems of large population agglomerations have to be addressed in their own right.

The implications of the above analysis are serious in the context of the next twenty to thirty years. Planning has to search for methods that will help cope with growth of cities rather than attempting to stifle this growth. The appreciable decline in the growth of level of urbanization and of urban population growth in the last decade is unlikely to continue. I have argued that the distribution of city sizes in India is relatively even and the historical record suggests that no dramatic changes should be expected in the next two to three decades. Thus, the problems of different-sized towns have to be dealt with according to their respective needs. It must be recognized that the largest of our cities, particularly Bombay, Calcutta, Delhi, and Madras, and increasingly, Bangalore, Ahmedabad, Hyderabad, and Pune, have to be regarded as national cities. While each of them is a regional centre, each performs national functions as well. They should be seen as performing a useful as well as productive role for the region and the country as a whole.

The Regional Pattern of Urban Growth and Economic Development

Indian urban growth is best analysed by disaggregating it by regions or states rather than by looking at the experience of different-size groups of towns and cities. Table 4.6 presents the growth-rates of urban and rural populations in major states between 1951 and 1991, and Table 4.7 the urbanization levels at the time of each census since 1951. What is noteworthy is the significant variation of urban growth between the different states of India. Data are provided for only the larger states which had at least 10 million total population in 1971, thus excluding smaller states whose urbanization levels and rates of growth are, in some cases, erratic.[2]

At the lower end of the urban scale are eastern states like Bihar and Orrissa with urbanization levels of just above 13 per cent and at the upper end are the industrial western states of Gujarat and Maharashtra with urbanization levels of around 40 per cent in 1991. The eastern states in India are comparable to some of the poorest and least-urbanized countries in the world, whereas the more developed western states exhibit levels of urbanization typical for middle-income countries. Thus, in terms of urbanization levels, India's states span the range of fifty or so countries from the least-developed up to the lower middle-income range of about US$500 per capita.

What are the emerging patterns from the variance found between states during the last forty years? Each decade has brought a somewhat different pattern.

It might have been expected that the more developed of the states

TABLE 4.6 Growth of Urban and Rural Population in India by State, 1951–1991 (per cent per year)[a]

State	Urban population				Rural population			
	1951–61	1961–71	1971–81	1981–91	1951–61	1961–71	1971–81	1981–91
Andhra Pradesh	1.5	2.9	4.0	3.6	1.5	1.7	1.6	1.7
Bihar	4.1	3.7	4.4	2.7	1.6	1.8	1.9	2.0
Gujarat	1.8	3.5	3.5	2.9	2.6	2.3	2.0	1.4
Haryana	3.1	3.1	4.8	3.6	2.9	2.8	2.0	1.9
Karnataka	1.7	3.1	4.2	2.6	2.1	1.9	1.7	1.6
Madhya Pradesh	4.0	3.9	4.6	3.7	1.9	2.3	1.8	2.0
Maharashtra	2.0	3.5	3.4	3.3	2.2	2.0	1.6	1.7
Orissa	6.5	5.2	5.3	3.1	1.6	2.0	1.4	1.6
Punjab	2.6	2.3	3.7	2.6	1.8	1.9	1.6	1.6
Rajasthan	1.1	3.3	4.6	3.3	2.6	2.3	2.4	2.2
Tamil Nadu	2.1	3.3	2.5	1.8	0.8	1.5	1.2	1.2
Uttar Pradesh	0.9	2.7	4.9	3.3	1.7	1.7	1.8	2.0
West Bengal	3.1	2.5	2.8	2.5	2.8	2.4	1.9	2.1
INDIA[b]	2.3	3.3	3.9	3.2	1.9	2.0	1.8	1.8

[a] Table includes all states with a total population greater than 10 million in 1971 except Assam and Kerala.
[b] Including all states except Assam and Jammu and Kashmir.

Sources: Mohan, 1985; and Census of India, 1991.

TABLE 4.7 Level of Urbanization[a] in India by State, 1951–1991[b]

State	1951	1961	1971	1981	1991
Andhra Pradesh	17.4	17.4	19.5	23.3	26.8
Bihar	6.8	8.4	10.0	12.5	13.2
Gujarat	27.2	25.8	28.1	31.1	41.0
Haryana	17.0	17.2	17.7	22.0	24.8
Karnataka	22.9	22.3	24.3	28.9	30.9
Madhya Pradesh	12.0	14.3	16.3	20.3	23.2
Maharashtra	28.8	28.2	31.2	35.0	38.7
Orissa	4.1	6.3	8.4	11.3	13.4
Punjab	21.7	23.1	23.7	27.7	29.7
Rajasthan	18.5	16.3	17.6	20.9	22.9
Tamil Nadu	24.4	26.7	30.3	33.0	34.2
Uttar Pradesh	13.6	12.9	14.0	18.0	19.9
West Bengal	23.8	24.5	24.7	26.5	27.4
INDIA[c]	17.6	18.3	20.2	23.7	26.1

[a] Urban population as percentage of total population.
[b] Table includes all states with total population greater than 10 million in 1971 except Kerala and Assam.
[c] Including all states except Jammu and Kashmir and Assam.

Source: Mohan and Pant, 1982; and Census of India, 1991.

(in terms of per capita state domestic product or level of industrialization) would exhibit the normal phenomenon of accelerating urban growth along with industrialization and development. This has not happened in India. In fact, of the four most industrialized and developed states at the time of independence in 1947, Gujarat, Maharashtra, Tamil Nadu, and West Bengal, the latter two have exhibited urban growth rather lower than the national average (see Table 4.8). Similarly, it would be expected that the least-developed states would urbanize slowly to begin with and then gradually begin to accelerate as overall income growth and industrialization picks up speed. In fact, in India, the least-developed states have exhibited urban growth, in terms of both urban population growth and change in level of urbanization, which has often been higher over the decades than the national average. Among the least-developed states, Madhya Pradesh has urbanized consistently faster than the national average, while Bihar, Orissa, Rajasthan, and Uttar Pradesh have also urbanized somewhat faster in at least two of the four decades. The variation in levels of urbanization has consequently declined over the decades, except during the 1980s.

TABLE 4.8 Selected Indicators of Economic Development in India by State, 1961–1991

State	Per capita net domestic product (constant 1970–1 prices in Rs)				Value added in factory sector[a] (current Rs per capita)				Agricultural productivity[b] (tons per person)			
	1961	1971	1981	1988–9	1961	1971	1981	1987–8	1959–62	1970–3	1980–3	1988–91
Andhra Pradesh	518	586	647	793	9	29	109	188	0.88	0.81	1.11	1.12
Bihar	390	418	441	527	14	31	72	221	0.71	0.71	0.62	0.71
Gujarat	697	845	913	1161	52	108	334	694	0.59	0.79	0.90	0.85
Haryana[c]	—	932	1060	1380	—	70	268	494	—	2.66	3.08	3.97
Karnataka	559	675	687	865	14	62	162	300	0.82	0.98	1.02	0.94
Madhya Pradesh	472	489	517	680	8	27	115	236	1.24	1.24	1.22	1.31
Maharashtra	769	811	970	1183	69	167	476	793	0.92	0.56	1.11	1.20
Orissa	392	541	529	625	6	27	75	150	0.98	0.97	0.99	1.20
Punjab	760[d]	1067	1383	1803	4	52	230	507	1.73	3.08	4.68	5.62
Rajasthan	500	629	535	722	5	26	98	160	1.10	1.29	1.23	1.40
Tamil Nadu	571	616	584	791	24	75	254	507	0.90	0.99	0.78	0.91
Uttar Pradesh	457	493	519	625	9	24	68	186	0.85	0.98	1.16	1.35
West Bengal	758	729	797	954	50	97	252	371	0.92	1.07	0.90	1.22
Mean	558	676	737	931	20	61	193	370	0.90	1.24	1.36	1.68
Coefficient of Variation	0.24	0.27	0.36	0.37	1.05	0.67	0.61	0.55	0.42	0.59	0.85	0.82
Ratio between richest and poorest state	2.0	2.5	3.1	3.4	17	7	7	5	2.9	5.5	7.5	7.9

[a] Value added in factory sector is the total value added in organized manufacturing in the state divided by total population.
[b] Agricultural productivity is defined as the total average annual output of food grains in the state divided by the total male agricultural labour-force.
[c] Haryana was part of Punjab until 1964.
[d] Undivided Punjab.

Source: Centre for Monitoring the Indian Economy, 1992.

During 1971–81 all the poorer states experienced accelerated rates of urban growth, whereas all the richer states had relatively slow growth. In some of the poorer areas—Orissa, parts of Bihar, eastern Uttar Pradesh, and eastern Madhya Pradesh, where current urbanization levels are low and towns far apart, there was a tendency for reclassification of large villages as towns, and the potential for the emergence of new towns seemed to be greater. However, this pattern has not repeated itself during 1981–91 and only Madhya Pradesh among the poorer states has exhibited an above-average growth-rate during the past decade. Among the richer states, Haryana shows above-average urban growth as does Maharashtra.

It is also useful to examine the pattern of rural population growth. The growth-rate of rural population has been remarkably steady over the last four decades, varying between 1.8 per cent and 2.0 per cent per annum during this period. There was a significant fall in rural population growth-rate in 1971–81, associated with the acceleration in urbanization, and hence an increased urban-rural growth differential (URGD). The surprise of the 1991 census is the maintenance of the 1.8 per cent annual rural population growth along with the significant deceleration in the urban growth-rate. What is disturbing is the tendency of rural population growth-rates to increase in the poorer states of Bihar, Madhya Pradesh, Orissa, and Uttar Pradesh. Among the richer states, West Bengal seems to have experienced a development inversion with declining urban growth accompanied by increasing rural population growth. Only one state, Gujarat, exhibited a significant fall in rural population growth over the past two decades. It is notable that rural population growth has tended to decline in the high agricultural productivity states of Haryana and Punjab while increasing in the low productivity states. It thus appears that it is not agricultural prosperity that is keeping people from migrating to urban areas, rather it seems to be rural poverty that has that effect. During 1971 to 1981 there was a perceptible decline in the proportion of male labour-force engaged in agriculture, from 69.9 per cent in 1971 to 65.6 per cent in 1981. This transformation has continued during the 1980s, as the proportion declined further to 62.9 per cent in 1991. However, whereas in the 1970s, the annual rural population growth-rate fell correspondingly from 2.0 per cent to 1.8 per cent, in the 1980s, the fall in the proportion of the agricultural labour-force was not accompanied by a similar fall in rural population growth. The shift in labour from agriculture to the tertiary sector was significant within rural areas. However, neither in the 1970s, nor in the 1980s, was rural population growth higher than urban population growth in any state.

How is the emerging pattern of urban growth to be interpreted? It is helpful to examine some rough indicators of economic development. Different indexes of economic development are available to explore the connection between urbanization and economic development. This analysis is restricted to the period since 1961 because consistent data are difficult to compile before then. Even since 1961 magnitudes of state domestic produce are only approximate since they are difficult to compute accurately in a country-wide common market. The Central Statistical Organization (CSO) publishes these estimates as compiled by the various state statistical bureaus but does not vouch for their consistency or accuracy. The data here are taken from those compiled from the official data by the Centre for Monitoring the Indian Economy, a private organization. These data are adequate for deriving broad patterns.

Table 4.8 provides three indexes of economic development at the state level: the per capita net domestic product as an overall proxy for development; value added per capita in the factory sector (organized manufacturing activity) as a proxy for state urban productivity, and per capita agricultural productivity as a proxy for rural productivity. Overall, there is an increasing trend in inequality between states in per capita net domestic product, though the increases in inequality slowed down during the 1980s. The ratio between the richest state and the poorest state has increased consistently from 2.0 in 1961, 2.5 in 1971, 3.1 in 1981, to 3.4 in 1991. Correspondingly the inequality between states in agricultural productivity has tended to increase since 1961 due mainly to the onset of the Green Revolution in north-western India in the late 1960s and early 1970s. There has been some reversal of this increasing inequality during the 1980s, though the difference between the most productive and least productive states continued to increase. In contrast, inequality in manufacturing productivity has fallen significantly over time and this tendency continued during the 1980s, along with a consistent reduction in distance between the most and least productive states.

At the time of independence the dominance of Gujarat, Maharashtra, Tamil Nadu, and West Bengal in manufacturing was absolute. Other states had very little manufacturing activity. Within those four states manufacturing was concentrated in the major cities—Ahmedabad, Bombay, Madras, and Calcutta. Consequently Indian policy-makers were consistently concerned with industrial dispersion. These data and other evidence (e.g. Udai Sekhar, 1983) indicate that there has been significant success in achieving this objective. In contrast, inequality between states with regard to agricultural productivity has become considerably worse. The fruits of the so-called 'Green Revolution' (the introduction of new agricultural technologies) have largely been confined to the wheat-producing areas of the country, although of late, progress in rice cultivation in other areas has also been reported. Large dry areas of central India which grow other food grains have yet to see any perceptible increases in agricultural productivity. Consequently, the Green Revolution has exacerbated income differences among states even though differences in the industrial performance shrank dramatically.

These data provide the basis for interpreting the somewhat puzzling pattern of urbanization observed between the states in India. The advanced industrial states of Gujarat, Maharashtra, Tamil Nadu, and West Bengal exhibit low levels of agricultural productivity. Increases in industrial productivity and production have had no connection with agricultural productivity in these states. Thus the industrialization in these states has little to do with their hinterlands (except for the sugar industry in Maharastra). Industrial development in these states has not contributed significantly to overall development, particularly in rural areas. Hence the acceleration in urbanization which might be expected to accompany industrialization and development has been muted. Haryana and Punjab, whose development has been led by agriculture, exhibit a more balanced development pattern and may be expected to undergo accelerated urbanization in the years to come as they further intensify their industrialization. Agricultural productivity growth brings with it demand for products and services which need to be produced in small and medium towns in close proximity to the areas of agricultural production. There is also then greater possibility of non-farm rural employment in response to such demands. This kind of process does not appear to have been set in motion in the more industrialized states of India. Other evidence suggests that they have, however, given more attention to industrial dispersion to the more backward areas within their own states. West Bengal is most striking as a case of de-industrialization since independence, with consistently low rates of urban–rural growth differential (URGD). Agricultural productivity had been stagnant until recently, whereas industrial productivity has come down to the national average. At the time of independence, industrial productivity of West Bengal was about 2.5 times the national average.

Now it is about the same as the national average. Urban growth has been correspondingly low. If agricultural prosperity shows significant and consistent improvement in coming decades we may expect increasing and more urban growth. The normal phenomenon of accelerating urban growth with industrialization seems to have been muted in the advanced industrial states in India because of the absence of corresponding improvements in agricultural productivity.

The poorer states exhibit a similar problem in the low degree of connectivity between their industrial development pattern and the agricultural base. This is reflected in the maintenance of high rural population growth-rates in these states in the past decade. The burst of urbanization experienced in 1971–81 has not continued in these states during the past decade, except in Madhya Pradesh. Manufacturing productivity in these states has continued to show significant improvement, but agricultural productivity levels continue to remain low. The lack of adequate urban infrastructure investment may have contributed to the slow-down of urbanization in these states during the 1980s. The general lack of organized sector industrial employment growth in the 1980s may have also contributed to slower urbanization during the period. Overall, it may be said that the acceleration in industrial growth rates in India during the 1980s was not accompanied by a corresponding acceleration in urban growth rates probably because of the low growth in industrial employment and a relative stagnation in agricultural growth rates in most areas of the country. It is only the north-western region which has significantly higher levels of agricultural productivity relative to the country as a whole. It is also a matter of conjecture whether low investment in urban infrastructure has itself contributed to the slow-down of urbanization in the 1980s. It may be that information on the lack of employment opportunities, the poor quality and non-availability of housing, and the poor state of urban amenities such as water and sanitation, in some urban areas is now being better transmitted to rural areas.

A major conscious strategy is needed to increase agricultural production and productivity across India along with an expansion in non-agricultural employment. The correction of various macroeconomic distortions in trade and industrial policies which are currently being pursued in the ongoing economic reform process in India may contribute to improvement on this score. The reduction in the overall protection of industry and the removal of various export constraints on agriculture will, for example, help in reducing the bias against agriculture embedded in such industrial and trade policies. Similarly, the reduction of high protection of capital-intensive industries is also likely to shift resources to more labour-intensive manufacturing sectors, thereby increasing non-agricultural employment. However, little attention is being devoted to the improvement of urban finance and management and to the provision of adequate shelter.

A positive approach to urbanization would imply a different strategy than that which has been followed in the past. The current rate of urban infrastructure investment is altogether inadequate to sustain the demand of a modernizing, industrializing, and increasingly urban economy. At the same time, the national fiscal crunch suggests that few discretionary resources will be available in the years to come. Hence, the key areas for attention in coping with city growth are financing and management of urban development.

Urban Local Governance

The 1991 census reports fifty-three cities with populations of over half a million each and 296 cities with a population of 100 000 each. Even with modest rates of urban population growth, these numbers will continue to increase in the foreseeable future. There is, thus, increasing need for investments in urban infrastructure: sanitation, drainage, sewerage, water supply, roads, electricity, and transportation. The requirements, though not well quantified, of both financial and physical resources are likely to be huge. The essential problem is that such infrastructure yields benefits over a long period of time while investments have to be made now.

The first necessity is the expansion of responsibilities and functions of local bodies along with strengthening of their finances. As a principle, it should be clear that rural areas should not suffer because of the needs of investments in urban infrastructure. Hence, urban areas need to be largely self-financing.

The current state of the management of urban development covers a very wide variety of experiences. Some large cities are still without any organized management worthy of note while others might well be over-managed. Most cities are governed by municipal corporations or municipal boards, but many of these have been suspended for varying lengths of time and are being administered by state-appointed officials.

In cities where there are major urban investment programmes, there are usually urban development authorities operating directly under the state government rather than the local authority. The reasons are essentially two-fold. First, local authorities have, for long, been regarded as inefficient, ineffective, corrupt, and too susceptible to local pressures. Second, the financing of urban development has been such that the funds have come from the central or state levels—hence it is thought prudent to entrust the execution of works to an agency which is directly under state control. The idea is then for the Urban Development Authority (UDA) to be responsible for capital investment works while the municipal corporation gets relegated to maintenance tasks. In some cases, the demarcation of responsibilities between the municipal corporation and the Urban Development Authorities is not entirely clear and leads to avoidable disputes. As mentioned, the general pattern is that the UDA is responsible for making the investments which are then taken over and maintained by the corporation. This often causes difficulties since there has typically been no connection between the investments and the local financing capabilities. Hence the local authority is often not able to raise resources adequate for the maintenance of the new infrastructure. Clearly, this problem will remain until the financing, planning, and management of urban development is co-ordinated such that local authorities can be held accountable.

Since the trend over the last two or three decades has been a steady erosion of the functions, powers, and prestige of local government, the prognosis for their regeneration has not been optimistic until the recent constitutional amendment for devolution of powers to local bodies. Yet there would appear to be limited choices available. Indeed, the rehabilitation of local authorities from the present state of neglect would need major shifts in the national and international techniques of financing urban development. Local authorities need not be seen as unnecessary stumbling blocks in programme formulation and execution. Urban development authorities could be subsidiaries of the local authorities rather than of the state governments but with appropriate checks and balances such that the local authority can indeed be held responsible for the funds placed at its disposal and for execution of programmes planned. The local authority, e.g. a municipal corporation, should then be integrated into the national and state capital budget process. Urban development planning exercises would then be subject to the discipline of national, state, and local resource constraints automatically. The resources would be related to the availability of financial resources whether they are borrowed from the state or centre, or raised locally, or raised nationally in the capital market. Such a procedure would improve financial discipline and induce the formulation of investments more in line with a city's paying capacity. Infrastructure investments would be made within the corporation's fiscal resources and tax capabilities.

At present the administrative, fiscal, management, and accounting capabilities of corporations are such that these ideas would be seen to be naïve. The case of Bombay has, however, shown the advantages of a strong local government once it takes hold and is recognized and respected by the state government. For a local government to be strong in the long run, it needs to be strong at the grass roots. Further, the collection of user and other civic charges requires a decentralized form of organization. Hence, the strengthening of local authorities could also include an expanded programme of intermediate forms of organization geared to urban community development through a participative process of planning, execution, financing, monitoring, and evaluation of projects and programmes. At present, in urban areas, there is no one between the citizen and the corporation or municipal board. In a participative sense, in the places where the local authority is suspended, the citizen really has no intermediary between himself and the state government. While there may be some argument about the efficacy of 'panchayats' (village local bodies), etc., there is at least an existing system of local participation in rural areas such that groups of 1000 to 10 000 people, depending on the size of the village, have some say in the running of their local affairs.

The recent constitutional amendments providing for 'panchayati raj' in the rural areas and 'nagar palikas' (municipal bodies) in the urban areas has given constitutional recognition for local government for the first time. Although there is general agreement in all political parties for this move favouring administrative and political decentralization, the actual mechanism by which this amendment will be implemented in practice remains to be seen. Optimism on this score would be misplaced at present: the amendment has not originated in a popular demand for this move. Rather, it has been a top-down action. There is also a crying need in urban areas for the existence of intermediate organizations which elicit the participation of people in decision-making. Only then would there be a closer connection between tangible works and resource raising. There needs to be widespread adoption of urban community development programmes.

Financing Urban Development

The main revenue sources of urban areas are property taxes and octroi duties (a tax on goods entering urban areas). There appears to be general agreement that urban local finances have not been as buoyant as might have been expected or as is desirable. In principle, local taxes should increase faster than national and state taxes because of increasing urbanization and industrialization. Property taxes and octroi are the main revenue sources for local bodies: both these sources should be very buoyant. When there is an increase in housing construction and existing properties are constantly appreciating in value, property taxes should be expected to lead the growth in revenues. Similarly, with increased urban activity, octroi can be expected to increase faster than other state and national revenues. In principle, therefore, financing urban development should not be a serious problem.

In practice, there are serious difficulties in the levy and collection of both these main sources of urban local revenue. Octroi, which has been generally more buoyant than the collection of property taxes, is a bad tax since it impedes movement of goods across the country. It is therefore being eliminated from most states as a source of revenue. Its substitutes are generally not as buoyant. The other main source, the property tax has not been buoyant because of the existence of rent control laws which make it difficult to revalue properties as they rise in value. Its administration in most cities has also been far from satisfactory.

If rent control laws are suitably amended thereby enabling regular property revaluation, increases in property values resulting from investments in public infrastructure would then be captured in the tax net and these revenues would help in maintenance expenditures as well as in debt-servicing of the capital expenditures incurred. At the same time, the provision of services such as water-supply, garbage collection, transportation, and the like, whose consumption is by identifiable households, should be based on the levy of user charges as far as is possible. In order to help the poor, such charges could subsume a system of cross subsidies.

Even if it is assumed that all these changes occur such that current revenues increase adequately to cover the financing of capital as well as maintenance charges of urban infrastructure and other investments, the problem of finding initial funds would still remain.

The brief review of local governments and the resources that they command has shown that urban management and financing needs a completely new strategy in India if urban growth is to be accelerated and if it is to occur in a balanced and healthy manner. This section puts forward some ideas which could ameliorate the current dismal situation in India. The success of the current economic reform process depends crucially on accelerated industrialization activity. Moreover, successful industrialization should bring in its wake a revival of growth in industrial and urban employment in the 1990s. This will face difficulty unless there is adequate investment in urban infrastructure and housing which would enable the urban labour-force expansion which results from increased demand for labour from industry.

The essential problem of urban infrastructure investment is that such infrastructure yields benefits over a long period of time whereas investments have to be made now. Given national priorities and the very serious fiscal constraints in the foreseeable future, it is difficult to see a radical step up in the allocation of budgetary resources for urban infrastructure investment. Yet demand for these investments will be difficult to resist and, therefore, willy-nilly, they will have to be made. It is important that urban infrastructure is made largely self-financing. Systems must be designed in such a way that their long-term viability is ensured. This would involve appropriate pricing of the facilities and services offered as well as reforms in the local tax structures since not all urban public services can be based on user charges. A number of interrelated fiscal, administrative and legal measures are needed in connection with the financing of urban development.

Urban infrastructure investments are characteristically long-term investments in the sense that the benefits which accrue do so over a long period of time. Thus sewerage, drainage, water-supply systems, and roads constructed today are expected to yield benefits for at least fifty years. Hence the finance needed for such investments should be long-term finance. And it is the lack of long-term sources of funds that has been a key constraint in the financing of urban development. The result has been that these investments have been neglected until absolutely unavoidable and then resort has been made to central funds supported by external financing, characteristically from the World Bank. This is not a new problem nor is it specific to India. Rapidly urbanizing countries even in the last century had to resort to large magnitudes of external financing for urban infrastructure investments (W. Arthur Lewis, 1978). The problem arises because urbanization is accompanied by or caused by increasing levels of industrialization.

Hence the demand for savings to finance what are usually termed 'productive' investments with quicker pay-off periods is also very high. It is therefore usually difficult to find the long-term finances necessary for the financing of urban infrastructure investment. Hence resort is typically made to external financing.

Different countries have solved the problem of urban infrastructure financing in different ways. In the USA, the standard method of financing is through the use of municipal bonds. The resource cost of raising funds through the floating of municipal bonds is reduced for the local bodies by making the interest on them free of income tax. Interest paid on them can therefore be lower than on other bonds. This is effectively a subsidy from the federal government to city governments. The well-developed financial market in the USA does the rest. Ratings of bonds floated by different authorities account for the different degrees of risk associated with bonds for different cities. Conversely, city authorities have an incentive to keep their fiscal house in order such that they can get and retain favourable ratings.

In the case of India, it is probably difficult to evolve a similar system for some time. Capital markets need to work better and local authorities need to be more creditworthy. Furthermore, in India, although the financial system is now gradually moving away from credit rationing, the bond market is yet to be developed. Municipal bonds would therefore seem a long way off. At the same time, the current problem is that the prognosis for increasing levels of international long-term concessional finance is gloomy. It is essential that measures be taken now such that long-term funds are increasingly available internally. This becomes all the more necessary as we have a larger number of towns moving into the 100 000-plus, the 500 000-plus category, and the 1 million-plus category. Even if there were the possibility of additional external finance it is unlikely that the volume available would be enough for all these needs. Moreover, the lack of internal finance has affected the smaller towns and cities more adversely since external finance finds it easier to go to the metropolitan cities.

It is in this context that discussion on the strengthening of local bodies—both organizationally and financially—assumes great importance. Only if local bodies can be held financially viable as well as responsible can a system of long-term finance be instituted for proving resources to local authorities. As an intermediate measure, it should be possible for an apex body at the national level to raise and distribute long-term funds.[3] Such a body could float government-guaranteed long-term securities such as twenty- to thirty-year Urban Development Bonds. These would then be marketable just like any other public-sector bonds. Similarly, if they are made tax free, the interest rate given could be lower and hence the cost of funds kept within prudent limits. Such funds can be passed down directly to local or, in some of the larger states, through state-level financing bodies. Such a system, involving better project design, project evaluation, and local fiscal responsibility, should promote the recovery of costs through the levy and collection of appropriate user charges.

Once such a system has been in place for some time, it will have helped in inducing autonomous fiscal responsibility in local bodies. At that point, it could be possible for some of the larger municipal authorities to raise public resources by directly floating municipal bonds. The ability of local bodies to raise funds—both directly as well as indirectly through the apex body—would depend crucially on their fiscal position and their capability to invest these funds in resource-raising investments. Municipal bodies must be held accountable for the investments they make. It is for this reason that it is a bad idea to separate investment responsibilities from maintenance responsibilities as has been done in many cities by founding urban development authorities. These authorities are state agencies made responsible for investment activities independent of the existing local municipal bodies. Completed investments are typically transferred to the municipal bodies for maintenance, while little is done to raise the resources for undertaking the maintenance responsibilities. If, however, such a system enforces better fiscal discipline, and local authorities do pay back the borrowed funds according to schedule, there would be an overall increase in the efficiency of utilization of resources. The establishment of such a body could also help in beginning a system of better urban-project evaluation and professionalization of project preparation, appraisal, and evaluation.

The foregoing should make clear the urgency of integrating local management and financial responsibilities in India in the interest of healthy urban development. Some of the problems of urban growth that have been encountered in the 1970s and 1980s have undoubtedly arisen from the lack of accountability in the system. There is simply no one who is really responsible for urban development. The central government is too remote from the concerns of specific cities and, moreover, lacks viable instruments. The state governments are also beset with state-level problems. Local governments do not really exist or are very weak. It is no wonder then that India is encountering problems in healthy urban growth and has even witnessed a deceleration during the 1980s.

Urban Housing

The problem that perhaps causes the most concern to a majority of urban dwellers is the problem of finding an appropriate place to live. The popular feeling that prices of shelter of all kinds have been rising excessively would indicate that housing investment has not kept pace with increases in housing demand. The national income accounts indicate that housing investment as a proportion of gross capital formation in the country has declined from about 30 per cent in 1950 to only about 13–15 per cent in the 1980s. To a large extent this is as should be expected in an economy undergoing considerable diversification with massive investments in industry. What is more worrying, however, is that the rate of growth of gross capital stock in housing (in terms of value) has been only about 1.5 to 1.6 per cent a year whereas total annual population growth has been about 2.2 per cent. At the same time the prices of construction materials appear to have been rising much faster than the general wholesale price index for quite some time. In rural areas the pressures of population growth and the ravages of deforestation also seem to have taken their toll on the availability of construction materials.

Census data indicate that the quality of shelter per capita has declined over the last thirty years as measured by indices of crowding. Between 1961 and 1971, for example, the average number of persons per room increased from about 2.6 to 2.8. Surprisingly, urban and rural trends are identical in this respect. The information available from the 1981 census indicates that, if anything, the situation had got worse during the 1970s. The pace of change in the quality of housing which was evident during the 1950s and 1960s seems to have been arrested in the 1970s. A new approach to the provision of shelter is, therefore, sorely needed, before conditions decline even further. Paradoxically, it must be more bold in providing for an expanded housing stock, as well as more cautious in what we expect to provide taking account of the existing levels and distribution of income in the country. Given the low levels of income and the low rate of growth it would be unrealistic to expect people to devote much greater proportions of income to housing. Hence the approach needs to change from the prescription of unrealistic norms to the provision of facilities and conditions that enable people to obtain the maximum quality of shelter according to their needs as well as their abilities.

The introduction of new housing finance institutions has now made possible the introduction of a widespread system of housing finance. At the national level, more realistic economic estimates need to be made of the likely demand for housing over the next decade. It is only then that appropriate measures can be taken to provide an institutional structure which permits easier investment in housing for all income groups. Given the kinds of price increases observed for construction materials, it will also be necessary to examine the supply situation of construction materials and to ensure their expanded supply. Since almost 90 per cent of housing investment is done in the private sector, measures need to be taken such that impediments to housing investment and finance are removed. Indeed the successful functioning of a housing-finance system depends on making housing more marketable. This means that the legal framework governing transfer of real property has to be reviewed and adapted to make the conveyancing of real property easier. Further, at the local level, various town planning and building by-laws affect housing investment in different ways. Adequate attention is still not given to the role of 'informal', or 'unorganized', or 'popular' housing which effectively provides shelter to the vast majority of the urban poor. The housing finance system needs to be adapted to the needs of the less well-off: the formal ways of providing housing finance need to be changed so that the poor can also have access to such funds. It must also be understood that a section of the poor simply cannot afford ownership housing, nor is it well suited to their needs. Hence special attention needs to be given to the rental needs of a large proportion of the population. Despite good intentions, it is unlikely that the public sector can directly provide shelter to the poorest of the urban population. It must, however, find ways to assist them in their own efforts.

The estimate of total (urban and rural) gross capital formation in residential dwellings was Rs75 billion in 1985–6. Although data are not yet available for the most recent years, gross capital formation in residential dwellings in 1990–1 may be estimated to be in the range of Rs125–140 billion. It is difficult to estimate the urban component of this housing investment. Annual demand for new urban housing investment amounts to roughly 5 to 6 per cent of current urban households.

Assuming a current urban population of about 44 million households, and a cost of about Rs50 000 (in 1990–1 prices) for an average standard urban house, total maximal annual demand for new housing investment in urban areas amounts to about Rs110–30 billion. The rural population may be assumed to comprise about 120 million households and the rural population growth-rate is about 1.7 to 1.8 per cent. Hence assuming a 1.5 to 2 per cent replacement demand, total rural demand would be for 3 to 3.5 per cent of rural households. If the 'weighted average standard rural house' costs about Rs17 500 (1990–1 prices), total rural demand works out to about Rs60–75 billion. Hence, total maximal rural and urban housing demand together amounts to about Rs170–205 billion. Total urban housing demand comprises about 65 per cent of the total in the country. Assuming the same proportion, actual gross capital formation in urban housing in 1990–1 would be about Rs80–90 billion. Hence the maximal demand of Rs110–30 billion is about Rs20–40 billion in excess of existing gross capital formation in urban housing. It would seem therefore that it is not too difficult a task to meet the new housing requirements thrown up by the increasing urban population. These rough calculations also suggest that a cumulative underinvestment of about 25 per cent every year would lead to serious housing and infrastructure deficiencies in urban areas. This has probably been happening during the 1980s thereby contributing to the urban slow-down.

It must be emphasized that the exercise outlined above is full of assumptions and approximations. But these have all erred on the side of exaggerating housing demand. However, the indication is that the easy availability of housing finance to all who desire it would probably increase total urban housing investment by no more than Rs20 to Rs40 billion—not too large an amount to be financed. Although the shortfalls are significant they are not unbridgeable, given sensible policies.

The main point of this exercise is to demonstrate that the issue of urban shelter is not an insoluble problem, that the increase in the availability of housing finance that may be required in any realistic assessment of savings and investments patterns is quite possibly within the reach of existing resources. It could well be that the level of saving could increase enough so that a radical rearrangement of investment priorities is not necessary. What will also happen as a result of the easier availability of housing finance is the monetization of some investments which were made earlier in kind.

Shelter for the Poor

Typically, the poorest of the urban dwellers, as many as 30 to 50 per cent in most cities, live in dwellings which have been constructed by themselves or with the help of neighbours, friends, and local artisans and construction workers. This would include squatters as well as other slum dwellers. The dwellings range from mud huts to 'pucca' (masonry) structures. The key point is that in this sector there is little use of architects, contractors, and engineers; there is little formal design; much of the investment is physical as opposed to financial; materials used are often waste materials not used elsewhere. The poor are thus able to construct dwellings at costs much lower than any system of public-sector construction can permit. A survey of Pune slums in the late 1970s, for example, indicated that the best shanty dwellings comparable to or better than any public-sector-provided housing for the poor, cost just over Rs2000 on average—in contrast to a minimum of Rs8000–10 000 at which the public sector was able to provide such housing (Bapat, 1981). These investments are substantial both in the aggregate as well as for each household. This sector may be termed the 'popular sector' and this kind of investment must be regarded as very useful, rather than as a nuisance of slum dwellers as is often the case. Indeed, since most of these dwellings are not able to meet the minimum standards laid out in most local building codes, they are thus unauthorized by definition.

The poor face specific constraints in their quest for shelter. The public sector should not attempt to provide standards that are higher than what the poor can afford and for which an adequate amount of subsidies cannot be found, given other needs in the economy. Innovative solutions have indeed been developed in different places in the country. One good example is the Hyderabad Urban Community Development Programme, another is the Bustee Improvement Programme in Calcutta (Planning Commission, 1986*b*). The Urban Community Development Programme in Hyderabad is now about twenty-five years old, but housing activities in the programme are of more recent vintage. The idea is essentially to give low-interest loans to dwellers *in situ* and assistance in the procurement of materials. Because of the community development programme this can be done for groups of people at a time so that whole communities get upgraded. The Bustee Improvement Programme in Calcutta focused on improvements in common services such as water supply, sanitation, and the paving of lanes. Other innovative programmes, some government sponsored, some community organized, and others initiated by voluntary organizations, can be found around the country.

The lessons common to these various programmes is that the government should eschew direct construction of houses for the poor. Public agencies need to concentrate on helping people with amenities that are beyond their reach, access to resources, and legal support:

(i) Provision of water supply, sanitation and roads;
(ii) land development;
(iii) housing finance in small amounts for the purchase of materials for new construction as well as repairs;
(iv) easy availability of materials that are typically used;
(v) security of tenure;
(vi) removal of legal impediments to such housing.

It is often argued that such a policy would merely perpetuate slums. This is true in the sense that people would continue to live in structures that are regarded as inadequate by most engineers and architects. But policies such as slum clearance and provision of public housing are worse because they can only provide for a fortunate few while others are totally neglected. It must be recognized that national resources are conserved when people are allowed to use waste and recycled materials to provide shelter for themselves while the public sector encourages these activities by helping in the provision of finances, materials, and security of tenure. Another key insight gained from the experience of different programmes around the country is that there is a great variety of approaches to the improvement of shelter for the poor. Thus public programmes must provide flexibility. A major problem in the implementation of programmes involving the disbursement of small loans to large numbers of people is that transaction costs are high relative to the loan amount. Non-formal methods for the disbursement of loans as well as the collection of payments can considerably reduce these high transaction costs. Again, different methods have been found for such activities—as by the Working Women's Forum in Madras, and by the Self-Employed Women's Association in Ahmedabad—for the administration of small loans for self-employed women. In each case, there has been a high degree of community organization.

The assumption in most government housing programmes is that subsidies are required for the less well-off to achieve access to better housing. The result is that the limited resources available finance the needs of a fortunate few. Interest subsidies have been the dominant mode of providing these subsidies. What is needed instead is an 'unbundling' of the housing package. Given the low levels of and high variance in earnings that many less well-off are subject to, it is easier for them to make small investments incrementally over a period of time than to make a large one-time investment in a fully finished house. They are also loath to enter into long-term fixed liabilities which may be difficult to honour. The need then is to match this kind of demand profile with an appropriate housing supply response. A good proportion of the poor would prefer easily available unsubsidized finance as against difficult to get subsidized finance.

Urban Land Development

The most serious impediment to housing investments in India today is probably the lack of an adequate supply of developed urban land. The result is enforced illegality of occupation for many of the poor. It is true, of course, that for the poorest, the only way they can achieve any access to shelter is through squatting. They can simply not afford any level of housing payments. It is difficult to devise any organized system for helping this section of the population. The best that the Government can do is to not thwart the poor in helping themselves. This is another method of unbundling the housing package in order to make it affordable. The poor acquire free land by squatting, then incrementally develop a structure as they can afford it. At this point the public authorities can help by legalizing the settlement and providing access to public services; and once the settlement is legalized, it will also be possible for the households to receive housing finance in small doses for further upgrading of the structure. This process is difficult to 'organize' since it is characteristically an 'unorganized process'. It is not feasible for the government to pronounce an open invitation to squatting, hence there must invariably be some tension between the authorities and squatters. At the same time, there could clearly be a policy of regularization once a settlement has been in existence for some time. This is indeed what happens in many places through the political process.

The central problem concerns the conversion of agricultural land to non-agricultural uses. In India, the problem has become more complicated with the promulgation of the Urban Land Ceiling Act, which makes it very difficult for any but public authorities to assemble and develop urban land. The current pattern is for public authorities to identify and, by public notification, schedule large tracts of land on the periphery of the city for urban use, often at the time of framing a new Master Plan. The notified price is usually the prevailing price for agricultural land. Poor farmers are the net losers when large tracts of land on the periphery of a city are notified, while largely better-off urban settlers get the benefit of obtaining land at effectively subsidized prices. The low rates of compensation also give rise to legal objections from the owners and, as a result, land acquisition can be held up for long periods of time. In any case, because of financial constraints, public authorities often acquire notified land many years later, and develop it even later. The private sector, on the other hand, is not allowed to develop the land. The result is an overall shortage of urban land accompanied by an unwarranted rise in urban land prices. Public ownership and the shortage of serviced land encourage large-scale squatting. It is imperative that innovative schemes be tried which have the effect of speeding up land development in an organized manner and at low cost to the public exchequer. One alternative to large-scale land assembly and development is land readjustment, a method that has been used successfully in Japan and Korea.

Coping with City Growth: The Way To Go

The urban slow-down of the 1980s has provided some breathing space for infrastructure investment to catch up with the expanding urban population in India. However, this cannot be expected to last long. Before long, rapid urban growth will be with us again. Even if overall population growth slows down substantially, rapid urban growth will continue (see Mohan, 1984). Slower overall population growth will reduce the pressure on agriculture for the provision of mass employment; the productivity of agricultural labour would then improve farther leading to accelerated rural income growth. This would generate increased demand for urban goods while simultaneously releasing more labour for urban pursuits so urban demand for labour needed for the production of urban goods will consequently increase, thereby causing accelerated urbanization. The effect of a slow-down in overall population growth on a slowing of urban population growth will be muted.

The most disturbing aspect of the results provided by the 1991 census is the stagnation in the proportion of labour-force employed in manufacturing. This has happened for the first time since 1931. The economy is not producing enough jobs in the productive non-agricultural sectors. The more detailed tabulations reveal that the major employment gains have been made in the categories of 'agricultural labour' and 'other service'—both being categories which are residual in nature. Anyone who is not a cultivator in rural areas has little choice but to become an agricultural labourer or take up 'other service' activities. In urban areas, if it is not possible to find employment in manufacturing, transport, or construction, there is little to do but to join retail trade or 'other service' activities. Thus the most critical task facing policy-making is to once again orient the industrial sector toward greater employment generation without compromising productivity. Explicit urban policies have little role here. Rather, a combination of appropriate trade, tariff, and financial sector policies is required. Programmes regarding the urban infrastructure can only be supportive of this larger endeavour.

The Gordian knot of financing urban infrastructure investment and of inadequate urban shelter with extremely limited public resources can be cut by innovative strategies. The response to urban population growth at our stage of development must be a welcoming one. Indeed we must seek out ways to accelerate it in the interest of improving human welfare in the country.

It is commonly assumed that it is not really feasible to provide adequate levels of urban investment to cope with the emerging demands of increasing urbanization. However, the extremely rapid urbanization experienced by some East Asian and Latin American countries during the last three to four decades has been managed in a reasonably competent fashion—though some countries have been more successful than others. The challenge in India (and in China) is one of magnitude: no system as large as India or China has yet undergone rapid urbanization. The innovation that is required in urban management, urban financing, and urban governance will have to be of a continuing and self-adapting nature if this unique challenge is to be faced successfully. It is essential therefore that effective decentralization take place since the required level of innovation is unlikely to occur in a centralized bureaucratic framework.

This paper has also documented the increasing levels of regional inequality. It is likely that this inequality will continue to increase. Accelerated industrial development in the future will probably be regionally concentrated as it has been in China in recent years. Social and political strains are therefore bound to emerge, and systems and institutions will be stretched to cope with those strains. Increasing interregional migration would be one response to the increase in regional inequality.

But such movement of large numbers of people will itself increase social tensions. Awareness of this prognosis for the future should be used to better prepare policies and programmes which assist in the management of such trends. Attempts to buck these trends would prove fruitless.

The transformation in employment structure from agricultural to non-agricultural pursuits is long overdue in India. The next thirty to forty years will certainly see a dramatic change in the employment structure. When this happens, urbanization will accelerate rapidly. The stagnation of the 1980s has provided a pause for urban policy-makers: the 1990s will probably make a beginning in clearing this backlog.

Policies toward urbanization have traditionally been negative in tone and intent. This must be changed so that the process of urbanization is regarded as a positive force in the development of the country. In particular, urbanization has to be seen as supportive and necessary for the development of the rural sector. Policies connected with urbanization and urban development must pay particular attention to increasing the access of the poor to urban incomes and amenities. The urban environment must become supportive of the needs of a rapidly increasing urban population. Self-sustaining urban institutions that are responsive to the needs of the individual must be created. They must also be flexible enough to cope with the demands of a changing environment.

Notes

1. The total number of towns in the 20 000 to 100 000 range is 1268 according to the 1991 census.
2. Assam has been excluded in these tables because it was not possible to conduct a census there in 1981. Kerala has been excluded because of some definitional problems in the classification of towns at the lower end (see Mohan and Pant, 1982). And Jammu and Kashmir has been excluded because its population was less than 10 million.
3. This was also recommended by the Indian Planning Commission (1986*a*) in the Seventh Five Year Plan.

References

Bapat, Meera (1981), *Shanty Town and City: The Case of Poona* (Oxford: Pergamon Press).

Census of India (1975) (India Series, 1971), *General Population Tables*, 1971, Series I, part ii (New Delhi: Government of India Press).

—— (1981), *Provisional Population Tables*, Paper 2 of 1981 (New Delhi: Government of India Press).

—— (1989) (India Series, 1981), *Report of the Expert Committee on Population Projections*, Paper 1 of 1989 (New Delhi: Government of India Press).

—— (1991) (India Series, 1991), *Provisional Population Totals*, Series I, Paper 2 of 1991 (New Delhi: Government of India Press).

Centre for Monitoring of the Indian Economy (CMIE) (1992), *Basic Statistics Relating to the Indian Economy*, ii (Bombay: Centre for Monitoring the Indian Economy).

Jain, M. K. (1977), *Interstate Variations in the Trends of Urbanization in India* (Bombay: International Institute of Population Studies).

Kelkar, Vijay and Kumar, Rajiv (1990), 'Industrial Growth in the Eighties: Emerging Policy Issues', *Economic and Political Weekly*, 25: 209–25.

Lewis, W. Arthur (1978), *The Evolution of the International Economic Order* (Princeton University Press).

Mills, Edwin S. and Becker, Charles M. (1986), *Studies in Indian Urban Development* (New York: Oxford University Press).

Mohan, Rakesh (1984), 'The Effect of Population Growth, the Pattern of Demand and Technology on the Process of Urbanization', *Journal of Urban Economics*, 15: 125–56.

—— (1985), 'Urbanization in India's Future', *Population and Development Review*, 11: 619–45.

—— and Pant, Chandrashekar (1982), 'Morphology of Urbanisation in India: Some Results from 1981 Census', *Economic and Political Weekly*, 17: 1534–40, and 1579–88.

—— and Thottan, Pushpa (1992), 'The Regional Spread of Urbanization, Industrialization and Urban Poverty', in Barbara Harriss, S. Guhan, and R. H. Cassen (eds.), *Poverty in India: Research and Policy* (Bombay: Oxford University Press), 76–141.

Planning Commission (1986*a*), *Report of the Task Force on Financing of Urban Development* (New Delhi: Government of India Press).

—— (1986*b*), *Report of the Task Force on Shelter for the Urban Poor* (New Delhi: Government of India Press).

Sekhar, A. Uday (1983), 'Industrial Location Policy: The Indian Experience', World Bank Staff Working Paper No. 620 (Washington, DC: World Bank).

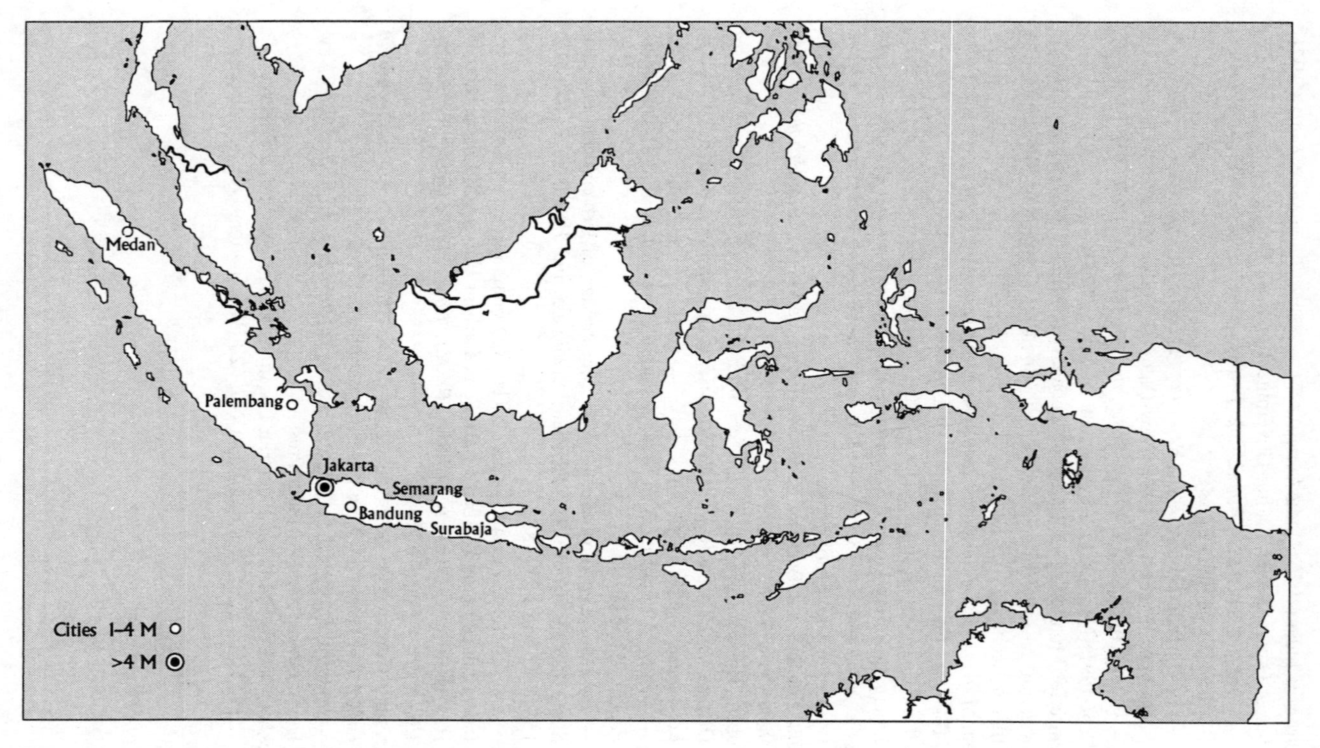

FIG. 5.1 Urban Agglomerations in Indonesia with more than 1 Million Inhabitants in 1990

5

Urbanization in Indonesia: City and Countryside Linked

GRAEME HUGO

INDONESIA, the world's fourth most populous nation, has long been characterized as one of the least urbanized countries with the vast majority of its population comprising a peasantry entirely dependent upon agriculture for its livelihood. Such a stereotype has never been totally accurate, but it holds little validity in the 1990s. Certainly, at the time that Indonesia declared its independence from its Dutch colonial rulers (1945), less than one in ten Indonesians lived in urban areas or relied upon the non-agricultural sectors for their livelihood and the domestic and export economies were overwhelmingly based upon primary production. However, Indonesia has subsequently experienced an urban transition, the rapid pace of which is reflected in the fact that by the early 1990s a third of Indonesians were urban dwellers, for the first time less than half of Indonesian workers were employed in the agricultural sector, and agriculture's share of Gross Domestic Product had fallen to 18.5 per cent (Hugo, 1993*a*).

In the present chapter we seek to clarify some of the salient features of Indonesia's contemporary urban transformation. Before focusing upon the process of urbanization, it is necessary to set the scene by outlining some of the major features of contemporary Indonesian development which are influential in shaping the level and nature of urbanization. In addition, it is not possible to understand urbanization in present-day Indonesia without reference to the historical development of that pattern.

The level of urbanization of a nation is conventionally defined as the proportion of that nation's population residing in urban areas. There have been several criticisms of the criteria used to define urban areas in Indonesia (e.g. Hugo, 1981*a*; NUDSP, 1985; Rietveld, 1988; Hugo, 1993*b*). What is especially important to bear in mind, however, is not so much the issue of where the boundaries should be drawn around

Indonesia's urban areas, but the fact that millions of Indonesians residing in areas classified for census purposes as rural spend much of their working lives in cities via processes of circular migration and commuting. Many rural residents work for extended parts of the year in cities but keep their families and their permanent place of residence in rural areas. Hence, the distinction between urban and rural populations in Indonesia, as elsewhere, is becoming blurred (Hugo, 1992*a*).

The Context of Urbanization

Indonesia is an enormously diverse nation consisting of more than 3000 islands (see Figure 5.2), more than 200 ethno-linguistic groups, and significant representations of each of the world's major religions, although 87 per cent profess Islam. One of the most serious perennial population issues in Indonesia relates to huge variations in population density between Inner (Java–Madura–Bali) and Outer (other islands) Indonesia. Population densities vary from 784 people per square kilometre in Java to five in Irian Jaya. Java comprises 6.9 per cent of the total land area of the nation but had 60 per cent of the population in 1990 (although this had declined from 65 per cent in 1961).

Like many of its Asian neighbours, Indonesia has recorded very high levels of economic growth over the last two decades as Table 5.1 shows.[1] In the 1970s and early 1980s economic development was fuelled by escalating oil export income, but in recent years there has been a rapid expansion of manufacturing through a major shift in economic policy whereby 'promotion of exports and private investments, both local and foreign, have replaced a policy of import substitutions marked by high tariffs and heavy government intervention in distributing capital, licenses and other means of production' (Schwarz, 1990: 40). This massive shift in government policy has not only led to an increase in the tempo of economic growth but had significant implications for the structure of employment in Indonesia and the extent and spatial patterning of urbanization. Between 1971 and 1990 there was only a slightly greater numerical increase in the number of employed people living in rural areas (17.527 million) than in urban areas (13.196 million) despite the fact that urban areas accounted for only 17.2 per cent of the total population in 1971 and 30.9 per cent in 1990 (Table 5.2). Hence, the proportion of all employed persons in Indonesia living in cities increased from 17.2 per cent in 1971 to 26.7 per cent in 1990 and employment

TABLE 5.1 Basic Economic and Social Indicators for Indonesia, 1970–1992

Indicator	Year	Value	Average annual change (%) 1970–80	1980–91
GNP Per Capita (US$)	1992	670	n.a.	4.0
GDP (US$m)	1992	126 364	7.2	5.7
Agricultural Production (%)	1970	45	4.1	3.1
	1992	19		
Industry Production (%)	1970	19	9.6	6.1
	1992	40		
Manufacturing Production (%)	1970	10	14.0	12.3
	1992	21		
Services Production (%)	1970	36	7.7	6.8
	1992	40		
TOTAL POPULATION	1992	184.3 m	2.4	1.8

Source: World Bank, 1994.

TABLE 5.2 Employed Persons by Industry, Indonesia, 1971–1990

	1971		1990		Increase (%) 1971–90
	Number	Per cent	Number	Per cent	
Urban					
Agriculture	653 508	11.8	1 839 687	9.7	181.5
Manufacturing	1 452 240	26.3	3 516 486	18.6	142.1
Services	3 424 866	61.9	13 521 788	71.6	294.8
Not Stated	520 386	0.0	369 122	0.0	0.0
TOTALS	6 051 000	100.0	19 247 083	100.0	218.1
Rural					
Agriculture	25 808 930	76.2	33 610 698	64.6	30.2
Manufacturing	2 957 640	8.7	5 437 874	10.5	83.9
Services	5 070 240	15.1	12 987 837	24.9	156.2
Not Stated	1 373 190	0.0	700 367	0.0	0.0
TOTALS	35 210 000	100.0	52 736 776	100.0	49.8

Note: In all percentage calculations the 'not stateds' have been omitted.

Sources: 1971 and 1990 Censuses of Indonesia.

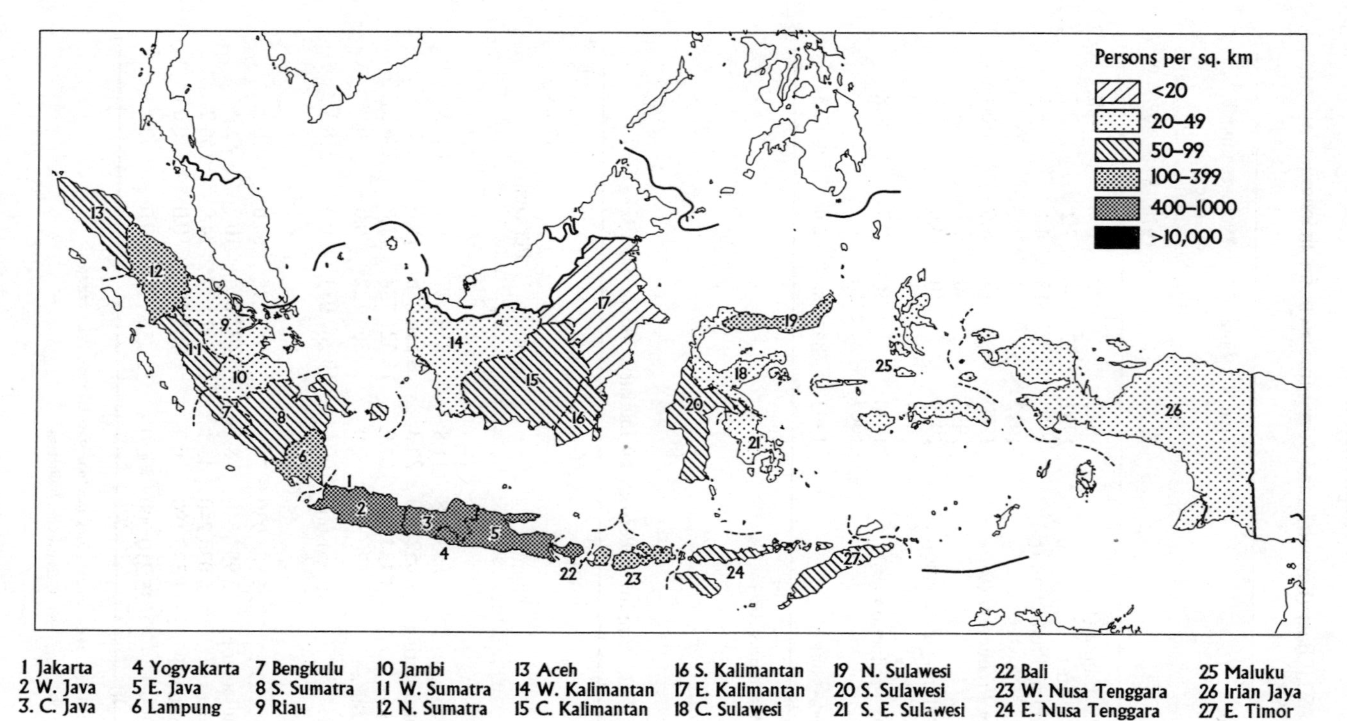

FIG. 5.2 Indonesia: Population Density, 1990

Source: 1990 Census of Indonesia.

increased almost three times as fast in urban as in rural areas over the period. The change in the mix of industry types in rural and urban areas is also significant. In rural areas, agriculture was the slowest growing employment sector, increasing by less than a third over the 1971–90 period, but the other sectors increased more than twice as fast so that agriculture's share of total employment of rural dwellers fell from 76.2 per cent in 1971 to 64.6 per cent in 1990. At the same time, the number of rural dwellers employed in manufacturing almost doubled and the proportion of the rural work-force in this sector increased from 8.7 to 10.5 per cent. In the services sector in rural areas, the numerical increase in jobs (7.9 million) was greater than the increase in agricultural jobs (7.8 million) and services increased their share of the total from 15.1 to 24.9 per cent. Hence, in Indonesian rural areas in 1990, less than two-thirds of employed persons worked in agricultural jobs. This is only partly a reflection of movement of industry into 'greenfield' rural areas and is more indicative of the involvement of rural dwellers in long-distance commuting and circular migration to work in urban-based non-agricultural enterprises.

Turning to the urban sector in Indonesia, the most striking pattern shown in Table 5.2 is the quadrupling of employment in the services sector over the two decades and its increase in dominance from 61.9 to 71.6 per cent of all jobs. This in part reflects the increased growth of the urban informal sector—an important element in the absorption of rural–urban migrants. However, it is also interesting to note that employment in agriculture among urban residents almost trebled, indicative of the blurring of the distinction between urban and rural areas in Indonesia.

Despite the rapidity of recent economic growth in Indonesia and its classification by the World Bank as one of Asia's eight 'highest flying' economies (Awanohara, 1993: 29), Indonesia is still squarely within the ranks of less developed countries. According to official estimates the incidence of poverty in Indonesia declined from 40.1 per cent in 1976 to 15.1 per cent in 1990, and the number of Indonesians living in poverty was halved from 54.2 to 27.2 million (World Bank, 1992). There is a relatively high degree of inequality with the top 20 per cent of income earners accounting for 42.3 per cent of expenditures in 1990 (World Bank, 1993: 296). Infant mortality rates remain high, but have declined from 145 per 1000 in 1955–60 to 124 in 1965–70 and 75 in 1986–90. The proportion of Indonesians aged 10 years and over who had never been to school declined from 27.5 per cent in 1980 to 16.1 per cent in 1990, while the proportion with a high-school education or above rose from 11 to 22.3 per cent. Total fertility rates have declined dramatically from 5.7 in 1955–60 to 5.1 in 1970–5, 3.5 in 1985–90, and 3.0 in 1988–91. This has been in part due to a highly successful family planning programme which has resulted in around half of currently married women practising a modern form of contraception. Population growth-rates have declined from 2.4 per cent per annum in the 1970s to current rates of around 1.7 per cent.

Indonesia has thus experienced massive economic and social change since Independence, but especially in the last two decades. These changes have impinged significantly on the nation's space-economy, especially upon the distribution, size, structure, functioning, and impact of urban centres. We will now turn to a brief description of the evolution of Indonesia's urban system before examining some of the major contemporary features of that system.

The Pre-Colonial and Colonial History of Urbanization

In pre-colonial times, mobility was greatly constrained for most Indonesians although some significant movements, especially of the agricultural colonization, seasonal, and trading types, occurred (Hugo, 1980: 97–100). The pre-colonial class structure, the rise and fall of inland kingdoms and coastal sultanates, the regular incidence of famine, the development of various trading patterns through the Indonesian archipelago, the spread of new types of agriculture and various environmental disasters, all shaped the patterns and levels of the movements which did occur. However, movement outside of the well-trodden local area was usually prevented by lack of transport infrastructure, the obvious difficulties of moving between the regions of Indonesia's separate ethnolinguistic groups, and the political constraints exerted by the control of élite groups in some areas. Nevertheless, a significant network of urban centres developed during the pre-colonial period (Reid, 1980; McGee, 1967). There were a few inland cities which McGee (1967) designates 'sacred cities'. These were the centres of kingdoms where élite groups settled, living off tribute appropriated from the rural hinterland they ruled over. They reached substantial sizes as testified to by magnificent temple complexes such as Borobodur and Prambanan which still exist. While there were some commercial activities in these cities, their principal function was the religious and military domination of their hinterlands. In contrast to the planned and substantial buildings of the sacred cities, the second type of pre-European urban development were 'market cities' which were not only functionally different but less permanent and more chaotic in appearance (McGee, 1967). Their function was overwhelmingly one of trade, and they were located at strategic trading points along the coasts of Java, Sumatra, Kalimantan, and Sulawesi. Figure 5.3 shows the location of the major market and inland cities which existed in Indonesia before European contact.

Reid (1980: 235) shows that trading cities existed in South-East Asia as far back as records allow us to go, but notes that the period 'from the late fourteenth to the seventeenth centuries, however, affords a picture of exceptionally rapid growth of trade, and of a network of indigenous cities necessary to sustain it'. While data are uneven and unsatisfactory, populations between 50 000 and 100 000 appear to have been reached by Demak in the early sixteenth century and by Aceh, Makassar, Surabaya and Banten in the seventeenth. Reid argues that, in relation to its population, South-East Asia was one of the most urbanized regions of the world at that time.

The Portuguese and Dutch at first fitted into the existing trading patterns and had an impact on the development of the port cities. However, as the Dutch gradually increased the strength and geographical extent of their control over the archipelago they reshaped the urban system to meet their needs. Prior to the nineteenth century the major Dutch impact was in Java. By 1619 the Dutch had taken control of the small port of Sunda Kelapa, renamed it Batavia (later Jakarta), and made it the base of their East Indian operations. Castles (1967: 155) points out that during this period the Dutch, partly for security reasons, did not encourage the hinterland's indigenous population to settle in or near Batavia. Raffles (1817, i. 64) wrote that initially the Dutch, 'having no confidence in the natives, endeavoured to drive them from the vicinity of Batavia, with the view of establishing round their metropolis an extensive and desert barrier'. The fleeing Sundanese left a vacuum in this area (known as the *Ommelanden*) so that, when the Dutch realized the possibilities of growing sugar for export in the vicinity of Batavia, slaves were brought in from elsewhere. Hence, the local population became a minority in the heterogeneous population of Batavia as can be seen in Table 5.3.

Raffles' census of 1815 indicated that the population of Batavia was 47 217. Since the populations of Batavia's closest rivals, Semarang and Surabaya, were 20 000 and 25 000 respectively, it was clearly Java's

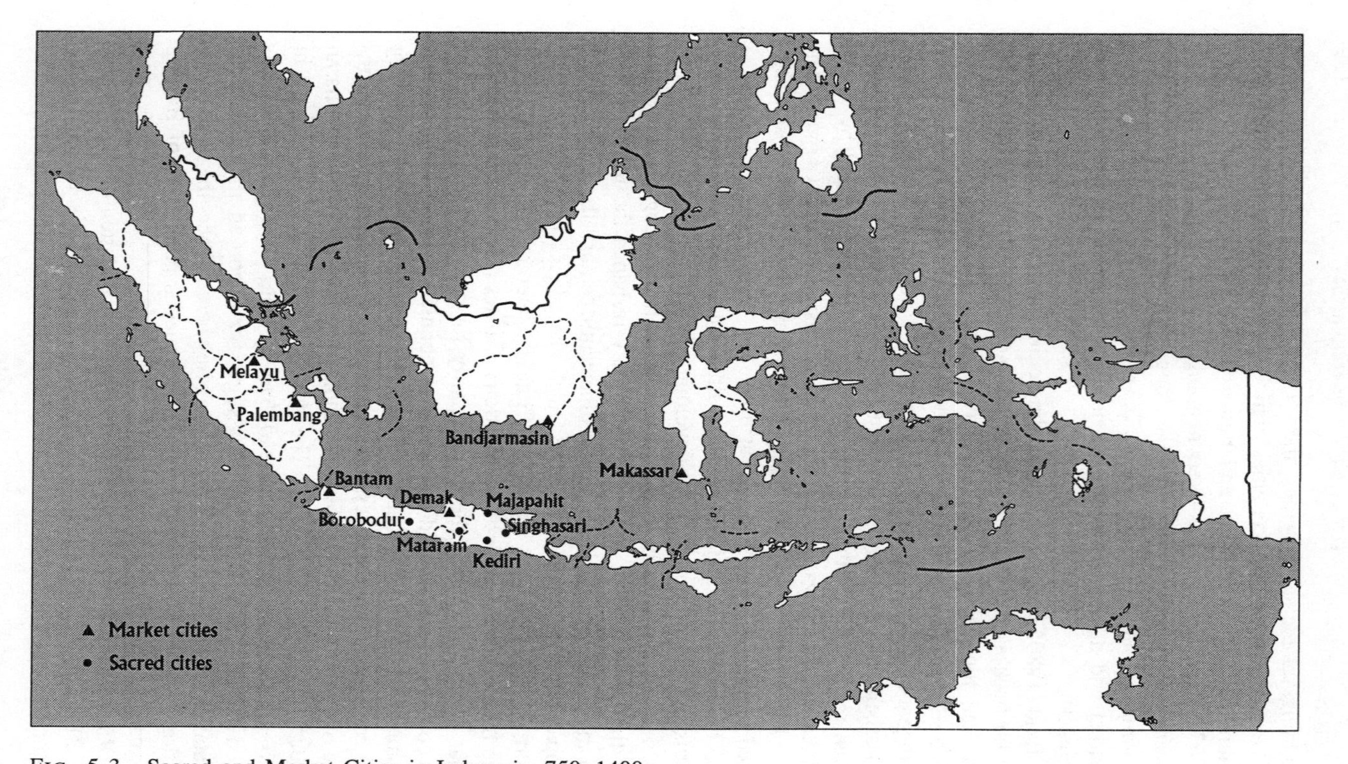

FIG. 5.3 Sacred and Market Cities in Indonesia, 750–1400

Source: Adapted from McGee, 1967: 35.

TABLE 5.3 Population of Batavia by Origin, 1673 and 1815

Origin	1673		1815	
	Number	Per cent	Number	Per cent
Europeans and part-Europeans	2750	8.6	2028	4.3
Chinese	2747	8.6	11 854	25.1
Mardijkers	5362	16.7	n.a.	n.a.
Javans	6339	19.8	3331	7.1
Balinese	981	3.1	7720	16.4
Malays	611	1.9	3155	6.7
Slaves	13 278	41.4	14 239	30.2
Others	—	—	4890	10.2
TOTAL	32 068	100.1	47 217	100.0

Sources: Castles, 1967: 157; Raffles 1817, ii. 246.

principal city—a status it was to lose during the nineteenth century. The recency of migration to the city in 1815 is reflected in the population's masculinity (sex ratio 1203 males: 1000 females) and ethnic heterogeneity (see Table 5.3). Slaves were still a substantial proportion of the total population. Most came from Bali and South Sulawesi, but at different times Sumbawa, Sumba, Flores, Timor, Nias, Kalimantan, and Pampanga (Luzon) made contributions. Castles (1967: 156) points out that the unhealthiness of Batavia necessitated constant replenishment of population from the outside such that during the third quarter of the eighteenth century 4000 slaves were being imported annually. The Sundanese thus remained a minority element in Batavia's population.

The nineteenth century saw a major change in Indonesia's urban system. In bringing their plundering of Java's raw materials to new heights of efficiency, the Dutch created a system of colonial cities and towns which functioned as key devices in expediting the production and delivery of raw materials and the subjugation of the local population. The three main elements in this urban development were first the large entrepôt city of Batavia which was the linchpin of the system—the point of contact between the colony and the home country. Secondly, there were a number of secondary ports such as Surabaya, Semarang, Makassar, and Banjarmasin which were smaller entrepôt cities. Thirdly, the Dutch created a hierarchical network of smaller centres which were the intermediaries through which the entrepôts' hinterlands were controlled and divested of their export crops. Much of this rapid urban growth was achieved by migration (Ranneft, 1916).

The expansion of Dutch colonialism assured the supremacy of the coastal cities which functioned as trans-shipping points for the agricultural products of their hinterlands. By 1905 Batavia's population had increased to 138 551 (Ranneft, 1916: 84) representing an annual growth rate of 0.74 per cent for the ninety years since 1815. This may seem a modest rate by contemporary standards, but in the light of Batavia's proverbial unhealthiness (Castles, 1967: 156) and high mortality rates (De Haan, 1935: 685, 702) this figure indicates that substantial replenishment of the population occurred from the outside. Indeed, Java's coastal cities had previously been small and ephemeral centres and the scale of the nineteenth-century development was a new phenomenon. This growth, however, was not sufficient to establish Batavia as Java's primate city because the other colonial cities of Surabaya (East Java) and Semarang (Central Java) were important exporting centres and recorded similar growth-rates.

Table 5.4 shows the population growth recorded by the twelve largest cities in Indonesia during the later colonial years and indicates that the period immediately preceding the 1930 census was one of particularly rapid growth, with all major centres, except the ancient capitals of Surakarta and Yogyakarta, recording growth-rates twice as large as those experienced by the population as a whole. One important point evident in Table 5.4 is that, unlike many other colonies, the urban hierarchy of the Netherlands East Indies was not dominated by a single primate port city. Indeed Batavia, the colonial capital, had a smaller population than the port of Surabaya until the early years of this century. The reasons for a 'more balanced' urban system are due to the fact that Indonesia's large size and archipelago nature meant that with the limited transport and communication technology available to them, the colonial Dutch were forced to operate through several scattered entrepôt centres. While Batavia was the colonial capital, the hinterland from which it collected raw materials for export was largely restricted to West Java; Surabaya dominated the eastern part of Java, Semarang drained central Java, and later Medan controlled northern Sumatra. With growing sophistication in transport and communication the capital's hinterland was greatly extended in both the later colonial and the Independence years, and the Indonesian urban system has shown a consistent tendency to move toward a primate city size distribution in which Batavia, now Jakarta, dominates the urban system. This is shown

TABLE 5.4 Population in Major Indonesian Cities, 1855–1930

City	Island	1855	1883	1895	1905	1920	1930	Average annual growth (%)	
								1855–1905	1905–1930
Batavia	Java	55 000	121 637	114 566	138 551	306 309	533 015	1.85	5.5
Surabaya	Java	88 527	121 637	124 529	150 198	192 190	341 675	2.14	3.35
Semarang	Java	n.a.	67 575	82 692	96 660	158 036	217 796	1.64	3.30
Bandung	Java	11 223*	—	46 326	47 400	94 800	166 815	2.92	5.15
Surakarta	Java	—	—	104 589	118 378	139 285	165 484	1.25	1.35
Yogyakarta	Java	43 000	—	58 299	79 569	103 711	136 649	1.23	2.18
Palembang	Sumatra	—	—	—	61 000	73 726	108 145	—	2.30
Malang	Java	—	—	—	30 000	42 981	86 646	—	4.33
Makassar	Celebes	—	—	—	26 000	56 718	84 855	—	4.83
Medan	Sumatra	—	—	—	—	45 248	76 584	—	5.42
Pekalongan	Java	—	—	—	—	47 852	65 982	—	3.28
Banjarmasin	Borneo	—	—	—	—	46 993	65 698	—	3.61

* 1846 figure.

Sources: Volkstelling, 1933–6; Milone, 1966; Ranneft, 1916.

TABLE 5.5 Four-City Primacy Index[a] for Indonesia, 1890–1990

Year	Index	Year	Index
1890	0.39	1961	1.17
1905	0.59	1971	1.16
1920	0.69	1980	1.25
1930	0.73	1990[b]	1.09
1955	0.87	1990[c]	1.49

[a] The four-city primacy index represents the population of the largest city (Jakarta), divided by the combined population of the three next largest centres (Surabaya, Semarang and Bandung up to 1961, thereafter Semarang was replaced by Medan as the fourth largest centre).

[b] Index based on official municipality and provincial boundaries, i.e. Jakarta 8.2 million, Surabaya 2.4 million, Bandung 2 million, and Medan 1.7 million.

[c] Index based upon contiguous urban area including overspill areas, i.e. Jakarta 13.1 million, Surabaya 3.2 million, Bandung 3.3 million, Medan 2.3 million.

Sources: Hugo, 1980; 1980 and 1990 Censuses of Indonesia.

clearly by the increase in the primacy index for Indonesia between 1890 and 1980. It will be noted in Table 5.5, however, that there was an apparent fall in the primacy index between 1980 and 1990 (Table 5.5). This, however, is a statistical artefact of the overspilling of urban development beyond official urban boundaries. If we take a definition of the four major metropolitan centres which not only includes urban population in the respective municipalities (or in the case of Jakarta-Special Province) used to calculate the first 1990 index in Table 5.5 but add urban development in the adjoining *kabupaten* to take into account this overspill, a very different pattern emerges. Hence, Table 5.5 indicates that if this more realistic urban definition is used, the primacy index is considerably higher than that prevailing at earlier times.

Another feature of the pattern of urban growth evident in Table 5.4 is the predominance of Java. It was not until the 1920s that an Outer Island city exceeded 100 000 in population and even then there were six Javan cities with much larger populations. The dominance of Javan cities was partly attributable to the concentration of colonial activity in Java. It is also noticeable in Table 5.4 that the fastest growing cities were those who owed their growth, and in some cases even their origin (e.g. Bandung), to the colonial system. The ancient indigenous cities of Surakarta and Yogyakarta recorded the lowest rates of growth of all the major cities.

There is no doubt that migration played a key role in the growth of urban centres. Apart from the fact that a significant proportion of urban dwellers were immigrants from overseas, immigrants made up a majority among the indigenous urban population. It was not until the eighteenth century that Indonesians became a majority in Batavia and their representation increased during the 1800s to reach 71 per cent in 1905 and 83 per cent in 1930. Following the reduction in slavery, many Sundanese from the immediate hinterland of Batavia and Banten migrated to the city to take up work in such occupations as household domestics (Milone, 1966: 229). Much of the Sundanese migration to Batavia was initiated by *mandur* (labour recruiters/foremen) who often brought whole villages of labourers to the capital, the group usually specializing in a particular type of activity (Milone, 1966: 236). Most indigenous people in Batavia were poor and lived in the older section of the city or in overcrowded *kampung* which were more rural than urban in character. The latter were encouraged by an influential element among the Dutch who 'felt that Indonesians should be discouraged from migrating to the city, for it was an unnatural way of life for them. If once there, however, they should be left to continue the style of life that they had in the village' (Milone, 1966: 555).

Several smaller towns developed during the nineteenth century at places selected by the Dutch as administrative centres, points for collecting raw materials, military garrisons, and hill stations (resorts in upland areas where colonial officers and their families took leave, especially during the monsoon season). These third- and fourth-order urban centres in the colonial urban hierarchy gained most of their migrants at the expense of adjoining regions.

Migration was equally important in the growth of cities in the Outer Islands during the early years of this century. In Medan, located on Sumatra's east coast and a service centre for plantations in that region, 60 per cent of the indigenous population were born outside of the city, while even higher proportions were recorded in smaller, more specialized colonial urban centres such as the coffee- and pepper-exporting city of Telukbetung (62 per cent) and the coal-mining town of Sawahlunto (80 per cent).

International migration was most important in the growth of Indonesia's cities during the peak colonial impact years. Table 5.6 shows that the Chinese and other non-indigenous Asian populations increased at a substantially faster rate than the indigenous population and by 1930 they comprised more than 2 per cent of the total population of the Netherlands East Indies. The colonial system tended to channel the Chinese and other non-indigenous Asian groups into employment as

TABLE 5.6 Indigenous and Non-Indigenous Populations of Indonesia, 1860–1930

Year	Indigenous	Europeans	Chinese	Other Asian
1860	15 409 944	43 876	221 438	8909
1900	34 666 659	91 142	537 316	27 399
1930	59 138 067	240 417	1 233 214	115 535
Average annual growth (%)				
1860–1900	2.05	1.86	2.23	2.85
1900–20	1.78	3.28	2.80	4.91

Source: Volkstelling, 1936, viii.

TABLE 5.7 Distribution of Indigenous and Non-Indigenous Population Between Urban and Rural Areas in Indonesia, 1930 (%)

	Indigenous	Europeans	Chinese	Other Asians
Netherlands East Indies:				
Large cities	3.7	63.7	27.5	36.2
Towns	2.7	11.8	16.6	21.0
Rural areas	93.7	24.5	55.9	42.8
Java:				
Urban	7.4	79.9	58.7	40.8
Rural	92.6	20.1	41.3	59.2
Outer Islands:				
Urban	4.0	58.1	31.0	41.4
Rural	96.0	41.9	69.0	58.6

Source: Volkstelling, 1936, viii.

foremen and white-collar workers in Western industrial enterprises (Skinner, 1963: 98), small- and medium-scale internal retail trade, and provision of internal credit facilities and community services (Vries and Cohen, 1938: 269), and other service industries (Aten, 1952–3: 19–27). Consequently, the non-indigenous Asian segment of the population was strongly concentrated in urban areas, as Table 5.7 shows.

The major colonial cities which developed in the nineteenth and early twentieth century took on a distinctive form and structure still evident today. The cities bore the stamp of the clear status-ranking which emerged during the nineteenth century and which had a racial-economic basis and was supported by legal distinctions. At the apex of the status pyramid were the Europeans, followed by Eurasians, Indonesian Christians, Chinese, Arabs, and Islamic Indonesians (Milone, 1966: 149). In Batavia, for example, by 1905 the European proportion had increased from 4.3 per cent in 1815 to 9.3 per cent, largely as a result of the modification of laws regulating the alienation of land in 1870 (Cobban, 1970: 2). The old coastal town was rejected by Europeans earlier in the century, and they established their homes and administration buildings some two kilometres inland at Weltevreden, which took on a dispersed garden-town character still evident today. The Chinese population in Batavia were restricted as to the areas where they could settle. Most lived in the Glodok area, which was the camp to which they were exclusively restricted in the mid-eighteenth century, in shop-houses in Weltevreden, and near the markets of Tanah Abang, Pasar Senen, and Pasar Baru (Milone, 1966: 210).

After 1900, the large urban areas like Batavia began to take on many of the features that characterize them today. Most of the indigenous inmigrants to the cities lived in 'autonomous enclaves of native settlement that stood out in their general air of backwardness from the air of light and propriety which prevailed in the European section of cities' (Cobban, 1970: 228). Particular urban *kampung* were favoured for settlement by persons from particular home areas. Thus inmigrants from Tanggerang and Banten were strongly concentrated in the older sections of the city, whereas those from Priangan tended to settle in the Weltevreden area (Volkstelling, 1933, i. 29). This followed in the tradition of the development of 'quarters' of particular ethnic groups in Indonesian cities which pre-dated European contact. Another important feature of early twentieth-century migration to Batavia still evident is the tendency for persons from a particular region or village to take up the same occupation. In fact 1930 census officials found the phenomenon so widespread that they make mention of it in their report (Volkstelling, 1933, i. 29).

Discussions of population mobility are frequently confined to more or less permanent types of movement. However, colonialism in Indonesia also had a significant impact on circular or temporary types of population movement. The meagre migration literature written during the colonial period reflects the importance of non-permanent forms of mobility. Scheltema (1926: 872–3) and Ranneft (1916), for example, produced a typology of population movement the most striking feature of which is its emphasis upon circular movements. It is clear that much of this temporary movement was rural–urban in nature. For example, Ranneft (1916) in describing some of the types of temporary migrations characteristic of the early years of this century in Java, stresses the significance of circular labour migration to cities such as that from Banten to Batavia, Demak to Semarang, and Gresik and Sidarjo to Surabaya. He suggests that one of the effects of this type of mobility was to prevent Java following Europe in developing a large urban proletariat. This being due to the fact that although the temporary migration was caused largely by the penetration of capitalism, temporary workers, by retaining such strong links with their villages of origin, remained largely 'traditional men'.

A substantial growth of the urban population occurred in the later colonial years. Between 1920 and 1930 the urban population of the Netherlands East Indies grew from 2 881 576 to 4 034 149, and the proportion of the total population living in urban areas increased from 5.8 per cent to 6.7 per cent. There was thus significant rural–urban migration but it could in no way be described as the 'great shaking loose of migrants from the countryside' posited by Zelinsky (1971: 236) as characterizing the second stage of the 'mobility transition'. A contemporary writer of the later colonial period, Ranneft (1929: 80), published an interesting discussion of why there was not a great exodus to the towns in Java, as had occurred earlier in Europe. He recognized that the limited urban growth reflected the important connection between job opportunities and rural to urban migration. The aim of the Dutch was to perpetuate Java as a 'country of various "subsidiary industries"'. The unprocessed and semi-processed raw materials extracted from the country supported secondary industry in the Netherlands—not in Indonesia. Cities in Indonesia were subsidiary, forwarding centres in which tertiary activities supporting this function increased, but most colonial employment creation was in *rural areas* through plantation development and later by expansion of the irrigated wet rice area. In addition, as Fisher (1964: 296) has pointed out:

> Dutch factory owners in Holland had no desire to create new rivals for the Indonesian market and until the 1930s Western style industry remained virtually limited to the initial processing of locally produced raw materials and the provision of maintenance facilities for railways, motor vehicles, shopping and the machinery used by mines and estates.

Clearly, large-scale industrialization could not occur in the Netherlands East Indies under colonial rule. In fact, existing handicraft industry declined under the impact of Western imports, and profits extracted from the agricultural sector were repatriated to the Netherlands and not invested in local industrial development.

Patterns of Urbanization Since Independence

The pattern of change in urban growth and urbanization in Indonesia over the last seventy years is shown in Table 5.8. The top row of the table gives some idea of the rate of *urban growth* in Indonesia. It can be seen that in each of the intercensal periods the rate of urban growth far outstripped that of the rural population. The urban population has increased almost twentyfold since 1920 while the rural population has less than trebled. If we examine *urbanization*, however, the gains over the last seven decades appear more modest. The third row of Table 5.8 shows that the percentage of Indonesians living in urban areas increased fivefold from 5.8 per cent in 1920 to 30.9 per cent in 1990.

The period since Independence in 1949 has seen three distinct phases in Indonesia's urbanization. The first phase includes the first two decades of Independence under President Soekarno who inherited a country which not only had limited urban, industrial, and infrastructural development, but also one in which a decade of Japanese occupation and independence struggle had devastated much of that development. The establishment of new bureaucratic structures, the rapid expansion of education, and attempts to develop industry to reduce dependency on industrialized nations saw rapid growth of cities in the 1950s (Table 5.8). Moreover, the repatriation of large numbers of people of Dutch and Chinese origin saw an Indonesianization of the cities. The economy of Indonesia, however, retained many of its colonial characteristics with a heavy dependency upon the exports of raw materials to industrialized nations and a large semi-subsistence agricultural sector absorbing not only the bulk of workers but also the increments to the work-force. The massive disruptions caused by the war with Malaysia and the breakdown of the Indonesian economy in the first half of the 1960s saw a downturn in urban growth with some urban dwellers being forced to move back to live temporarily with their rural-based relatives. McGee (1991: 4) explains that 'the political process, characterized by colonial devolution and vigorous nationalism, of the post-Independence era was affecting the large cities of the region. In this period, the political élites were mainly interested in establishing the

TABLE 5.8 Distribution of Population Between Urban and Rural Areas in Indonesia, 1920–1990

Characteristic	Census year						Average annual growth (%)				
	1920[a]	1930	1961	1971	1980	1990	1920–1930	1930–1961	1961–1971	1971–1980	1980–1990
Urban population	2 881 576	4 034 149	14 358 372	20 465 377	32 845 829	55 391 171	+3.42	+4.18	+3.61	+5.40	+5.36
Rural population	46 418 424	56 693 084	82 660 457	98 874 687	114 485 994	123 808 052	+2.02	+1.22	+1.79	+1.64	+0.79
Total population	49 300 000	60 727 333	97 018 829	119 140 064	147 331 823	179 199 223	+2.11	+1.52	+2.08	+2.39	+1.98
Per cent urban	5.8	6.7	14.8	17.2	22.3	30.9	+1.45	+2.59	+1.51	+2.93	+3.32

[a] Source of 1920 statistics: Milone (1966). An inaccuracy coefficient of 5 per cent has been suggested for these figures.

Sources: Hugo *et al*., 1987: 89; 1990 Census of Indonesia.

"custody" of their countries, and economic goals were less important.' He also points out that due to

> the rapidly accelerating rates of population growth in the 1950s, slow agricultural growth and thus limited employment prospects in rural areas, along with continuing political instability based on regional, political or ethnic discontent, there were considerable flows of population to the cities at the time. The economic structures of these cities were ill-equipped to cope with this influx for there had been very little growth of manufacturing. As a consequence, new migrants crowded into low-income service occupations and often settled in poorly serviced squatter settlements. This reinforced the social and economic dualism of the cities. (McGee, 1991: 5)

The New Order era under President Soeharto extends from the late 1960s to the present and can be divided into two distinct periods. The first includes the years up to the mid-1980s and included the 'oil boom'. Economic growth in the 1970s was at an average annual rate of 8.1 per cent and per capita GDP grew by 6.3 per cent in real terms. Large increases in export income from oil, especially after 1973, were important in this rapid growth. The government adopted a strategy of import substitution and, with the substantial export revenues from oil, industry grew rapidly (at around 16 per cent per annum), albeit from a very small base. Also employment in the bureaucracy, in education, and in the service sector generally expanded very rapidly. Thus the average annual growth-rate of Indonesia's urban population increased from 3.6 per cent in the 1960s to 5.4 per cent in the 1970s. Even allowing fully for boundary changes, the general consensus is that the Indonesian urban population grew at around 4 per cent per annum in the 1970s—more than double the rate in the rural sector (World Bank, 1984; NUDSP, 1985).

The fall in oil prices in the mid-1980s threatened the basis of Indonesia's rapid economic growth. Annual GDP growth slowed to 4.3 per cent per annum between 1981 and 1988, per capita income in dollar terms fell from $530 in 1980 to $480 in 1988 and the World Bank reclassified Indonesia from a 'lower middle' to a 'low' income country (Booth, 1992: 1). The government responded with a major shift in its economic strategy toward liberalization, dismantling import substitution strategies, encouraging foreign and local investment, and encouraging exports, especially manufactures. This strategy obviously favoured urban over rural areas. Hence, while urbanization probably slowed down in the middle part of the 1980s, there can be no doubt that in the latter part of the decade and in the 1990s rapid urbanization has resumed. In the 1980s, the rate of urban growth was more than six times greater than that of the rural population, and the absolute growth of the urban population (almost 23 million) was twice that of the rural population (11.8 million).

Indonesia's urban areas have not only recorded massive population gains during the 1980s but there also has been a huge increase in the lateral extent of urban areas. This lateral extension of Indonesia's urban areas has tended to occur in corridors, along major transport routes radiating out from (and linking) major urban areas (McGee, 1991; Firman, 1989, 1991, 1992). This phenomenon, together with a rapid increase in rural to urban circular migration is producing a new form of diffuse urbanization in Indonesia, especially in densely settled Java. The overlapping of urban and rural populations and areas is producing a blurring of distinctions. It is most intense in the areas around Jakarta (JABOTABEK[2]), and around Surabaya (GERANGKERTO-SUSIDO[3]), and along the transport corridors linking major cities (especially Jakarta–Bandung,[4] Jakarta–Cirebon, Surabaya–Malang, and Yogyakarta–Semarang). Such a pattern of diffuse urbanization surrounding a major metropolitan centre may even overlap national boundaries with the development of SIJORI, i.e. overspilling of Singapore's industrial development into the adjoining Malaysian state of Johore and the Indonesian province of Riau. In the latter case the rapid urban growth on the island of Batam is very much an extension of Singapore (Hugo, 1991).

An issue which is much discussed in Asia relates to the distribution of the sizes of cities. We are hampered in our analysis here because at the time of writing we do not have available 1990 census data for *functional* urban areas which were prepared for 1971 and 1980 by the National Urban Development Strategy Project (1985) and have had to revert to using unsatisfactory administrative definitions. Figure 5.4 shows the rank-size distribution of Indonesia's major cities in 1990, 1980, 1930 and 1850. The diagram also shows the equivalent 'idealized' rank-size curves which represent the city distributions which would occur if the sizes of all other cities were related to the size of the largest city (Jakarta) according to the rank-size rule. The latter in its simplest form states that in any nation or region, the population of a given city tends to be equal to the population of the largest city divided by the rank of the given city (Haggett, 1975: 358). It is apparent from the diagram that in 1980 and 1990 the idealized rank-size

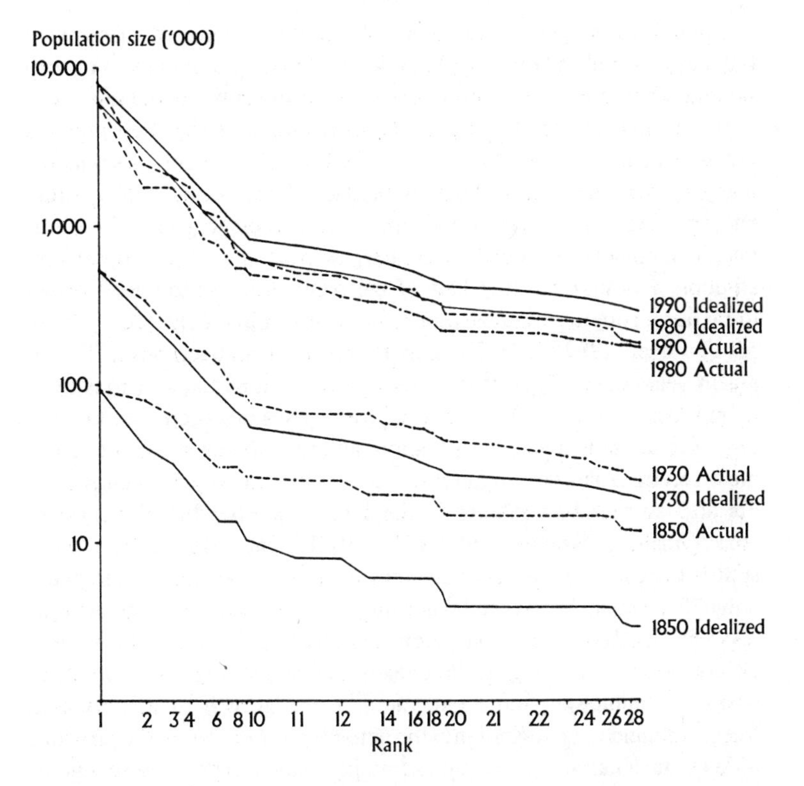

FIG. 5.4 Actual and Idealized Rank-Size Distributions of Indonesian Cities in 1850, 1930, 1980, and 1990

Sources: Hageman, 1852; Volkstelling, 1933–6; NUDSP, 1985; 1990 Census of Indonesia.

distribution overestimates the size of all twenty-eight other cities. This reflects a concentration (or, in terms of the rank-size rule, an 'over-concentration') of the urban population in the nation's largest city.

A primacy index for Indonesia which divides the population of the largest city (Jakarta) by that of the three next largest (Surabaya, Bandung, and Semarang) is presented in Table 5.5. The index progressively increases from 0.39 in 1890 to 1.25 in 1980, and it has increased further since, at least when a realistic urban definition is used to compute the index. While it is true that this level of primacy is not high when compared with many countries (Renaud, 1981), it should be pointed out that the countries with higher levels of primacy tend to be small- or medium-sized in population terms and their national space comprises a contiguous land area. The present degree of primacy in Indonesia is both historically large in the Indonesian context and substantial when compared with countries of comparable population size.

In the post-Independence period, notwithstanding the deep political and social change which has occurred, Indonesia's economy has retained many of the 'dependent' features of the colonial years with a stress on exporting raw materials and importing processed goods. The limited development of secondary industry associated with the import substitution policy of the first half of the New Order period also tended to favour a pattern of concentration in the urban hierarchy (Azis, 1989: 56; Gardiner, 1992: 283). The export-oriented manufacturing policy of recent years encourages urban growth in Jakarta and its environs even more (Azis, 1989: 59). These developments have favoured Jakarta moving more towards a primate situation, although this is masked because most recent urban development in the capital actually is occurring in the area surrounding Jakarta in the Jabotabek extended metropolitan area (Ginsburg, Koppel, and McGee, 1991). The extent of this overspill is indicated by the fact that whereas at the 1980 census there were 456 625 residents of West Java who gave their previous province of residence as Jakarta, this had more than trebled to 1 558 641 in 1990. Of course many settling in the overspill areas are migrants from elsewhere in Indonesia. Some indication of the growth of Jakarta as a 'megacity' and the overspill into the adjoining *kabupaten* in the province of West Java can be gained by examining some trends in those adjoining areas.

The true geographical extent of the functioning megacity of Jakarta is shown in Figure 5.5. Table 5.9 shows growth in the Jakarta metropolitan region over the last two decades and indicates how unrealistic 'official' data on the population of the city are. The first part of the table shows growth in the administrative region of Jakarta giving a population of 8.2 million in 1990 and intercensal population gains of around a third. The second part shows growth in those parts of the adjoining three overspill *kabupaten* defined as urban at the three censuses. This gives a metropolitan region population of 13.1 million in 1990 and very rapid growth rates partly because of reclassification of formerly rural areas as urban between censuses. The final part of the table includes the fixed boundaries of the entire Jabotabek region and shows a more than doubling of the population between 1971 and 1990 (while the national population increased at less than half this rate) and gives a metropolitan region population of 17.1 million. In functional

TABLE 5.9 Population in Jakarta Metropolitan Area According to Different Definitions, 1971–1990

Definition	Population			Increase (%)	
	1971	1980	1990	1971–1980	1980–1990
Special Capital City District (DKI) of Jakarta	4 546 492	6 071 748	8 227 766	33.5	35.5
DKI Jakarta and Urban Bogor, Bekasi and Tanggerang *kabupaten*	4 838 221	7 373 563	13 096 693	52.4	77.6
DKI Jakarta, Kotamadya, Bogor, and *kabupaten* Bogor, Bekasi and Tanggerang (Jabotabek)	8 374 243	11 485 019	17 105 357	37.1	48.9

Sources: 1971, 1980, and 1990 Censuses of Indonesia.

terms the latter figure is certainly more realistic than that for DKI Jakarta. The transformation of the population of the areas adjoining DKI Jakarta is reflected in the fact that the number of residents in these areas who were employed in non-agricultural occupations increased from 583 450 in 1971 to 2 359 832 in 1990. It should be noted that there was already substantial commuting out of the region to Jakarta in 1971 but the growth of non-agricultural employment in the next two decades was 304.5 per cent compared with an increase in total population of 131.9 per cent.

There have been several post-Independence developments which have assisted in extending Jakarta's dominance in Indonesia. Jakarta has become the focus and symbol of Indonesian national unity. The Independence period has seen a greater centralization of activities in the national capital than was the case under the Dutch. Not only did Jakarta maintain its function as the major intermediary with other countries, but it became the major focus of national commercial, industrial, administrative, and political development. Widespread improvements

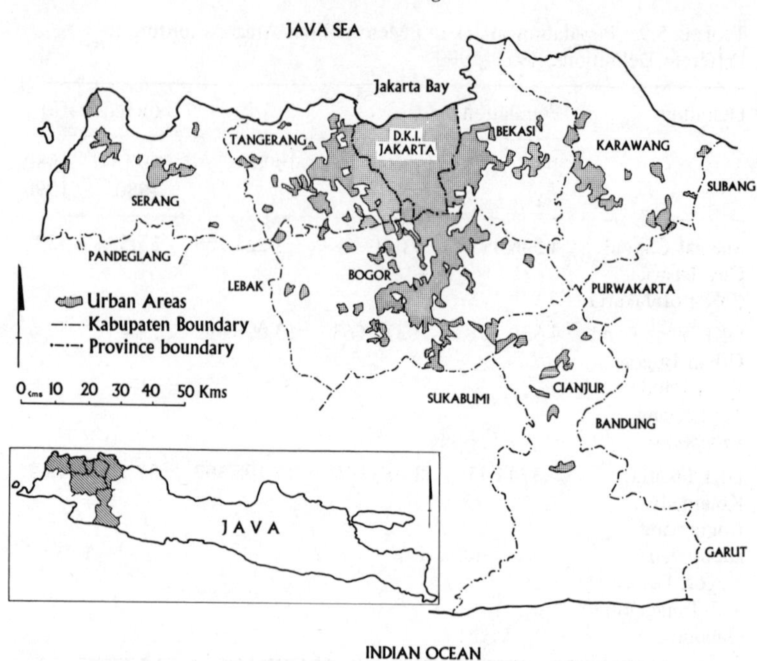

FIG. 5.5 The Functioning Urban Region of Greater Jakarta

in transport and communications made possible a much greater degree of centralization of administrative and commercial activity in the national capital than was the case during Dutch rule. Jakarta's growing primacy in Indonesia is reflected not only in population size but also in the fact that in 1990, although it had 4.6 per cent of the national population, it accounted for 37 per cent of all passenger cars and 38 per cent of telephones.

In a nation as large, heterogeneous, and geographically fragmented as Indonesia it is not surprising that there have been significant regional differentials in the extent and pace of the urban transition. Table 5.10 shows that Java (especially West Java) has remained the most urbanized region of Indonesia with more than a third of its residents in urban areas. The growth of the urban population of West Java was particularly rapid with its urban population more than doubling between the 1980 and 1990 censuses. Much of this growth was associated with the overspill of Jakarta into the surrounding West Java areas of Bogor, Tanggerang, and Bekasi (Figure 5.5). By 1990 there were almost as

TABLE 5.10 Distribution of Population Between Urban and Rural Areas in Indonesia by Province, 1980 and 1990

Region	1980		1990		Average annual growth (%)		Per cent Urban 1990
	Urban	Rural	Urban	Rural	Urban	Rural	
Java	22 926 377	68 290 593	38 335 297	69 182 666	5.28	0.00	35.6
DKI Jakarta	6 071 748	408 906	8 222 515	0	3.08	0.00	100.0
West Java	5 770 868	21 678 972	12 208 176	23 170 307	7.78	0.67	34.5
Central Java	4 756 007	20 611 337	7 694 539	20 822 247	4.93	0.10	27.0
DI Yogyakarta	607 267	2 142 861	1 294 056	1 618 555	7.86	(2.77)*	44.4
East Java	5 720 487	23 448 517	8 916 011	23 571 557	4.54	0.05	27.4
Sumatra	5 481 488	22 514 439	9 293 747	27 128 739	5.42	1.88	25.5
Lampung	576 872	4 047 366	747 327	5 256 782	2.62	2.65	12.4
Bengkulu	72 492	695 496	240 192	938 759	12.73	3.04	20.4
South Sumatra	1 267 009	3 360 710	1 839 492	4 438 453	3.80	2.82	29.3
Riau	588 212	1 575 684	1 047 454	2 233 592	5.94	3.55	31.7
Jambi	182 846	1 261 630	432 727	1 581 327	9.00	2.28	21.5
West Sumatra	433 120	2 973 012	807 983	3 190 694	6.43	0.71	20.2
North Sumatra	2 127 436	6 223 514	3 638 832	6 613 479	5.51	0.61	35.5
Aceh	233 501	2 377 027	539 740	2 875 653	8.74	1.92	15.8
Kalimantan	1 441 300	5 275 596	2 506 657	6 596 249	5.69	2.26	27.5
West Kalimantan	416 923	2 067 968	642 989	2 592 377	4.43	2.29	19.9
Central Kalimantan	98 257	855 919	245 249	1 150 612	9.58	3.00	17.6
South Kalimantan	440 901	1 622 326	702 950	1 893 697	4.78	1.56	28.0
East Kalimantan	485 219	729 383	915 469	959 563	6.55	2.78	48.7

TABLE 5.10 Distribution of Population Between Urban and Rural Areas in Indonesia by Province, 1980–1990 (*Cont.*)

Region	1980		1990		Average annual growth (%)		Per cent Urban 1990
	Urban	Rural	Urban	Rural	Urban	Rural	
Sulawesi	1 654 190	8 746 358	2 761 021	9 750 142	5.26	1.09	22.1
Central Sulawesi	115 472	1 169 056	281 134	1 422 196	9.31	1.98	16.4
North Sulawesi	354 607	1 760 215	564 795	1 913 151	4.76	0.84	22.8
South Sulawesi	1 096 075	4 963 489	1 685 443	5 295 146	4.40	0.65	24.1
South-East Sulawesi	88 036	853 598	229 649	1 119 649	10.06	2.75	17.0
Other Islands	1 342 474	9 659 008	2 494 449	11 150 256	6.39	1.45	18.2
Bali	363 336	2 106 388	734 237	2 043 119	7.29	(0.30)*	26.4
West Nusa Tenggara	383 421	2 340 257	582 180	2 789 519	4.26	1.77	17.3
East Nusa Tenggara	205 457	2 531 531	372 242	2 895 677	6.12	1.35	11.4
Maluku	152 944	1 255 507	352 438	1 498 649	8.71	1.79	19.0
Irian Jaya	237 316	869 975	395 131	1 233 956	5.23	3.56	24.1
East Timor	0	555 350	58 221	689 336	n.a.	2.18	7.8
TOTAL INDONESIA	32 845 829	114 485 994	55 391 171	123 808 052	5.36	0.79	30.1

* Negative values

Sources: Biro Pusat Statistik, 1980 and 1990 Censuses of Indonesia.

many urban-based residents (20.4 million) as rural residents (23.2 million). In Sumatra there is a wide variety of levels of urbanization among the provinces with North Sumatra the most urbanized. There are still some areas in the Outer Islands such as East Timor, East and West Nusa Tenggara, Central and South-East Sulawesi, Central Kalimantan, Lampung, and Aceh, which have very low levels of urbanization.

The broad positive relationship between economic and urbanization levels in Indonesia's provinces is evident in Figure 5.6. The main outliers from the pattern are firstly the small, largely urban provinces of Jakarta and Yogyakarta and, secondly, the provinces in which there are large-scale oil, gas, and mineral exploitation activities (Riau, East Kalimantan, and Aceh).

Population Movement to Urban Areas

Internal migration has played a major role in the growth of Indonesia's urban areas. Before examining some aspects of the rural to urban population movement we need to touch on a pattern of internal migration of particular interest in Indonesia—that between densely populated Java–Bali of 'Inner' Indonesia and the Outer Islands. Since the early years of this century, successive governments have had active programmes to 'transmigrate' rural people from Java–Bali to the other islands. However, transmigration is only one element (and not the largest one) in a substantial migration from Java to the other islands. Table 5.11 shows that the number of people living in the Outer Islands but who had migrated there from Java increased by 73 per cent between 1971 and 1980 while the number who had moved in the opposite direction increased by only 15 per cent. Overall, there was a net migration loss to Java of 2.35 million people. During the 1980s, however, there has been a distinct change. There was net increase of migrants from Java residing in the Outer Islands in the 1980s, a similar magnitude to that recorded in the 1970s, but a percentage increase substantially lower. The most striking change, however, in Table 5.11 is in the number of migrants from the Outer Islands residing in Java—this doubled between 1980 and 1990. As a result there was only a comparatively small increase in the net migration loss from Java, from 2.35 million in 1980 to 2.71 million in 1990. Two-thirds of migrants from the Outer Islands who have settled in Java lived in urban areas (Table 5.12).

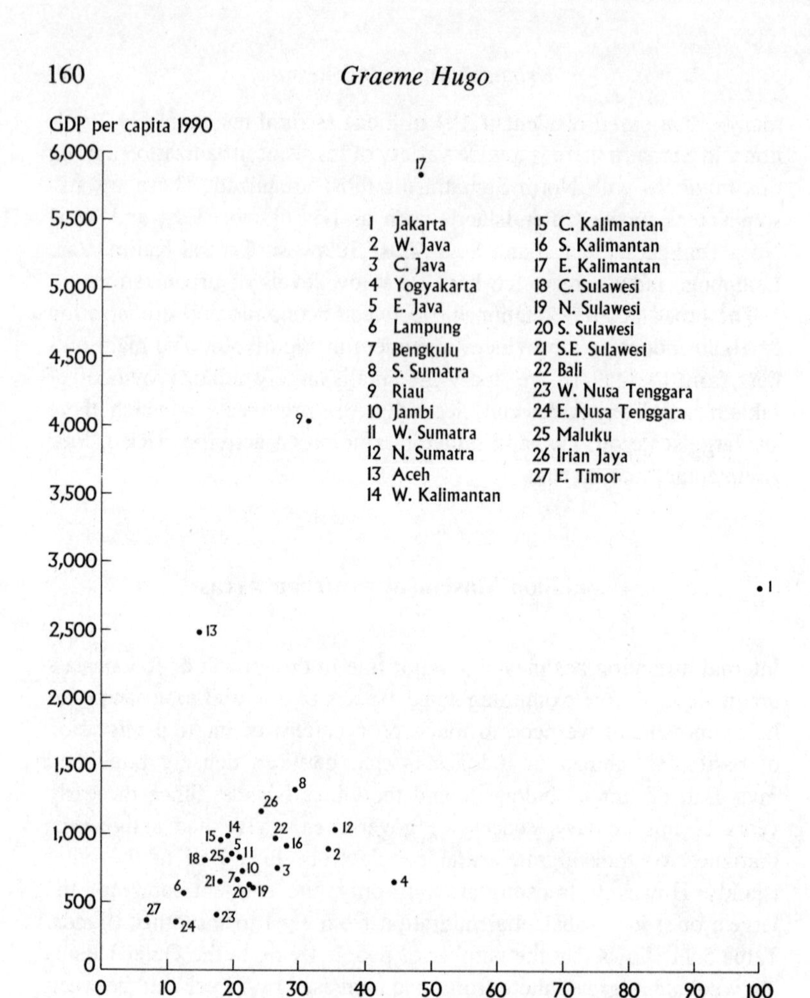

FIG. 5.6 Indonesian Provinces: Relationship between Percentage of Population Living in Urban Areas and Gross Domestic Product Per Capita, 1990

Source: Biro Pusat Statistik, Jakarta.

TABLE 5.11 Migration Into and Out of Java, 1971, 1980, and 1990

	1971	1980	1990	Increase (%)	
				1971–80	1980–90
Total Outmigrants	2 062 206	3 572 560	5 149 470	73	44
Total Inmigrants	1 067 777	1 225 560	2 434 719	15	99
Net Outmigration	−994 429	−2 347 000	−2 714 751	136	16

Note: Based upon 'most recent' migration data, using census question on province of previous residence.

Sources: 1971, 1980 and 1990 Censuses of Indonesia.

TABLE 5.12 Migration Into and Out of Java by Urban/Rural Destination, 1990

	All migrants		Migrants aged 5+ arriving 1985–90		
	Urban	Rural	Urban	Rural	Total
Total Outmigrants	1 233 874	3 915 596	361 857	601 483	963 340
Total Inmigrants	1 600 333	834 386	464 739	309 050	773 789
Net Migration	+366 459	−3 081 210	+102 882	−292 433	−189 551

Note: Based upon 'most recent' migration data, using census question on province of previous residence.

Source: 1990 Census of Indonesia.

Indeed, when only urban-destined migrants between Java and the other islands are considered, Java records a net gain.

On the other hand, more than three-quarters of Java-born people living in the other islands in 1990 resided in rural areas. Hence, while urban-destined migrants from Java to the other islands are by no means insignificant in number (1.2 million) they are outnumbered 3 to 1 by rural-destined movers and the net rural flow is heavily in favour of the Outer Islands. If only migrants who moved in the five years before the census are examined, Table 5.12 shows that the net migration loss to Java over the 1985–90 period was quite small, especially in relation to the gross volume of movement. This reflects the shift in Indonesia's development policy in the late 1980s and 1990s which has focused upon attracting investment for industrial development. This has resulted in new employment growth being concentrated in Java, especially in its urban areas, particularly in and around Jakarta. Hence, the net redistribution from Java to the Outer Islands has been greatly reduced in recent times.

Most of the inhabitants of Indonesian cities are close to their rural migrant origins. At the 1990 census, only 17.6 per cent of urban dwellers in Indonesian cities were migrants from another province. Since intra-provincial migrants outnumber their inter-provincial counterparts by up to five to one (Hugo, 1975), it is apparent that migrants or their off-spring make up the majority of Indonesian urban dwellers. Moreover, field surveys have indicated that not only has migration to cities increased in tempo in Indonesia, but rural–urban interaction has become more complex with greater incidence of commuting and other forms of circulation between urban and rural areas. At least 25 per cent of rural households on Java in the early 1980s had one or more family members working for part of the year in urban areas (Hugo, 1991). This implies that at least 3.75 million people are involved in this form of migration on Java, equivalent to 16.5 per cent of the reported 1980 urban population. The effect on the urban labour-force is much greater since virtually all such movers are either employed, or looking for work—the figure of 3.75 million is equivalent to just over half of recorded urban employment in Java at the 1980 census. Of course, since migrants are only working in the cities for part of the year the average effect is less than this, but quite likely about one-sixth of the average daily urban work-force consists of temporary migrants.

Studies in the early 1970s showed that in West Java circular migration was an important part of rural-based families' survival strategies whereby family members were allocated to labour markets in large cities (Hugo, 1975, 1982). Most work in the informal sector, the flexible working regime of which permits them to return to their village on a regular basis. Such strategies allow movers to retain their village-based work (which often varies between seasons) and to maximize the benefits of their work by 'earning in the city but spending in the village'. These rural–urban movements have greatly increased over the last two decades. This was evident in a 1985 study undertaken in the cities of Surabaya (East Java), Surakarta (Central Java), Den Pasar (Bali), Ujung Pandang (South Sulawesi), Padang (West Sumatra), and Palembang (South Sumatra) which showed that while circular migration is most common in Java, it is significant around all Indonesian provincial centres (Molo (ed.), 1986). It overwhelmingly involves low-income families and individuals with few skills who participate in the informal sector at the urban destination, 24.3 per cent owned no land at all and only 28.7 per cent had more than 0.75 ha of agricultural land. Circular migration plays a major role in strengthening urban–rural linkages and blurring distinctions between urban and rural populations (Hugo, 1992*a*).

Long-distance daily commuting to Indonesian cities, especially Jakarta, is also assuming greater significance since it was identified in the 1950s (Heeren, 1955) and studied as an important phenomenon in the early 1970s (Hugo, 1975; Koentjaraningrat, 1974, 1975). More recent studies have shown that commuting to Jakarta by train, bus, minibus, bicycle, and walking extends even beyond the Jabotabek area (DKI Jakarta, 1984, 1985; Koestoes, 1992). Commuting is significant around provincial cities as well. For Yogyakarta, Mantra (1981) showed that there was substantial long-distance commuting in the 1970s, as did Castles (1967) in several cities in East and Central Java in the 1960s, Yunus (1979) in Ujung Pandang in South Sulawesi, and Leinbach and Suwarno (1985) in Medan. The largest study, of rural villages within a 50 km radius of the cities of Banda Aceh, Palembang, Bandung, Malang, and Ujung Pandang not only demonstrated the increasing significance of commuting with improvements in transport but revealed some important differences between commuters and circular migrants (Bakir (ed.), 1986). White-collar, professional, and formal-sector workers were much more common among commuters than among circular migrants. Obviously, commuting is more amenable to being combined with a formal fixed work-time regimen than circular migration. It is apparent then that census data of populations resident in Indonesian urban areas severely underestimate the actual day-time populations of these cities and the numbers who gain their livelihood in those cities. Non-permanent migrations to Indonesia's cities are significant in scale and have a major social and economic impact on both cities and the countryside.

One major change in rural to urban movements is the increasing participation of women (Hugo, 1992*c*). For much of the post-Independence period there has been significant rural to urban migration of women to work in Indonesian urban areas, especially as domestic servants but also to some extent as petty traders. Structural changes in Indonesia in the last two decades have had a substantial impact on female migration patterns. First, the effects of the changing international division of labour and capital have seen the movement off-shore of the production processes of highly industrialized countries to low-wage nations like Indonesia. In particular, many factories established in the Jakarta region have sought to employ young unmarried women (Anaf, 1986: 20; Wolf, 1984) because of their relatively high degree of efficiency, lower wages, and more malleable nature relative to other segments of the labour-force (Lim, 1984). In addition, the increasing commercialization of agricultural production and processing, and the increasing proletarianization of the agricultural work-force, have had a significant impact on the migration of women. Over the last two decades there have been sweeping changes in agriculture in Java involving increased mechanization and changes in production processes which have displaced a considerable amount of labour in agriculture and initiated population movements (Hugo, 1985). Many of these changes have impinged disproportionately upon women. To take just one example, Collier *et al.* (1974: 120) estimated that the introduction of mechanical hullers replacing hand-pounding of rice by women in the early 1970s resulted in an annual loss of 125 million woman days of wage labour.

Another set of influences which have a complex two-way relationship with increased female migration in Indonesia are related to wider social and economic changes over the last quarter century which have reshaped the roles and statuses of women in the nation and increased their propensity to move. These include the transformation effected by increased levels and changing patterns of education, increased mass media exposure, and the transportation revolution (Hugo, 1981*b*).

Alatas (1987) has analysed the work-force participation of the 40 per cent of the females in Jakarta who were migrants. She found that migrants tended to be better educated than non-migrants. Female migrants had a higher level of participation in the work-force than non-migrant women and showed a much greater tendency to be employed in service activities. Some 34 per cent of all migrant females in Jakarta in 1980 were domestic servants, compared with only 3.6 per cent of non-migrant women, and most of them had been in the city less than two years reflecting the high 'turnover' migration of this group. After services, trade and manufacturing industry were the main areas of employment among female migrants in Jakarta. A third of migrant women worked more than sixty-four hours a week compared with 13 per cent of non-migrant women workers.

Studies of the effects of female rural–urban migration in Indonesia are few. Changes wrought by migration do not necessarily lead to an improvement in conditions for women. Migration has the potential to enhance the status of women because it usually is associated with a separation of the nuclear family from other kin and traditional power structures and hence a greater reliance by spouse and children on the nuclear family, a weakening of wider kinship relationships, and a consequent widening of the roles of nuclear family members, especially women (de Gonzales, 1961: 1274). On the other hand migration may lead to a multiplicity of new roles placing heavy burdens on women and in some cases separation from the support systems of the place of origin exposes them to greater risks of exploitation. Wolf (1984) notes, for example, that young women working in factories were rarely aware of their rights in the work situation, although Anaf (1986: 68) found that this was less so in Pasar Rebo, Jakarta. This vulnerability is, of course, also present among women involved in illegal activities such as prostitution.

Contrasts within the Urban System

The metropolitan region of Jakarta is the major focus of rural–urban migration in Indonesia. Unfortunately, the true scale of this movement cannot be assessed since Jakarta has spilled beyond its boundaries into the adjoining *kabupaten* of Bogor, Tanggerang, and Bekasi. However, it is useful to examine census data relating to migrants in the capital city district of Jakarta.

Table 5.13 shows that the number of migrants in Jakarta has increased substantially over the last three decades with the number of lifetime[5] migrants increasing by 114 per cent, while the total population increased by 183 per cent. Some 3.23 million of Jakarta's 8.23 million residents in 1990 had lived outside of the capital city district.

Jakarta has an extensive migration field, attracting significant numbers of migrants from all over Indonesia. As Jakarta has grown, the Jakarta-born have increased their share of the city's population, but even in 1990 almost 40 per cent were born outside of the capital. However, the distribution of origins of migrants has changed dramatically. In particular, the number of Jakarta residents born in adjoining West Java has changed little in thirty years and the West Java-born's share of the population fell from 26.9 to 10.5 per cent so that by 1980 there were more Central Java-born people in the capital, despite the greater distance. Indeed, in recent years there has been a net migration loss to West Java. This reversal is a function of several factors:

TABLE 5.13 Immigrants in the Capital District of Jakarta, 1961–1990

Type of migrants	1961	1971	1980	1990	Increase (%)		
					1961–71	1971–80	1980–90
Lifetime Migrants[a]	1 483 231	1 821 833	2 496 128	3 170 215	22.8	37.0	27.0
Most Recent Migrants[b]	n.a.	1 866 635	2 532 791	3 230 092	n.a.	35.7	27.5
Five Year Migrants[c]	n.a.	n.a.	718 619	833 029	n.a.	n.a.	15.9
TOTAL POPULATION	2 906 533	4 546 492	6 071 748	8 227 766	56.4	33.5	35.5

[a] Lifetime migrants are persons living in a province other than their province of birth.
[b] Most recent migrants are persons who have ever lived in another province.
[c] Five year migrants are persons who moved to their present province of residence from another province within the five years preceding the census.

Sources: 1961, 1971, 1980, and 1990 Censuses of Indonesia.

- The adoption of circular migration and commuting strategies of West Javans wishing to work in Jakarta, *instead* of permanently shifting to the capital.
- A high rate of return migration among West Java migrants to Jakarta so that there is a high 'turnover' of migrants. This is a feature of all migration to Jakarta, but especially marked for West Java migrants.
- The suburbanization of West Java migrants back into the West Java parts of Jabotabek.

Jakarta is becoming a truly Indonesian city, the only part of the nation where no one single ethnic group is dominant, and where there is substantial representation of all the major ethnic groups in the nation. Jakarta has come to symbolize and embody Indonesian national unity.

With respect to the characteristics of migrants to Jakarta we can make a number of generalizations:

- In terms of gender, permanent migration has become more balanced over the years. In 1961 the sex ratio[6] among migrants was 110, by 1971 it had fallen to 105, by 1990 to 99. Females dominate among recent migrants: there are only 82 males for every 100 females among migrants who arrived between 1985 and 1990. This is due to a greater turnover of female migrants, particularly those employed as domestic servants, prostitutes, and petty traders who often spend only a few years in the city before returning to their village.
- The dominance of females among recent migrants declines with distance from Jakarta of migrants' provinces of origin.
- Males are dominant among circular migrants and commuters to Jakarta although the participation of women in this form of movement is increasing.
- As with other migrations, permanent migration to Jakarta is dominated by people in the young adult cohorts. There is also some tendency for older female widows to migrate to Jakarta to join their children (Hugo, 1992*b*). In contrast, the age distribution of commuters and circular migrants, while wholly concentrated in the working ages, is less clustered in the young adult ages.
- There is considerable positive educational selectivity in migration to Jakarta, and the extent of this selectivity strengthens with increasing distance of the province of origin from the capital. This is closely associated with occupational differentials: people in the

lower-paying occupations tend to be drawn from provinces closer by.

- Permanent migration to Jakarta is selective of white-collar and more skilled workers and of those involved in government or other formal sector occupations.

Turning to other Indonesian cities, Table 5.14 shows considerable variation in their growth-rates. In fact, the growth-rates for the last decade are problematic in the absence of 1990 data for functionally defined urban areas. Still, a general pattern of smaller Indonesian cities growing more slowly than the largest agglomerations is evident. Some resource-based cities in the Outer Islands are growing very rapidly as is Batam, the city which is receiving overspill development from Singapore in the so-called 'Growth Triangle' involving Malaysia, Singapore, and Indonesia. But long-established centres which provide essentially service and administrative functions, especially those in Java, have grown slowly (Hugo *et al.*, 1987; NUDSP, 1985; World Bank, 1984). Despite the wide variation in growth experiences, some generalizations can be made about provincial cities:

- Net migration has generally played less of a role in the growth of smaller cities than it has in Jakarta. Natural increase and realignment of urban boundaries are the dominant elements of urban growth and urbanization.
- Migration tends to be predominantly from within the province in which the city is located. Hence, in contrast to Jakarta, these cities tend to be dominated by a single ethnic group, although they are more heterogeneous than the adjoining rural areas. Thus Bandung is largely a Sundanese city, Surabaya Javanese-Madurese, Yogyakarta Javanese, Medan Batak, Ujung Pandang Bugis-Makasarese, etc.
- Migration turnover in provincial cities is greater than in large cities like Jakarta, Bandung, and Surabaya. This is due to step migration and a significant 'floating population' made up of persons employed by government agencies or private enterprise who are transferred to regional urban centres for limited periods (Evers, 1972).
- Circular migration and commuting are highly significant in these centres (Hugo and Mantra, 1983).
- Migration to provincial cities seems to be less selective of females than is the case for that to the largest cities, especially Jakarta.
- There is a tendency for migration to be selective of young adults,

TABLE 5.14 Population and Population Growth-Rate of Urban Centres with More Than 100 000 Inhabitants in 1980 for 1920–1990

Centre	Population						Average annual growth (%)			
	1920	1930	1961	1971[e]	1980[e]	1990	1920–61	1961–71	1971–80	1980–90
Java										
Jakarta	306 309	533 015	2 973 052	4 084 950	6 071 750	8 227 746	5.4[a]	4.6[a]	4.5[a]	3.1
Surabaya	192 190	341 675	1 007 945	1 308 630	1 737 020	2 473 272	4.2[a]	4.4[a]	3.2	3.6
Bandung	94 800	166 815	972 566	1 310 910	1 744 520	3 349 993[c]	5.1[a]	2.1	3.6	6.7
Semarang	158 036	217 796	503 153	566 380	820 140	1 389 227[c]	2.8	2.5	4.2	5.4
Malang	42 981	86 646	341 452	365 280	491 470	1 130 034[d]	4.6[a]	2.2	3.4	8.7
Surakarta	134 285	165 484	367 626	438 490	539 980	503 827	2.2	1.2[b]	2.3	−0.7
Yogyakarta	103 711	136 649	312 698	347 260	460 170	412 059	2.4	0.9[b]	3.2	−1.1
Bogor	45 595	65 431	154 092	330 160	884 985	2 195 207[c]	2.9	2.4	9.5[a]	9.5
Kediri	43 222	48 567	159 918	130 500	176 260	469 179[c]	2.8	1.2[b]	3.4	10.3
Cirebon	33 051	54 079	158 299	195 360	265 720	868 330[c]	3.4	1.2[b]	3.5	12.6
Madiun	31 593	41 872	123 373	135 660	169 920	246 802[c]	2.9	1.0[b]	2.5	3.8
Tasikmalaya	14 216	25 605	125 525	146 620	192 270	370 555[d]	4.5[a]	0.8[b]	3.1	6.8
Jember	16 491	20 222	94 089	147 350	171 280	428 495[c]	4.0[a]	2.7	1.7[b]	9.6
Pekalongan	47 852	65 982	102 380	191 590	249 170	432 807[c]	1.7[b]	0.9[b]	3.0	5.7
Magelang	36 213	52 944	96 454	126 800	152 950	290 998[c]	2.2	1.4[b]	2.1	6.6
Tegal	34 687	43 015	89 016	229 590	326 800	723 539[c]	2.2	1.7[b]	4.0	8.3
Sukabumi	23 533	34 191	80 438	160 840	215 290	453 714[c]	2.4	1.8[b]	3.3	7.7
Kudus	42 045	54 524	74 911	114 840	154 480	358 222[d]	2.0	2.0	3.4	8.8
Garut	14 063	24 219	76 244	113 320	145 620	269 859[d]	1.7	3.9	2.8	6.4
Bekasi	n.a.	n.a.	32 012	61 320	188 668	1 152 883[c]	n.a.	6.5[a]	11.9[a]	19.8
Purwokerto	12 584	33 266	80 556	113 160	143 790		5.2[a]	3.4	2.7	
Cilacap	18 991	28 309	55 333	85 470	127 020	307 435[c]	3.7	4.3[a]	4.5[a]	9.2
Depok	n.a.	n.a.	n.a.	44 640	126 690		n.a.	n.a.	12.3[a]	
Pasuruan	n.a.	37 081	63 408	87 020	119 090	408 697[c]	2.1	2.1	3.6	13.1
Klaten	9373	12 039	33 400	97 550	117 560	384 885[c]	4.5[a]	10.7[a]	2.1	12.6
Cianjur	11 955	20 812	62 546	81 430	105 660	241 839[c]	3.8	2.6	2.9	8.6
Salatiaga						98 012				

TABLE 5.14 Population and Population Growth-Rate of Urban Centres with More Than 100 000 Inhabitants in 1980 for 1920–1990 (*Cont.*)

Centre	Population						Average annual growth (%)			
	1920	1930	1961	1971[e]	1980[e]	1990	1920–61	1961–71	1971–80	1980–90
Sumatra										
Medan	45 248	76 584	479 098	892 130	1 265 210	1 730 052	5.3[a]	2.9	4.0	3.2
Palembang	73 726	108 145	474 971	504 340	757 490	1 144 047	3.1	2.1	4.6[a]	4.2
Tanjung Karang	14 980	25 170	133 901	166 010	357 690		5.2[a]	4.0[a]	8.9[a]	
Padang	38 169	52 054	143 699	188 940	296 680	631 263	3.3	3.2	5.1[a]	7.8
Jambi	11 311	22 071	113 080	94 350	155 760	339 786	5.2[a]	3.4	5.7[a]	8.1
Pekanbaru[f]	n.a.	10 000	70 821	126 090	186 200	398 621	6.5[a]	7.4[a]	4.4[a]	7.9
Pematang Siantar	9460	15 328	114 870	143 420	172 910	219 316	5.3[a]	1.2[b]	2.1	2.4
Kalimantan										
Banjarmasin	46 993	65 698	214 096	224 060	330 130	480 737	3.6	2.8	4.4[a]	3.8
Pontianak	28 731	45 196	150 220	182 190	276 670	466 108	4.0[a]	3.8	4.8[a]	3.7
Samarinda	6879	11 086	69 715	90 770	182 470	407 174	6.1[a]	7.0[a]	8.1[a]	8.4
Balikpapan	n.a.	29 843	91 706	97 370	208 040	344 147	2.8	4.1[a]	8.8[a]	5.2
Sulawesi										
Ujung Pandang	56 718	84 855	384 159	423 560	638 800	944 372	4.1[a]	2.7	4.7[a]	4.0
Manado	17 062	27 544	129 912	151 490	217 090	320 600	4.6[a]	2.7	4.1	4.0

Other Islands										
Mataram	n.a.	n.a.	n.a.	155 100	210 490		n.a.	n.a.	3.5	
Ambon	11 120	17 334	56 037	64 700	111 910	275 888	3.9	3.6	6.3[a]	9.4
Den Pasar	8501	16 639	56 780	72 830	159 230		4.2[a]	1.5[a]	9.1[a]	

[a] More than twice the national population growth rate.
[b] Below the national growth rate.
[c] *Kotamadya* and *kota kabupaten* (i.e. municipalities and other regency capitals).
[d] Urban kabupaten only.
[e] The 1971 and 1980 figures are as defined by the National Urban Development Strategy Project for boundaries of functional urban areas which differ in some cases from administrative definitions of urban areas.
[f] The 1930 population figure for Pekanbaru is the upper limit of the estimate made by census officials (Withington 1963, 242). The annual growth rate for Pekanbaru is for 1930–71, not 1920–71.

Sources: Hugo 1981*b*: 65; NUDSP 1985: 12–13; 1990 Census of Indonesia.

but it appears that young families rather than young adult singles and couples without children predominate in the movement to Jakarta.
- There are substantial educational differentials in permanent migration to provincial urban areas—to a greater degree than in Jakarta (Hugo and Mantra, 1983).

Overall it would seem that there are significant differences between population movements to the largest metropolitan centres in Indonesia (Jakarta, Surabaya, Bandung) and those to provincial cities and towns. These small cities and towns are not recording substantial net gains of migrants, and the selective and 'step-by-step' nature of migration up the urban hierarchy has meant that they are losing their best educated, their most skilled, and their entrepreneurially oriented young people to the largest cities. There are thus not only wide disparities in social and economic conditions between urban and rural areas but also between the largest cities and other urban areas, disparities that generate concern among Indonesia's planners.

Official estimates of poverty in Indonesia indicate that it has halved in the last fifteen years. However, it is interesting to observe that, according to some estimates of poverty in Indonesia, the incidence of poverty in rural Indonesia fell below that in urban areas in the late 1980s. In 1990, one in three Indonesians living in poverty resided in urban areas, whereas in 1976 fewer than one in five were urbanites. Poverty then is increasingly becoming an urban problem, especially on Java.

It has been argued that rural–urban migration has contributed to the improvement in the situation in rural areas because it relieves the pressure on rural job opportunities and because of the flow of remittances from urban–based migrants to their rural-based families. The World Bank (1992: 114), for example, has estimated that rural–urban migration 'by moving people from lower productivity activities (measured by lower incomes) to higher productivity ones contributed about 18 per cent of total per capita income growth in 1990. Within rural areas, the impact of outmigration to cities was estimated to contribute as much as 30 per cent of the growth of rural per capita incomes.'

Indonesia has had relatively low levels of open unemployment (around 2 per cent) but high levels of underemployment (around 40 per cent) (Hugo, 1993*a*). However, the 1990 census indicates that open unemployment has increased substantially and is being increasingly felt in urban areas. One dimension of this problem is 'the increasing number of school and university graduates, who either cannot find a job or attain employment commensurate with their expectations. An estimated 62 per cent of young people aged between 15 and 19 with a high school education were looking for work in 1986. The fear is that this group, concentrated as they are in urban areas, could pose a threat to stability' (Vatikiotis, 1988: 39).

Conclusion

This chapter has concentrated largely upon the demographic aspects of urbanization in Indonesia. By the year 2000, a third of Indonesians will be living in cities, and many more will be living outside those cities but gaining their livelihood there via commuting and circular migration. The population movement between village and city in Indonesia is not only increasing the size of Indonesia's urban population but also strengthening linkages between urban and rural areas with an increasing flow of information, goods, money, and people. This blurs the distinction between urban and rural and underlines that neither urban nor rural planning should be undertaken separately from each other.

Indonesia's cities are consistently depicted as having chronic shortages of housing, limited accessibility to fresh water, sewerage, garbage collection, and other infrastructural services, unacceptably high levels of air, noise, and water pollution, high levels of unemployment and underemployment, etc. Yet they are continuing to draw people from villages on a permanent or temporary basis in increasingly large numbers. This is because the 'unattractive' picture of cities so often painted by outsiders belies the fact that, some estimates of the incidence of poverty notwithstanding, urban Indonesians on average are better off than their rural cousins, on almost every indicator of social and economic well-being. Moreover, the current development strategy will encourage much more job creation in urban than in rural-based activities. In addition, there are signs that new developments in the agricultural sector will produce a significant displacement of labour within the next decade or so. Hence any suggestion that urbanization is intrinsically bad for Indonesia and Indonesians must be rejected. Despite the recent rapid urban growth, Indonesia will still be one of the world's least urbanized major nations in the year 2000. Of the nineteen nations with more than 50 million residents, only Bangladesh, China, India,

Nigeria, Thailand, and Vietnam have lower levels of urbanization than Indonesia. The scope for further urbanization is clearly there, and the key issue is *where* and *how* these increases be absorbed not whether they should or should not occur.

Jakarta must loom large in any discussion of Indonesian urbanization. The United Nations (1993) currently rank Jakarta as the nineteenth largest city in the world and estimate that by the end of the century it will be in the top twelve. However, it has been argued here that these assessments are based on an incorrect definition of the functioning city of Jakarta which has at least twice its 'official' number of residents. Some commentators expect Jakarta to have a much larger population by the end of the century. Douglass (1988*a*: 28), for example, states that 'By the year 2000 the greater Jakarta metropolitan region is expected to have as many as 30 million people, placing it ahead of Los Angeles as the seventh largest urban agglomeration in the world.'

While the argument that metropolitan regions can grow so large as to experience serious diseconomies lacks unequivocal supportive evidence (Jones, 1988: 142), it is generally accepted that the development and equity goals of Indonesia's government would be more effectively pursued if there were some decentralization of urban development away from Jakarta. The means by which this can or should be achieved, however, is the subject of some debate, although there is agreement about the futility of closed-city policies which have been rejected by subsequent governors (*Jakarta Post* 18 June 1988: 1). There would also appear to be agreement that it is generally more effective and desirable to target policies not so much toward individuals and/or family units as was the case with the 'closed-city' strategy but more to units and organizations which create employment (Fuchs and Demko, 1981). The overwhelming evidence from a wide range of studies on population movement in Indonesia is the predominance of employment-related motives in shaping how many people move, who moves, where they move from, and where they move to (Hugo *et al.*, 1987).

An emerging consensus in the literature suggests that less expensive, *indirect* strategies which work largely through sectoral policies may be more effective means of achieving population redistribution goals. This approach essentially involves:

- First, identification of the 'biases in macro and sectoral policy that unnecessarily promote the growth of large cities' (Jones, 1988: 192).

- Secondly, developing strategies to remove the 'blockages' which are presently making it disadvantageous for employment-generating enterprises to locate in medium-sized cities.

Indonesia has the major advantage of having a well-developed urban hierarchy with a substantial number of established medium-sized cities which can readily absorb accelerated growth and obviate the expense of the establishment of new growth centres. Moreover, many of these cities have a recent history of substantial growth which suggests that they have the potential to absorb and sustain economic growth.

Such strategies might include the following (NUDSP, 1985; Hamer *et al.*, 1986):

- Improvement of transport and communication systems so as to increase the access of intermediate cities to domestic and international markets as well as to their hinterlands.
- Reducing the degree of government regulation of business activities so as to eliminate the incentive to locate businesses near major administrative centres.
- Expanded use of industrial zones and estates in areas that have well-documented growth potential.
- Improvement of access to capital and financial advice for existing small- and medium-sized enterprises in intermediate cities.
- Improved provision of public services and utilities in medium-sized cities.
- Fiscal reform in intermediate cities away from reliance on the central government to a combination of grants, loans, increased local tax efforts, and user charges which will give the city administration control over their budgets and flexibility in planning.
- Greater local autonomy and a devolution of power to regional administrations. Although there are strong regional interest groups in Indonesia, since Independence political power and decision-making have become increasingly centralized in Jakarta and 'policy-making is essentially a top-down process in which the central government plays a predominant role' (MacAndrews, Fisher, and Sibero, 1982: 84). Regional development authorities (BAPPEDA) 'have little room to manœuvre in planning' and 'to a large extent they are simply agents of the centre implementing centrally determined policies' (Booth, 1981: 4).
- A corollary of the last point is that greater independent financial responsibility be given to the regions.

- Housing has been shown to be the most influential non-employment factor in migration which is amenable to policy intervention (Fuchs and Demko, 1981). While Indonesia's capacity to provide public housing in urban areas is limited, it should be directed at the poor and located in provincial urban centres rather than Jakarta.
- Intermediate city growth is likely to be enhanced, if it is part of a wider development strategy encompassing the hinterlands of those cities which stresses the expansion, diversification, and reform of agriculture, increases in farm employment, the development of agricultural services and processing, and better provision of services and utilities (Douglass, 1988*b*: 31–4).
- More attention should be paid to the role of the urban informal sector in intermediate cities in absorbing workers to ensure that existing policies do not work against this role but rather enhance it.

Even if decentralization strategies have a measure of success in Indonesia, Greater Jakarta will continue to grow rapidly. This raises a number of challenges for policy-makers and planners, not least because there is very little experience to draw on to manage cities with more than 20 million inhabitants:

- There can be no doubt that the already huge pressure upon housing, water supply, transport, garbage disposal, utility provision, health and education services in Jakarta will escalate. Much has been achieved in these areas in the capital over recent years, but there is still considerable scope for adoption of additional, flexible, accommodationist measures (Laquian, 1981) which fully recognize and support the contribution of the informal sector in employment creation, housing construction, and service provision and emphasize the provision of basic needs for all residents rather than quality provision for the relatively well-off. The elaboration and extension of existing programmes such as the Kampung Improvement Program and the development of new initiatives along these lines is crucial.
- Employment creation must be a priority goal in Jakarta. Here there is a need to work on many fronts. The productive utilization of the large number of unemployed graduates and the stimulation of the informal sector are two of the more promising and important avenues.

- Greater attention needs to be paid to the environmental consequences of Jakarta's growth. For example, the excessive drawing of water from underground aquifers has lowered the water-table producing an encroachment of seawater into much of the aquifer and causing building subsidence. Air pollution is at levels above WHO standards in some areas. Environmental health considerations must be recognized and given legislative and financial back-up.
- There is scope in Jakarta for raising revenue through local taxation and user-pays approaches to service provision. In this respect, Jones' recommendation 'to gear the standard of services provided to the ability to pay of both the city administration and the individual urban dwellers' (1988: 151) is important.
- Greater attention to the housing problems of the poor is sorely needed. This may necessitate investigation of low-cost housing alternatives which pay less attention to unrealistic 'minimum housing standards' and aesthetic niceties and more to involvement of future occupants in construction, sites-and-services schemes, and the availability of realistic financing. Rather than insisting upon minimum standards at the outset, allowance has to be made for families to upgrade the size and quality of their dwelling units as they become more established.
- The lateral expansion of Jakarta is of concern in that it is taking prime agricultural land out of production and creating environmental problems of major dimensions. The need for careful zoning in the Jabotabek and Jabopunjar regions is pressing (Douglass, 1988*b*). The potential for land-use conflicts is enormous and should be anticipated. Much of the region into which Jakarta is expanding is administered by regulations not appropriate to, and by officials with no experience of, metropolitan regions. There is clearly a need for a regional authority which has the power to control and co-ordinate land-use development in the region.

Notes

1. Indonesia's GDP grew by 7.5 per cent in 1989, 7.1 per cent in 1990, and 6.6 per cent in 1991 (*Far Eastern Economic Review*, 18 March 1993). In 1992 the rate was 6.3 per cent, and the 1993 rate was forecast to be 6.5 per cent (McBeth, 1993).

2. An acronym made up of Jakarta and the three adjoining West Java *kabupaten* of Bogor, Tanggerang, and Bekasi.
3. This incorporates Surabaya and the adjoining *kabupaten* of Gresik, Bangkalan, Mojokerto, and Sidoarjo.
4. This corridor has been given the acronym of JABOPUNJAR—Jakarta, Bogor, Puncak, and Cianjur.
5. Lifetime migrants are persons living in a province other than their province of birth at the time of the census.
6. Males per 100 females.

References

Alatas, S. (1987), *Migran Wanita Di Dki Jakarta Dan Peranannya Dalam Ketenagakerjaan: Studi Hasil Sensus Penduduk* [Female Migrants in Jakarta and Their Role in the Labour-Force: A Study Based on Population Census Results] (Jakarta: Universitas Indonesia).

Anaf, A. (1986), *Female Migration and Employment: A Case Study in Kecamatan Pasar Rebo*, Indonesian Institute of Science Monograph Series (Jakarta: National Institute of Economic and Social Research).

Aten, A. (1952–3), 'Some Remarks on Rural Industry', *Indonesia*, 6: 19–27, 193–216, 330–45, 411–22, 536–9.

Awanohara, S. (1993), 'The Magnificent Eight—World Bank Seeks Lessons from East Asia', *Far Eastern Economic Review*, 22 July: 79–80.

Azis, J. J. (1989), 'Key Issues in Indonesian Regional Development', in H. Hill (ed.), *Unity and Diversity: Regional Economic Development in Indonesia Since 1970* (Singapore: Oxford University Press).

Bakir, S. Z. (1986) (ed.), *Studi Mobilatas Ulang Aik Di Lima Kola Besar Di Indonesia* [Study of Commuting to Five Large Cities in Indonesia] (Jakarta: Ministry of Population and Environment).

Biro Pusat Statistik (BPS) (1963), *Sensus Penduduk 1961* (1961 Population Census) (Jakarta: Biro Pusat Statistik).

—— (1975), *Sensus Penduduk 1971: Penduduk Indonesia* (1971 Population Census: Population of Indonesia), Series D (Jakarta: Biro Pusat Statistik).

—— (1983), *Penduduk Indonesia: Hasil Sensus Penduduk 1980* (Population of Indonesia: Results of the 1980 Population Census), Series S, No. 2 (Jakarta: Biro Pusat Statistik).

—— (1992), *Penduduk Indonesia: Hasil Sensus Penduduk 1990* (Population of Indonesia: Results of the 1990 Population Census) (Jakarta: Biro Pusat Statistik).

Booth, A. (1981), *Financing Agricultural Development*, Institute of Local Government Studies, University of Birmingham (Birmingham).

—— (1992), 'Introduction', pp. 1–38 in A. Booth (ed.), *The Oil Boom and After: Indonesian Economic Policy and Performance in the Soeharta Era*, (Singapore: Oxford University Press).

Castles, L. (1967), 'An Ethnic Profile of Jakarta', *Indonesia*, 3: 153–204.

Cobban, J. L. (1970), 'The City on Java: An Essay in Historical Geography', unpublished Doctoral thesis, University of California, Berkeley.

Collier, W. R., Colfer, J., Sinarhadi, and Shaw, R. (1974), 'Choice of Technique in Milling: A Comment', *Bulletin of Indonesian Economic Studies*, 10/1: 106–20.

De Gonzales, N. L. S. (1961), 'Family Organisation in Five Types of Migratory Wage Labour', *American Anthropologist*, 63/6: 1264–80.

De Haan, F. (1935), *Oud Batavia* (2 vols.) (Batavia: G. Kolff).

Douglass, M. (1988*a*), 'Urbanization and National Urban Development Strategies in Asia: Indonesia, Korea and Thailand', University of Hawaii, Department of Urban and Regional Planning Division, Paper No. 9.

—— (1988*b*), 'Land Use and Environmental Sustainability of the Extended Metropolis: Jabotabek and Jabopunjur Corridor', paper presented to Conference on the Extended Metropolis in Asia, East–West Center, Environment and Policy Institute, Honolulu, 19–23 September.

DKI Jakarta Biro Bina Kependudukan dan Lingkungan Hidup (1984), *Pola Pengendalian Tingkat Mobilitas Kependudukan Daerah Khusus Ibukota Jakarta, 1983* [Management Patterns of Population Mobility Levels in the Special District of Jakarta, 1983] (Jakarta: Demographic Institute, University of Indonesia).

—— (1985), *Pola Pengendalian Mobilitas Penduduk Penglaju BOTABEK-Jakarta, 1984–1985* [Management Patterns of Commuting Between Bogor-Tangerang-Bekasi and Jakarta, 1984–1985] (Jakarta: Demographic Institute, University of Indonesia).

Evers, H. (1972), 'Preliminary Notes on Migration Patterns of a Sumatran Town', *Sumatra Research Bulletin*, 2/1: 18–23.

Firman, T. (1989), 'Pembangunan Kota-Kota Baru di Wilayah Metropolitan Jabotabek' [Development of New Cities in the Jabotabek Metropolitan Region], *Prisma*, 18/6: 49–60.

—— (1991), 'Penataan Koridor Antar Kota' [Managing Inter-City Corridors], *Kompas*, 11 January.

—— (1992), 'The Spatial Pattern of Urban Population Growth in Java, 1980–1990', *Bulletin of Indonesian Economic Studies*, 28/2: 95–109.

Fisher, C. A. (1964), *Southeast Asia* (London: Methuen).

Fuchs, R. J. and Demko, G. J. (1981), 'Population Distribution Measures and the Redistribution Mechanism', in United Nations, *Population Distribution Policies in Development Planning* (New York: United Nations), 70–85.

Gardiner, P. (1992), 'Urban Population in Indonesia: Future Trends', in T. J. Kim, G. Knaap, and I. J. Azis (eds.), *Spatial Development in Indonesia: Review and Prospects* (Aldershot: Avebury).

Ginsburg, N., Koppel, B., and McGee, T. G. (1991) (eds.), *The Extended Metropolis: Settlement Transition in Asia* (Honolulu: University of Hawaii Press).

Hageman, J. (1852), *Geschiedenis, Aardrijkskunde, Fabelleer en Tijdrekenkunde van Java* [History, Geography, Mythology and Chronology of Java], Part II (Batavia: Lange).

Haggett, P. (1975), *Geography: A Modern Synthesis* (New York: Harper and Row).

Hamer, A. D., Steer, A. D., and Williams, D. G. (1986), *Indonesia: The Challenge of Urbanization*, World Bank Staff Working Paper No. 787: 1–11, 33–61.

Heeren, H. J. (1955) (ed.), 'Urbanization of Djakarta', *EKI*, 8: 696–736.

Hugo, G. J. (1975), 'Population Mobility in West Java, Indonesia', unpublished Ph.D. thesis, Department of Demography, Australian National University, Canberra.

—— (1980), 'Population Movements in Indonesia During the Colonial Period', in J. J. Fox, R. G. Garnaut, P. T. McCawley and J. A. C. Mackie (eds.), *Indonesia: Australian Perspectives* (Canberra: Australian National University, Research School of Pacific Studies), 95–135.

—— (1981*a*), 'Levels, Trends and Patterns of Urbanisation', in ESCAP, *Migration, Urbanisation and Development in Indonesia* (New York: United Nations), 57–80.

—— (1981*b*), 'Road Transport, Population Mobility and Development in Indonesia', in G. W. Jones and H. V. Richter (eds.), *Population Mobility and Development: Southeast Asia and the Pacific*, Monograph No. 27. (Canberra: Australian National University Development Studies Centre), 335–86.

—— (1982), 'Circular Migration in Indonesia', *Population and Development Review*, 8/1: 59–83.

—— (1985), 'Structural Change and Labour Mobility in Rural Java', in G. Standing (ed.), *Labour Circulation and the Labour Process* (London: Croom Helm), 46–88.

—— (1991), 'Recent Developments in Indonesian Migration', paper presented to International Conference on Migration in Asia, University of Singapore, January.

—— (1992*a*), 'Migration and Rural–Urban Linkages in the ESCAP Region', in ESCAP, *Migration and Urbanization in Asia and the Pacific: Interrelationships with Socio-Economic Development and Evolving Policy Issues*, Selected Papers of the Pre-Conference Seminar, Fourth Asian and Pacific Population Conference, Seoul, 21–25 January (New York: United Nations), 91–117.

—— (1992*b*),'Ageing in Indonesia: A Neglected Area of Policy Concern', in D. R. Phillips (ed.), *Ageing in Newly Industrializing Countries of East and Southeast Asia* (London: Edward Arnold), 207–30.

—— (1992*c*), 'Women on the Move: Changing Patterns of Population Movement of Women in Indonesia', in S. Chant (ed.), *Gender and Migration in Developing Countries* (London: Belhaven Press), 174–96.

—— (1993*a*), *Manpower and Employment Situation in Indonesia 1992* (Jakarta: Indonesian Department of Labour).

—— (1993*b*), 'Urbanisasi: Indonesia in Transition', *Development Bulletin*, May 27: 46–9.

—— and Mantra, I. B. (1983), 'Population Movement To and From Small and Medium-Sized Towns and Cities in Indonesia', *Malaysian Journal of Tropical Geography*, 8: 10–32.

—— Hull, T. H., Hull, V. J., and Jones, G. W. (1987), *The Demographic Dimension in Indonesian Development* (Kuala Lumpur: Oxford University Press).

Jones, G. W. (1988), 'Urbanization Trends in Southeast Asia: Some Issues for Policy', *Journal of Southeast Asian Studies*, 19/1: 137–54.

Koentjaraningrat (1974), 'Mobilitas Penduduk sekitar Jakarta' [Population Mobility around Jakarta], *Masyarakat Indonesia*, 1/2: 45–60.

—— (1975), 'Population Mobility in Villages around Jakarta', *Bulletin of Indonesian Economic Studies*, 11: 108–20.

Koestoes, R. H. (1992), 'Residential Choice and Journey-To-Work in Desa-Kota Region, *Demografi Indonesia*, 37: 51–74.

Laquian, A. A. (1981), 'Review and Evaluation of Urban Accommodationist Policies in Population Redistribution', in United Nations, *Population Distribution Policies in Development Planning* (New York: United Nations), 101–12.

Leinbach, T. R. and Suwarno, B. (1985), 'Commuting and Circulation Characteristics in Intermediate Sized City: The Example of Medan, Indonesia', *Singapore Journal of Tropical Geography*, 6/1: 35–47.

Lim, Lin Lean (1984), 'Towards Meeting the Needs of Urban Female Factory Workers', in G. W. Jones (ed.), *Women in the Urban and Industrial Workforce of Southeast and East Asia*, Monograph No. 33 (Canberra: Australian National University Development Studies Centre), 129–48.

MacAndrews, C., Fisher, H. B., and Sibero, A. (1982), 'Regional Development Planning and Implementation in Indonesia: The Evolution of a National Policy', in C. MacAndrews and C. L. Sien (eds.), *Too Rapid Rural Development: Perceptions and Perspectives from Southeast Asia* (Athens, Oh.: Ohio University Press).

Mantra, I. B. (1981), *Population Movement in Wet Rice Community: A Case Study of Two Dukuh in Yogyakarta Special Region* (Yogyakarta: Gadjah Mada University Press).

McBeth, J. (1993), 'Show of Resilience', *Far Eastern Economic Review*, 11 November.

McGee, T. G. (1967), *The Southeast Asian City* (London: Camelot Press).

—— (1991), 'The Emergence of Desa Kota Regions in Asia: Expanding an Hypothesis', in N. Ginsburg, B. Koppel and T. G. McGee (eds.), *The Extended*

Metropolis: Settlement Transition in Asia (Honolulu: University of Hawaii Press), 3–26.

Milone, P. D. (1966), *Urban Areas in Indonesia* (Berkeley: University of California Press).

Molo, M. (1986) (ed.), *Studi Mobilitas Sirkuler Penduduk ke Enam Kota Besar di Indonesia* [Study of Circular Mobility to Six Large Cities in Indonesia] (Jakarta: Ministry of Population and Environment).

NUDSP (National Urban Development Strategy Project) (1985), *NUDS Final Report* (Jakarta: Directorate of City and Regional Planning, Department of Public Works).

Raffles, T. S. (1817), *The History of Java* (2 vols.) (London: Black, Parbury and Allen).

Ranneft, J. M. (1916), 'Volksverplaatsing op Java', *Tijdschrift voor het Binnenlandsch Bestuur*, 49: 59–87, 165–84.

—— (1929), 'The Economic Structure of Java', in B. Schrieke (ed.), *The Effect of Western Influence on Native Civilizations* (The Hague: G. Kolff), 71–84.

Reid, A. (1980), 'The Structure of Cities in Southeast Asia, Fifteenth to Seventeenth Centuries', *Journal of Southeast Asian Studies*, 11/2: 235–50.

Renaud, B. (1981), *National Urbanization Policies in Developing Countries* (Oxford University Press).

Rietveld, P. (1988), 'Urban Development Patterns in Indonesia', *Bulletin of Indonesian Economic Studies*, 24/1: 73–95.

Scheltema, A. M. P. A. (1926), 'De Groei van Java's Bevolking', *Koloniale Studiën*, 10/2: 849–83.

Schwarz, A. (1990), 'A Miracle Comes Home', *Far Eastern Economic Review*, 19 April: 40–4.

Skinner, G. W. (1963), 'The Chinese Minority', in R. T. McVey (ed.), *Indonesia* (New Haven: Human Relations Area Files), 91–117.

United Nations (1993), *World Urbanization Prospects: The 1992 Revision* (New York: United Nations).

Vatikiotis, M. (1988), 'Worrying About Idle Minds', *Far Eastern Economic Review*, 13 October: 39.

Volkstelling (Population Census) (1933–6), *Definitieve Uitkomsten van de Volkstelling 1930* (8 vols.) (Batavia: Department van Landbouw, Nijverheid en Handel).

Vries, E. de and Cohen, H. (1938), 'On Village Shopkeeping in Java and Madura', *Bulletin of the Colonial Institute of Amsterdam*, 1/4: 263–73.

Wolf, D. L. (1984), 'Making the Bread and Bringing it Home: Female Factory Workers and the Family Economy in Rural Java', in G. W. Jones (ed.), *Women in the Urban and Industrial Work-force of Southeast and East Asia*, Monograph No. 33 (Canberra: Australian National University Development Studies Centre), 215–34.

World Bank (1984), 'Economic and Social Development: An Overview of Regional Differentials and Related Processes', main report from *Indonesia: Selected Aspects of Spatial Development*, Report No. 4776-IND (World Bank: Country Programs Department, East Asia and the Pacific Regional Office).

—— (1992), *Indonesia: Agricultural Transformation: Challenges and Opportunities* (Washington, DC: World Bank).

—— (1993), *World Development Report 1993: Investing in Health* (New York: Oxford University Press).

—— (1994), *World Development Report 1994: Infrastructure for Development* (New York: Oxford University Press).

Yunus, D. (1979), ' "Pagaden": Suplaier Terbesar Kebutuhan Kota' ['Pagaden': Major Supplier of Urban Needs], *Pedoman Rakhyat*, 33/30: 1–2.

Zelinsky, W. (1971), 'The Hypothesis of the Mobility Transition', *Geographical Review*, 41/2: 219–49.

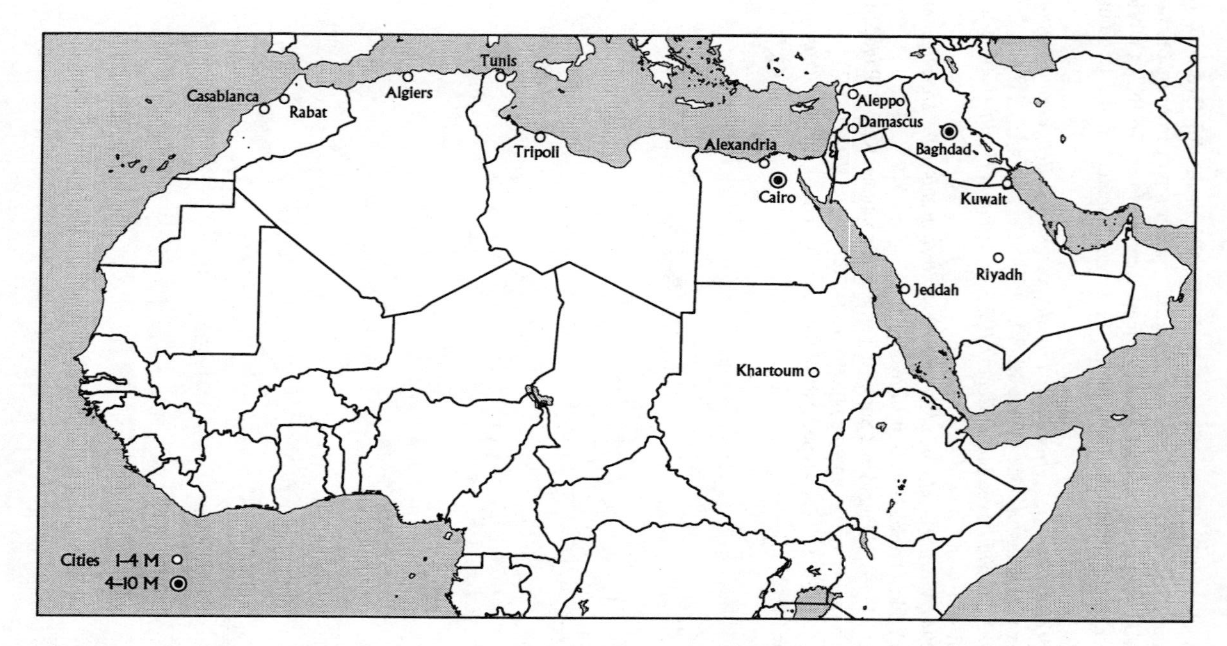

FIG. 6.1. Urban Agglomerations in the Arab World with more than 1 Million Inhabitants in 1990

6

Urbanization in the Arab World and the International System

JANET ABU-LUGHOD

IN a recent article intended to evaluate the 'State of the Art in Studies of Middle Eastern Urbanization' (1991), I concluded that most works on this topic were disappointing because they had been done either by historians with little understanding of the nature of cities (i.e. those lacking expertise in urban studies in general) or by atheoretical empiricists engaged in pure description, whether historical, spatial, or statistical.

Two things central to the scientific task of explaining change in urban communities in the Arab World were missing from virtually all studies. First, there was no explicit theoretical framework for seeing the city as the outcome of larger social, economic, and political processes within society. And second, there was little awareness of the extent to which global forces impinge not only on Third World societies in general but, quite specifically, on how Arab cities have developed and on the problems they face.

The first deficiency is probably due to the lack of social science urban expertise on the part of most scholars who have written about Middle Eastern cities, whether historically or from a contemporary perspective. At least, there is a clear correlation between the quality of a work and the degree to which the writer understands general urban

This chapter has been adapted and updated from my (1990), 'Dependency and Middle Eastern Cities', in Dan Chekki (ed.), *Research in Community Sociology* (Greenwich, Conn.: JAI Press, i. 35–52.) My comments are confined to cities of the Arab World, the largest component in the Middle East, an otherwise arbitrary aggregation of regional subunits. I include North Africa but do not cover Turkey and Iran because they were never directly colonized and are linguistically and culturally distinct. Furthermore, since my own research in the region has been limited to the cities of the Arab world, it would be presumptuous for me to go beyond them. I have also excluded Israel/Palestine, because that country has a unique current colonial situation that would require separate treatment. Despite these omissions and the narrower geographic scope, it is, as we shall see, difficult to generalize about urbanization patterns, even in the smaller unit of the Arab World *per se*.

processes. It is encouraging to note that this deficiency is increasingly being overcome as social scientists and architects, including many from the Arab world itself, begin to examine the cities of the Middle East from a broader and more processual point of view.[1]

The present chapter will not address the first type of deficiency. Instead, it concentrates on the second, namely, the tendency for analysts of Middle Eastern urbanization to ignore the theoretical and practical implications of the international system and the role of Middle Eastern countries within it. This, it seems to me, is a more serious defect because it has insulated Middle Eastern urban studies from the strides that have recently been made in understanding urbanization in other parts of the Third World.[2] Many studies of Middle Eastern cities are weak in explanatory power because their authors wear blinkers that exclude crucial events and forces that lie beyond the urban environment being studied. This is a particularly serious fault when one wishes to explain cities in the Arab World, and the error is progressively compounded as one approaches the present time.

To explain the rise and fall of cities in the period of Islamic greatness (roughly between the seventh and twelfth centuries), one needs to know not only what was happening in the cities themselves, but what was happening to world trade routes and to the rise and fall of other cities and empires.[3] Similarly, to understand the sources of today's problems in Arab cities, one needs not only to comprehend the after-effects of imperialism and colonialism (both internally and with respect to the roles cities are called upon to serve in the international division of labour), but also the lightning-rod role this region continues to play in international political struggles. Despite this obvious need, very little of the literature on Arab cities is guided by any theoretical understanding of outside systems and how they impact upon urban growth and problems.

This is in sharp contrast to the urban literature on other parts of the Third World. Dependency theory, generated in Latin America and applied with great power to the analysis of urban problems there, has been virtually unapplied to the Middle Eastern context. Imperialism—and the Middle East has suffered from it as much as any other region—is seldom invoked to explain the type of parasitic urbanization that evolved in the Arab World, although no student of African (or, by today, even Indian) urbanization is ever allowed to forget it. Colonialism *per se* is usually ignored, except to note in a bland and accepting way that it created 'dual cities' throughout the region. (For exceptions, see

Abu-Lughod, 1980 and Wright, 1991.) Nor do urban studies on the Middle East pay much attention to the way such cities articulate either with the Arab region around them or with the economic and political world system that defines their role internationally.

In short, the field of Middle Eastern urban studies has been slow to catch up with the conceptual sophistication evinced in studies of other parts of the Third World and, therefore, has been essentially left out of comparative studies of urbanization. There are two possible explanations for this. One, which seems unlikely, is that scholars working in the field are simply less intelligent and more tunnel-visioned than those working on other regions. Certainly, mastering difficult languages can consume precious time and brainpower which among those working in other fields, such as Latin American studies, could otherwise be channelled to conceptual and empirical work. But that handicap would also apply to work on Chinese urbanization, a field noted for its level of sophistication. A contributory factor might be the undeveloped state of indigenous scholarship in the Arab world which has failed to generate the kind of creative thinking that English-speaking scholars of Latin America have been able to take from their local colleagues; it is just harder to be 'parasitic' on Arab scholars, although this is now changing.

A second explanation, however, may account for most of the variance. The Middle East is more fragmented than China and subject to more conflicting forces than Latin America. Like the peninsulas and islands of Pacific Rim countries, regional complexities make it hard to generalize. Special circumstances, such as the presence of petroleum resources in places that lack other bases for development and the widening gap between rich and poor Arab states, introduce variations that are more extreme than in other regions. Thus, studies of the Arab World cannot simply adapt the explanatory theories generated elsewhere, but must create new ones more relevant to the region.

In my own attempt to address this problem (see, for example, Abu-Lughod, 1984), I subdivided the twenty countries that make up this complex area into five different types whose patterns of urbanization and whose urban problems vary significantly from one another. In only a few cases do we find neo-colonial dynamics similar to those in Africa and Latin America which have been the inspiration for most theories of Third World urbanization. There are no cases that parallel the so-called NIC's which have recently forced revisions in urbanization theory. On the other hand, seven out of the twenty cases have no real parallels elsewhere. Finally, the entire region has been embroiled for half a century or more in a localized conflict which has had widening ramifications for the adjacent states. These are some of the reasons why the Arab World has defied easy comparison with other parts of the 'developing' world. Because of these factors, one cannot ignore the historical specificities that have shaped the countries in the region and their cities.

More than in Africa and Latin America, the people of the Arab World view themselves, at least on the ideological level, as a cultural, linguistic, and religious entity. Despite the hardening of political frontiers, the divergence in their economic and political interests, and even intraregional conflicts (of which the recent Iraq–Kuwait conflict was perhaps the most egregious), Arabs consider themselves part of the same 'people', sharing the memory of a common 'Golden Age' and a core of common values. However, the current states that constitute the Arab League, whose borders were primarily defined by Western powers in the early twentieth century, have a more ambivalent view of their identities. On the level of rhetoric they emphasize unity; on the level of praxis they can rarely agree upon a common policy, especially when that policy might have practical implications for a redistribution of privileges and scarce resources.

Originally, the ideology of Arab unity had anticipated the removal of all economic and, eventually, political barriers. But as the distribution of resources and wealth among the states became increasingly unequal over the ensuing decades of independence, citizenship barriers solidified, migration across state lines became increasingly difficult, and as a result, the rational transfers of the capital and labour that would be required for the economic development of the entire region became increasingly difficult to effectuate.

This fragmentation has had significant effects on the cities of the region. It has generated the five somewhat distinctive sub-types of urbanization that will be described in this chapter. Furthermore, it perpetuates urban problems that are unlikely to be solved unless there is a radical transformation of the regional and global systems in which they are embedded. Therefore, simple dependency theory, as it now stands, has little to offer in explanatory power for the evolution of urban problems in the Arab World nor do autarky and/or import substitution offer a useful set of prescriptions for their correction. On the other hand, it may be that the type of regional co-operation called for in the Middle East could even offer a model for the removal of dependency in the other regions of the world (Africa, Latin America) where dependency theory has offered a more powerful explanation of the past.

The Transformation of Cities in the Arab World Since 1950

The past forty years have witnessed a radical transformation of the Arab World and its cities and a complete rearrangement of the relative positions of the various Arab states in terms of their wealth, their modes of production, and their ways of life. It is useful to review this process before examining the present, and then the immediate future, which may represent a crucial turning point for the coming decades.

In the 1950s, as in the 1980s, the Arab World was in turmoil, even though it appeared at that earlier time that the situation was a temporary nadir out of which a more promising future might develop. The war in Palestine had left Palestinians without a homeland and many without a home, but even in other states where the effects of the 1948 defeat were not so direct or disastrous, the event set deep changes in motion. The early 1950s were marked by significant political changes in many states of the region. The most important of these was the overthrow of the Egyptian monarchy, followed by Gamal Abdul Nasser's rise to the head of this rapidly decolonizing state. The 1950s, then, were a time of hope in that country, when it seemed that Egypt might regain some of its former importance. Indeed, by 1955 Egypt had become one of the leaders of the non-aligned block and, once Israeli, French, and British forces were forced to withdraw from the Canal Zone after their 1956 invasion, the country began a long-overdue process of guided economic development which seemed destined to gain it greater autonomy and prosperity.

In other parts of the Arab World there were similar hopeful signs. In 1956 the French colonial administrations were forced out of Morocco and Tunisia, and after a fierce struggle, were finally expelled from Algeria in 1962. The decolonization of North Africa was presumably accomplished. In 1958 Iraq mounted a revolution that overthrew its corrupt monarchy. In the euphoria of these independence movements, it seemed that, almost everywhere, there were encouraging signs. One impulse of the time was toward greater unity, toward erasing the boundaries introduced by the colonial occupations. Federations between Iraq and Jordan, and then between Egypt and Syria in the late 1950s were initially welcomed optimistically as the first steps toward the creation of a larger regional unity, although they were to prove brief and abortive. A second encouraging sign was a genuine attempt to break dependency upon the former colonial powers; self-determination was sought in the economic as well as the political arena.

In the 1950s the Arab World was still only modestly urbanized. In most countries only 30 to 40 per cent of the population lived in cities, although the range was wide. City-states such as Kuwait and Qatar, because they were so restricted in area, had the highest percentages of their population in 'the city', but among the more urbanized countries were Egypt, Lebanon, Iraq, Syria, Jordan, and Tunisia. The populations in the remaining countries remained largely rural, with Libya, what was to become the United Arab Emirates, Algeria, and Morocco having 20–30 per cent in cities, and Saudi Arabia, Oman, Sudan, and Yemen having even less.

At that time only Egypt seemed plagued with what might be called classic Third-World 'urban problems'—although one now looks back on those as the halcyon days, before Cairo broke down under the ponderous weight of her bloated population. Jordan, of course, was an especially tragic case. If one defines 'urbanization' as an increase in the number of urban residents, Jordan had 'urbanized' significantly, but the sudden growth of its capital, Amman, from a tiny town to a metropolis generated enormous problems, because it had come largely from an influx of hundreds of thousands of impoverished Palestinian refugees.

But throughout the region, older forms of city-building persisted and cities had not yet been entirely engulfed by a building 'boom' of peripheral squatter settlements and modern 'downtowns'. And except for the Palestinians who had no alternative, populations tended to remain within their own borders, which were proving to be more impermeable than early hopes for unity had anticipated. Nor had regional integration made much headway. The economic interrelationships among the various Arab countries remained as minimal as during the colonial period. To travel by air from North Africa to the Arab countries of West Asia, one still had to go through European airports.

All this was to change. Just as 1948 had sent shock waves of change through adjacent countries, so the even more devastating 1967 Arab defeat by Israel was to alter conditions drastically. In Egypt, which had borne the brunt of the fighting, the economy collapsed, although Cairo continued to grow, in part because it had to absorb half a million refugees from the destroyed Suez Canal zone. Amman also grew, as some 300 000 refugees from the recently occupied West Bank and Gaza sought refuge there. And in Iraq, Saddam Hussain came to power, in part as a response to the sense of vulnerability that pervaded all the Arab nations after the defeat. Throughout the region, countries were determined to strengthen their armies and to equip them with the means of modern warfare, regardless of the cost in scarce foreign exchange.

If the 1960s were years of rapid population and urban growth, the 1970s were destined to bring even greater changes. The way in which things changed indicated the extent to which the separate Arab economies—relatively independent of one another up to then—had become inextricably intertwined. The period 1973–4 was, of course, a critical turning point. It was then that the so-called oil states began to reap unprecedented riches and to develop ambitious plans for economic and social development—plans for which their human and other resources were completely insufficient.

The optimism of the oil states, however, was scarcely paralleled in the 'confrontation states' (those bordering Israel), where conditions deteriorated considerably. It was then that the states on the 'front-line' began to live increasingly on outside subsidies—some state-to-state, some in the form of investment inflows, and then, as migration to the oil countries of the Gulf for employment became increasingly common, in the form of remittances sent by workers abroad. Essentially, the infusion of wealth generated by the high price of petroleum between 1973 and 1985 bought time for the Arab World, but, as we shall see, at a terribly high opportunity cost.

Trends in urbanization followed these shifts in economic organization. Table 6.1 cross-classifies Arab countries by their Gross Domestic Product or Gross National Products per capita *circa* 1975–6 and 1990s and their relative levels of urbanization in 1950, 1980, and 1990. Over time one can observe the oil-wealthy states converging into the cell of high Gross Domestic Product per capita and high level of urbanization, while the older-established countries, previously among the relatively more urbanized, slip to the middle levels of urbanization. Only a few countries remain throughout the entire period in the cell of low income and low urbanization. This table clearly shows how the relative ranks of the Arab countries altered after 1950.

TABLE 6.1 Relationship between Economy and Urbanization in Arab States 1950, 1980, and 1990

Level of urbanization	GDP/capita		
	High	Medium	Low
(a) *Urbanization level in 1950 by GDP/capita in 1976*			
High (30%+)	Kuwait	Bahrain	Egypt
	Qatar	Iraq	Tunisia
		Lebanon	Syria
			Jordan
Medium (20–30%)	Libya	Algeria	Morocco
	UAE		
Low (below 20%)	Saudi Arabia	Oman	Sudan
			Yemen
(b) *Urbanization level in 1980 by GDP/capita in 1976*			
High (60%+)	Kuwait	Bahrain	
	Qatar	Iraq	
	UAE	Lebanon	
	Saudi Arabia	Algeria	
Medium (40–60%)	Libya		Morocco
			Tunisia
			Syria
			Egypt
			Jordan
Low (below 40%)		Oman	Sudan
			N. Yemen
			Dem. Yemen

	GDP/capita			
	Very high	High	About average	Low
(c) *Urbanization level in 1990 by GDP/capita in 1990*				
Very high (75%+)	Bahrain		?Lebanon	
	Kuwait			
	Qatar			
	Saudi Arabia			
	UAE			
High (60–75%)		Libya		Jordan
		Iraq*		
About average (40–60%)		Algeria	Syria	Egypt
			Tunisia	Morocco
Low (below 30%)		Oman		Yemen**
				Sudan

* Prior to the destruction of 1990–1.

** North Yemen and Democratic Yemen reunited.

Types of Urbanization[4]

By the mid-1980s the countries of the region could be classified into five fairly distinctive types which had very different economies and suffered from very different kinds of urban 'problems'.

1. Economically marginal countries of the 'Fourth World', such as the Sudan and Yemen,[5] which lie outside the international economic system and have had very low rates of urbanization.
2. Neo-colonial economies in which political dependence upon a Metropole was transmuted, after independence, into economic satellite status. Morocco and Tunisia are examples of this.
3. The confrontation states—Egypt, Jordan, Syria, Palestine, and Lebanon—and eventually, to a certain extent, Iraq, whose economies had been shattered by war and who, with the exception of Iraq, increasingly depended on outside subsidies.
4. The diversified land-based agricultural states with some oil production, of which Algeria and Iraq were the chief examples. Iraq was to move, after the 1990–1 war, from the fourth category to the third, and in the post-1985 period of petroleum-price retrenchment, Algeria's position would also deteriorate.
5. The *nouveau riche* rentier states whose oil-based investment economies linked them, for better at first but then increasingly for worse, to the world economy. These include, of course, Libya, Saudi Arabia, Bahrain, Qatar, Kuwait, the United Arab Emirates, and in the most recent period, Oman as well. Yemen may eventually join them.

In each of these subgroups, political position and economic role combined to set the level of urbanization, the relative rates of urban growth, and the nature of the urban problems to be faced.

Fourth World

The simplest to describe, but the least significant in terms of urbanization, are the Fourth World countries, which have stagnating agriculture, relatively little industrialization, and provide surplus labour to other nations in the region. In the Sudan there appeared to be sufficient farmers left so that agricultural production was not held back (although the current civil war there is destroying much food production), but in Yemen the situation has been more serious. North Yemen was a major exporter of young men to Saudi Arabia and the magnitude of this emigration interfered with indigenous developments. According to estimates I made from Saudi data, in the early 1980s it appeared that more than half of the able-bodied men in the Yemen Republic were working abroad, leaving their fields untended or in the hands of women. The remittances from these labour migrants, while relatively valuable, were insufficient to fuel national development schemes because most were poorly paid unskilled labourers. Inability to produce viable urban economies, low rates of urban modernization, and a tendency for urban degradation to outstrip urban improvements were the characteristic problems of these countries.

In the 1990s some changes have occurred. The discovery of oil in Yemen is serving to improve conditions in that recently unified country, and the forced repatriation during the 1990–1 Gulf War of Yemenis who had been working in Saudi Arabia is restoring needed manpower. In contrast, the continuing civil war in the Sudan has depressed conditions even further; little can be done until the refugees from the south are restored to a viable life. Their plight intensifies all problems, urban and rural.

Neo-Colonial Economies

Somewhat similar blockages to development exist in Morocco and Tunisia, although their economies, albeit briefly redirected toward the oil countries, are still mostly tied to Europe. In both these neo-colonial countries, the urban consequences of dependency have been serious. Both inherited from the pre-colonial era numerous cities of pure Arab-Islamic character which had retained their surrounding walls until they were torn down in the present century, and which even today have kept their cellular organization into quarters, their narrow streets, and their linear *aswaq* (bazaars). On top of this extant urban system the French colonizers had imposed an urban hierarchy based upon a colonial extractive economy whose major link was to the Metropole. And beside pre-existing cities of historic significance the French had grafted new towns for their exclusive use. Although ostensibly the dual spatial structure paralleled a dual economic structure, neither was what it purported to be. Rather than separate but equal facilities, the dual system had been an efficient mechanism for draining surplus value from the natives (Abu-Lughod, 1980).

Decolonization had remarkably little immediate impact on this system, although once independence was achieved, the beneficiaries of the drained surplus were expanded to include a native élite. This indigenous group inherited both the vacated administrative posts in government and modern commerce and commodious urban quarters that had formerly been the exclusive preserve of the Europeans. The squatter settlements ringing the major cities were also inherited, and these, in the post-independence period, became the fastest growing parts of cities.

After abortive attempts by early ministers of economy to disengage their countries from the European system, both Tunisia and Morocco settled down to increasingly involuted and dependent roles in the international economy. The current heads of state favour joint ventures with foreign capitalists, expansion of state jobs, subsidies to industrialists engaged in import-substitution or piece-work for others and, thanks to the World Bank and French experts, an emphasis upon tourism as a hidden export.

The result has been to turn both countries into satellites of Europe, to which they export their labour, from which they import their manufactured goods and their tourists, and from whom they receive the semi-finished products which they then process and re-export, in a new form of the 'putting out' system. These strategies have not brought them universal prosperity, but have indeed stretched the internal gaps between rich and poor and threaten to stretch the social fabric to the breaking point.

These strategies have also affected the cities. Some amount of preservation is being done in the precious historical quarters or medinas (Tunis, Fez, etc.), but more with an eye toward the tourist trade than to creating a healthy environment and viable livelihoods for the poor who crowd into these districts. New construction in other quarters is chiefly in the form of villas and apartment houses for the wealthy. The growing populations of poor and middle-class families are increasingly thrown to the periphery where self-built housing, often on land not suitable for construction, has burgeoned. The old caste lines that formerly divided the dual cities are blurring, but largely through a degradation of the 'old' modern quarters adjacent to the medinas.

Heartland-Confrontation States

Egypt, Palestine, Syria, Lebanon, and to a much lesser extent, Jordan, were countries that were always more urbanized than most. They occupy the heartland not only of the Middle East but of the world. Their central location was never an unmixed blessing, however, since it made them the object of others' ambitions—starting with the Crusaders and ending with the Israelis. In today's world this has imposed especially heavy burdens—which unfortunately have not been alleviated by oil.

At present, in these countries (except Lebanon where disorder has obscured accurate information), cities have been growing at 4–5 per cent per year, with two-thirds of the increase coming from an excess of births over deaths, the rest from immigration from the countryside. Even higher rates would have prevailed, had growth not been deflected for a while by massive emigration for employment. The economy of each state is very weak; these countries are sustained only by virtue of significant subsidies from outside. These are the states I have classified as living by the 'charity mode of production'. It is clear that, regardless of how generated—whether through international state subsidies or private remittances—the funds that support these countries are not generated from within their indigenous economies.[6]

This fact has important consequences for the type of urbanization that occurs. First, since the cities are really supported from outside, primacy runs rampant. The largest city is the preferred destination for most internal migrants. Amman, for example, now contains more than half the population of Jordan, even though it has neither natural resources nor a real productive base, and Greater Cairo contains about one-fourth of the population of Egypt. Second, the inflow of funds from abroad has moved disproportionately not into productive facilities but into land as a capital 'bank'. This has resulted in rampant land speculation (which makes rational city planning impossible), grossly inflated land values, and in some cases, overbuilding of luxury flats. Inflation in the cost of all goods is inevitable when money flows in but the supply of goods does not expand commensurately, *viz.* Egypt's sky-rocketing inflation. Third, there are important social consequences of the fact that families have become so dependent upon remittances from workers abroad. Not only have families frequently been separated but there has been selective depletion of the local labour-force, particularly of the skilled and professional workers so much in demand in the oil countries. Nowhere did this pattern become as extreme as in Jordan where, according to the estimates of Birks and Sinclair (1980) made in the late 1970s, as much as half of the available labour-force was working abroad; immigrants from 'lower order' societies (such as Egypt, Sri Lanka, and Thailand) had to be imported to fill the gaps.

Nor should one underestimate the social and cultural effects that heavy dependence upon outside employment can have on the moral fibre of a society. Saad Eddin Ibrahim (1982) has written sensitively about the way 'seconding abroad' has undermined commitment to work and advancement at home; the 'luck' of finding employment in the Gulf substituted for more systematic ambitions.

The impact of all this on cities has been severe. The steady growth of the primate cities has overtaxed urban facilities, particularly when combined with a weak economy and an inadequately funded public sector. There has been inadequate housing built for the poor and those not receiving remittances. And because these cities have, with the exception of Amman, enjoyed long and illustrious histories, they contain large areas of historically significant urban heritage which, while they cry out for modernization and preservation, actually receive cruel abuse and overcrowding.

Recent events have revealed the vulnerabilities engendered by dependence upon remittances. In the wake of the Iraqi invasion of Kuwait, the major victims were the Palestinian and Egyptian expatriate communities in Kuwait. Driven out or fleeing, hundreds of thousands of guest workers returned to Jordan[7] and Egypt, leaving their jobs and savings abroad, ending their remittance-sending abilities, and crowding into their cities of origin. Their repatriation has significantly taxed local facilities at a time when their home economies can least afford to deal with new burdens.

The 'Semi-Oil' States

Algeria and Iraq previously stood between the impoverished, subsidized and/or still dependent countries, on the one hand, and the oil-wealthy states on the other. Both countries were plagued with an inherited problem of overconcentration of population in their respective capitals of Algiers and Baghdad but both had sought to address this through judicious policies of decentralization and new-town building in resource-extraction areas.

Algeria underwent perhaps the most radical social change of any Arab country, moving within only two decades from a war-torn oppressed nation to independence and a planned economy. Her transformation was aided considerably by the oil resources whose exploitation and marketing were government controlled, and whose extraction and shipping generated considerable multipliers in economic growth. However, her progress was handicapped by the difficulties with which she began.

Furthermore, what tenuous progress had been made in the 1970s was undermined in the 1980s when the price of oil dropped. As the economy contracted, dissatisfaction with the party in power led to a resurgence of protest movements which have recently taken a religious turn.

Iraq, in many ways, began with stronger possibilities and, until its eight-year war with Iran, made rapid strides to mobilize the capital provided by oil to generate electric power, irrigate farms, and fuel a diversified industrial base. The border war caused these plans to be set aside, and may have precipitated the harrowing events of 1990–1 that have left Iraq with insuperable problems. The destruction of much of the infrastructure of Baghdad and Basra through Allied bombing, the embargo on Iraq's ability to market her oil, and the still festering internecine conflicts between the central government and dissident Shi'ites in the south and Kurds in the north, have left Iraq one of the most distressed countries of the Middle East.

Thus, the two countries which, up to the early 1980s, had appeared to be making the most progress in development and independence, experienced the greatest setbacks in recent years.

The Oil States

Often, when contemporary urban developments in the Arab world are discussed, the cases that attract the most interest are the suddenly wealthy states which until little more than a few decades ago were sparsely populated by pastoralists, only lightly urbanized, and mostly poor. In these states, urbanization and social change occurred at lightning speed. It is important to remember, however, that they constitute only a tiny and atypical part of the Arab World. Despite population increases that were fed by international migration, they still account for fewer than 10 per cent of all Arabs. These small numbers, however, are not matched by small incomes. Over half of the GNP created in the entire Arab World goes to these states—an issue of crucial significance in the recent crises in the region.

The juxtaposition of small populations, large income, and the rapid rate at which the latter mounted—together with dependence upon a single nonrenewable resource, oil—yields a striking *in situ* experiment of instant urbanization and hothouse-forced social change. If social knowledge is advanced by the study of deviant cases (sociology *in extremis*), then there have been few better opportunities to see the processes of urbanization and social change laid bare.

The first thing to be noted has been the very rapid rate of population increase and its almost exclusively urban destination. With the exception of Libya and Saudi Arabia which have multiple cities, some agriculture, and a relatively extensive terrain, the rest of the oil states are essentially city-states with 80–90 per cent of the total population living in a single capital city. The second striking fact is that significant proportions of the residents are foreign nationals with very limited rights. In 1980 foreigners constituted over half of the populations of Kuwait, the UAE (United Arab Emirates), and Qatar and about a quarter of the population in the remaining states. Almost none of these 'guest workers' had a chance to become citizens, and their rights to remain in the countries were strictly limited. Even after the expulsion of many foreigners from Kuwait and Saudi Arabia during the Gulf War, the economies of these states remain dependent upon imported labour and, in its absence, things have already begun to fall apart.

A third fact is that the starting point of development for each of these societies during pre-oil days was quite low, as measured by education, national infrastructure, public services and the like. There were a few cities—many of them now capitals and grown to great size—but these were relatively tiny, often built of primitive materials, and associated with limited commerce, political/military functions, religion, and perhaps pearling and fishing. Therefore, just as the technical personnel needed for modernization had to be imported, so the new cities that were to burgeon could not be developed out of existing methods and modes but, in the last analysis, had also to be imported.

We witnessed in the newly rich Arab countries an unprecedented case of instant urbanization propelled not only by the collection within the capital city of virtually all the citizens of the state, previously scattered, but also by the importation of labourers to service both the productive economy and the citizens. What were the implications of this kind of urbanization, not only for the cities that resulted but for the social changes that occurred?

First, cities served primarily as sites for consumption rather than as centres for production. They became the arenas in which patrimonial state *largesse* was distributed. Cities, therefore, exercised an irresistible attraction for citizens. As centres of consumption they also attracted the lion's share of migrants. And as much as 75 per cent of the labour-force in each country was engaged not in production but in providing urban services.

Second, given the inhospitable terrain, cities were the only way that the greatly increased population could be absorbed. The result was that 80–90 per cent of the populations of Bahrain, Kuwait, Qatar, and the UAE had come to live in cities. While the percentages were lower in Libya and Saudi Arabia, because these countries control larger territories, their situations could prove to be only temporary, unless agriculture and herding alternatives can be rescued from their declining spiral.

Third, because of the extremely rapid rates at which the major cities grew, an incredibly high percentage of current urban residents are classified as migrants. In 1985, of the ten cities in the Arab world having the highest percentage of migrants, nine were located in the Gulf, where the proportion of migrants often reached as high as 85–90 per cent of the total city population.

The prevalence of migrants gives rise to a fourth characteristic of these cities, namely, their spatial patterns of ethnic segregation. Migrants have been of three basic types. First, there were local citizens who tended to reside in their own quarters. Second, there were migrant workers, mostly from the Arab countries, who were accompanied by their families and who viewed their sojourn as long-term. They tended to occupy flats in apartment houses rented from nationals and to sort themselves on the basis of specific nationality. In some areas early on hospitable to Asians (such as Dubai), Pakistani and Indian families also occupied their own districts. The third type of immigrant, the single male worker on temporary contract for a specific construction/industrial project or assigned to agricultural production, is even more isolated from the host society. Only the minority who work as domestic servants are not residentially segregated, but that is only because they are housed on the premises.

These four characteristics have yielded the new type of city that has appeared in these countries of instant urbanization. The fact that cities were chiefly consumption centres—and often of conspicuous consumption at that—gives them the appearance of gigantic display cases for goods. Not only were vast areas given over to shops where imported goods were repetitively displayed, but even the new buildings were built as displays. Set next to wide boulevards (the urban design least suited to the climate), the new structures, shiny, and almost bejewelled on their façades, symbolized advertisements to buy.

In addition, because the cities, like some giant vacuum cleaner, drew populations of decidedly different backgrounds, they had a distinctly uneven and unfinished quality. Construction materials clogged the ports or lay in untidy piles on almost every vacant lot. The dwarfed remnants of the Gulf's traditional cities—graceful constructions of warm-toned pressed brick and mudwash—terminated without grace at arbitrary points where new highways interrupted their flow. Often, the local population abandoned these areas to move into modern bungalow quarters; their places were taken by Pakistani or other Asian families. At the outskirts there were other unfinished zones, temporary shack-towns, some containing sedentarized bedouins, others containing Filipino labourers, or only partially demobilized soldiers from Korea.

And finally, because of the system of multiple-status labour migration and the increasingly uneasy relations between foreign workers and citizens, caste-like residential segregation characterized the spatial pattern. Ethnicity was the dominant criterion differentiating neighbourhoods. With time, these ethnic ghettos became increasingly institutionalized. Not only did they develop their own social and commercial services, but their residents were held accountable by the authorities for their collective good conduct. Misconduct of any kind was sufficient grounds for immediate expulsion. This was clearly demonstrated in Kuwait in the aftermath of the Iraqi invasion, when Palestinians, who numbered over 400 000 before the war, were expelled, leaving whole quarters of the city deserted.

It is still too early to predict the full range of social changes the oil boom set into motion through increased dependence upon foreign labour, or to predict the long-term effects of instant and saturation urbanization. Up to 1990 the region was remarkably free of anticipated difficulties, suggesting that it is easier to adjust to affluence than to poverty. But clearly, even before the Iraq–Kuwait war, basic erosions were already apparent. The older mechanisms of social control that held these societies together were severely strained, even though expanding wealth had been able to 'oil' these cleavages.

Indefinite continuation of that situation is, however, unlikely. The social structures that survived a rapid increase in resources were challenged by the war, which revealed basic weaknesses. These have been compounded by recent declines in the price of oil. Too much of the future of both states and cities had been committed to long-term, nonproductive but costly projects; it remains to be seen whether these societies can be placed on a course of deeper development and greater social justice. Thus far, the needed reforms have not been forthcoming.

Predictions of a Coming Crisis[8]

The above discussion demonstrates how interdependent the various Arab countries became in the past few decades. Thus, any economic recession in the oil countries must affect not only those states with the resource, but many other Arab countries as well. The ripple effects of any changes in the oil states cannot but be felt throughout the region. Each of the economic types, however, will be affected in somewhat different ways and degrees.

The countries called 'Fourth World' will be least affected by any difficulties in the oil countries; they benefited least from the expansion and therefore have the least to lose from a contraction. But in such economic basket cases, even modest losses can return them to pauperdom. Had oil not been discovered in Yemen, what would have happened in 1990 when the young men were sent home from Saudi Arabia or prevented from emigrating there? Clearly, the rate at which they returned and what they found at home determined the ease with which their labour could be absorbed into and mobilized by the society. With progressive governmental policies and advanced planning, disaster and turmoil can be avoided. In Yemen and the Sudan, preparations must be made in rural areas where agriculture can still yield a livelihood. Enforced channelling of all remittances *away* from petty commerce in urban areas and *into* small-scale industry and agriculture would have been the best way to prepare for the future.

The neo-colonial countries, also, are not likely to be strongly affected by reversals in the Gulf since, as we have shown, their dependence is still chiefly on Europe as an outlet for labour. In these places it is not hard to predict that radical changes in political regimes—and even revolution, in the case of Morocco—are likely, but the impetus for these changes will be found internally, rather than in the world system. The way to avert total upheaval is equally clear. Exploitation of their still untapped rich mineral resources, a break with the international 'putting out' system, and greater concern for the masses rather than the tourists are the measures these countries can take to avert the coming crisis. Here, however, the prognosis is not encouraging. Morocco has shown no, and Tunisia little, tendency to move in the needed directions.

The semi-oil states are in for difficult times but, if the Iraq boycott is lifted and political stability can be maintained in North Africa, both Algeria and Iraq are capable of weathering the economic retrenchments that will be required by the decline in oil prices. Here again, only serious attention to agriculture, labour-intensive industry, and decentralized 'growth pole' policies of urban development can manage to relieve pressures on Algiers and Baghdad and yield a balance between city and countryside. In some ways, retrenchment may even have a positive effect, since it will require (although unhappily not guarantee) a more rational husbanding and allocation of the resources that do exist.

But it is in the pauper economies of the confrontation states, now so heavily dependent upon steady and massive transfusions to remain alive, and in the oil-states themselves that the most dramatic effects of the contraction will be felt. Let me take up the latter first, since I want to end with Egypt and the other core countries whose futures seem bleakest.

The drop in the price of oil in the mid-1980s has already begun to affect the countries of the Gulf and Libya which had hitherto treated their resource as an unfailing supply of limitless funds. In the long run it is hard to imagine that the tiny principalities will not decline, one after the other, as it becomes more and more expensive to capture marginal supplies of the depletable resource. A precedent exists in Bahrain which early ran out of oil, even before the boom period. She was able to sustain her economy only by becoming the 'gateway' for goods, services, and communications coming to the region from outside. Her prosperity is essentially dependent upon what happens in the other economies. She will decline if they do.

Should this happen, what will be left of these 'instant' cities? One shudders to think. Tall apartment buildings constructed with foreign-made materials and serviced by foreign-produced utilities will need maintenance, repair, and reconstruction. But the foreign reserves squandered so lavishly on them will be in short supply. Elevators will break, plumbing lines clog, air conditioners, without which the structures will prove uninhabitable, will burn out their compressors. At first, servants will be sent home, then maintenance workers, and finally higher paid professionals and semi-professionals. If nationals are sufficiently trained, they can take over many of the essential functions now performed by guest workers. But a society cannot sustain itself by offering services to one another; there must also be something that generates material worth. At present, the industries that exist are run on large subsidies in the form of virtually free energy, just as the cars run on virtually free fuel. But when the resource becomes more essential for export, these subsidies will cease. While we are unlikely to reach the doomsday scenario of entire ghost towns standing uninhabited, sand dunes encroaching on the cities of Arabia, the likely situation will be working toward that direction. One thinks of what the alternative course could have been—a united Arab World where the great bonanza of the 1970s could have been invested in rich farmlands and prosperous industries in other parts of the region, away from the deserts of oil extraction.

Libya and Saudi Arabia face less desolate futures—or could, if needed measures are taken. It is still not too late to develop better agriculture and herding, the raising of domestic animals, and forms of industry other than petroleum cracking—industries that could give productive employment to supplement continued income, albeit on a reduced scale, from petroleum.

As noted before, however, the greatest difficulties will be experienced in the confrontation states that have already begun to lose the unhealthy glow of false prosperity and must now pay the price for their failure to become self-sufficient. Lebanon has already paid the price in blood and fratricide. From a thriving if parasitic commercial emporium it was reduced to a war-torn ruin, although it now seems to be recovering from its 1982 trauma. Jordan has already been overstressed by the return of many hundreds of thousands of citizens (mostly of Palestinian origin) from the Gulf countries, whom it has yet to integrate in a healthy economy. In and around Amman, in the boom years, there was enormous land speculation as expatriate families bought up building sites with an eye to eventual construction 'when they returned' or with a look towards quick profit. Half the present population of two million in Jordan already lives in the capital and enough land around the capital has been subdivided to accommodate the rest. But what will they all live on, after remittances stop? And will the Israelis really grant autonomy and open borders to the West Bank? Jordan's economy hangs on a narrow thread that could be snapped by any or all of these imponderables. Syria is also not out of danger, although agricultural resources are better developed and there is much less dependence upon workers abroad.

It is Egypt, however, which continues to face the greatest challenges in this new phase of history. Despite the fact that subsidies and remittances still flow in, and indeed have been increased as a reward for Egypt's position in the Gulf War, hardships are certainly present.

Inflation has severely eroded the buying power of the poor who cannot but resent the flashy Mercedes and videos that *infitah* (literally, economic 'opening') has brought. Middle-class workers hold several jobs and still cannot make ends meet. Even the wealthy, who have profited so much from the new projects, stand to lose something when they taper off.

By 1985 Egyptian labour emigrants were already coming home, radios, stereos, TVs, and stoves in their luggage. They were joined in 1990 by their less-fortunate brethren, escaping penniless and possessionless from Kuwait. It is estimated that at the peak of the oil boom, three million Egyptians were working abroad, mostly in the Gulf and Saudi Arabia. Most have now returned. While not all of them originated in Cairo, most descended on that already over-stressed city. Have they brought capital to invest in small enterprises in the smaller cities, towns, and villages of Egypt? Or have they returned only with heightened expectations, some consumer goods, but to no jobs? What will happen then, when the economy is bankrupt, when the expensive highways and flyovers and projects provided through foreign aid have to be paid for, and when funds will be lacking to keep them repaired and working? Then, we will have time to consider what could have been—massive investments in productive facilities decentralized to all parts of the country, investments in human resources to make the new person, reclamation of land from the desert, etc.

Just before his tragic death, Malcolm Kerr wrote, as the opening line to his last article discussing the Middle East, that the 1970s would later be known as the time when the Arabs 'missed the boat'. He had, I am sure, just this idea in mind. Arab unity could have brought the capital resources created by oil together with the land and human resources of the rest of the region to create viable, thriving, autonomous, economic development which would have averted the crises that have begun to wrack the region.

Is the Arab Experience Relevant to Urban Developments in Other Parts of the Third World?

This chapter has presented a somewhat detailed and close-grained analysis of the relationship between global, regional, and country-specific factors to demonstrate how they impinge upon the changing nature of cities in the Arab World. It is worth pointing out that in the case of only one sub-type did we find it particularly useful to apply an analysis that might be termed 'dependency'. In all other cases, although we utilized a political economy perspective which took into account the effects of the external system upon internal developments, we paid much closer attention to the details of such political and economic forces than is commonly done in 'nomothetic' science.

Does this mean that theories about Third World urbanization need to be revised to take account of regional variations? My answer is an unreserved, 'Yes'. Theories are descriptions of processes which may or may not prove valuable in illuminating specific cases. Even though theories derive from empirical examples in the last analysis, they are not generalizations from which specific cases and their outcomes can be deduced. Rather, theories are analytical tools for dissecting events. Thus, they cannot be judged as either correct or incorrect; a hammer or a screwdriver is not 'true' or 'false'. The value of any theoretical abstraction about process is determined more by its appropriateness to the case at hand than by its 'empirical truth'. And, in any case, the tool is *not* the product. It is merely one of the ways to make a product.

Just as the field of Middle Eastern urban studies has suffered from its lack of theoretical sophistication and its insulation from recent intellectual currents in the study of the Third World, so too often the critical evaluation of new trends in Latin American and African urbanization has suffered from a slavish and mechanical application of dependency theory. While such approaches gave us powerful insights into the past, they may be less relevant to the study of the present and future.

The entire global system has altered significantly within the past generation. And in many continental regions that formerly shared characteristics growing out of their common role in the international order, sub-types similar to those I have identified for the Middle East have been developing. It may be that we now require a finer-tuned analysis that distinguishes between a Nigeria and a Chad, a Brazil and a Bolivia, etc. One should not expect that the 'tool' of dependency theory, which gave us such valuable insights into the immediate post-colonial period, will retain its full power, now that various countries of the Third World (never a very helpful empirical category) are diverging in their political, economic, and urban developments. Perhaps the analysis contained in this article can demonstrate a new way to study urban problems comparatively.

Notes

1. For many years, the standard source on the Islamic city was Gustave von Grunebaum's 1995 article; he was *not* an urbanist. The works by Hakim (1986) and al-Hathloul (1981) on the relationship between law and city building during earlier Islamic times and the recent study by Celik (1986) on the modernization of Istanbul are vast improvements. So also are Brown's (1976) study of Salé; AlSayyad's review of early Arab cities (1991) and the collection edited by Germen (1983).
2. It is not without significance that when I was compiling material for my reader (Abu-Lughod and Hay, 1977, reprinted 1979), which was intended to cover the new theories as applied to Africa, Latin America, Asia and the Middle East, I was unable to locate any articles on Middle Eastern urbanization that utilized the new approaches. The book, therefore, is conspicuously lacking in references to the Middle East.
3. In my *Before European Hegemony: The World System* AD *1250–1350* (1989) I look at global connections among 'world market' cities in the thirteenth century. Even as early as this, it is clear that some of the forces affecting cities were purely local, some depended upon changes in the regions of which they were a part, and a surprising number depended upon the roles they played in a changing global system of production and commerce. To ignore political and economic factors at any of these three levels would be folly.
4. Parts of the following section have been adapted from Abu-Lughod (1984).
5. North and South Yemen, temporarily separated, were subsequently unified. The 1950 data are from the period before separation. Data for 1990 are after reunification.
6. The fragility of a system that depends upon the 'charity mode of existence' has recently been demonstrated in many ways. Lebanon's economic collapse and recent turmoil are attributable to a withdrawal of all forms of external subsidy, although ambitious plans to rebuild Beirut, via Arab capital investments, are currently underway. Jordan, in the wake of the Gulf War, has also reached a precarious state of existence, due to a withdrawal of American subsidies to punish her for siding with Iraq and to the loss of remittances that had formerly come from workers in Kuwait and Saudi Arabia. After Jordan signed an agreement with Israel US subventions were resumed, but remittances have never recovered.
7. By 1992, almost 400 000 Palestinians, out of the 450 000 who formerly lived in Kuwait, had been expelled. Hawali, an apartment-house quarter inhabited almost exclusively by Palestinians, became deserted.
8. This section was essentially written before the 1990–1 war in the Gulf. I have modified it only slightly since it is still too early to determine the long-range outcomes of that political realignment.

References

Abu-Lughod, Janet (1980), *Rabat: Urban Apartheid in Morocco* (Princeton University Press).

—— (1984), 'Cultures, "Modes of Production", and the Changing Nature of Cities in the Arab World', in John A. Agnew, John Mercer, and David E. Sopher (eds.), *The City in Cultural Context* (Boston: Allen & Unwin), 94–119.

—— (1989), *Before European Hegemony: The World System* AD *1250–1350* (New York: Oxford University Press).

—— (1990), 'Dependency and Middle Eastern Cities', in Dan Chekki (ed.), *Research in Community Sociology* (Greenwich, Conn.: JAI Press), i. 35–52.

—— (1991), 'The State of the Art in Studies of Middle Eastern Urbanization', in Earl Sullivan and Jacqueline Ismail (eds.), *The Contemporary Study of the Arab World* (Alberta, University of Alberta Press), 115–26.

—— and Hay, Jr. Richard (1977) (eds.), *Third World Urbanization* (Chicago: Maroufa Press; reprinted 1979 by Methuen).

AlSayyad, Nezar (1991), *Cities and Caliphs: On the Genesis of Arab Muslim Urbanism* (Westport, Conn.: Greenwood Press).

Birks, J. S. and Sinclair, C. A. (1980), *International Migration and Development in the Arab Region* (Geneva: ILO).

Brown, Kenneth (1976), *The People of Salé* (Cambridge, Mass.: Harvard University Press).

Celik, Zeynep (1986), *The Remaking of Istanbul: Portrait of an Ottoman City in the Nineteenth Century* (Seattle: University of Washington Press).

Germen, A. (1983) (ed.), *Islamic Architecture and Urbanism* (Dammam: King Faisal University).

Grunebaum, Gustave von (1955), 'The Structure of the Muslim Town', Memoir 81 of the American Anthropological Association, Ann Arbor.

Hakim, Basim (1986), *Arabo-Islamic Cities: Building and Planning Principles* (London: Routledge and Kegan Paul).

al-Hathloul, Saleh (1981), 'Tradition, Continuity and Change in the Physical Environment: The Arab Muslim City', Ph.D. diss. MIT.

Ibrahim, Saad Eddin (1982), *The New Arab Social Order: A Study of the Social Impact of Oil Wealth* (Boulder, Colo.: Westview Press).

Wright, Gwendolen (1991), *The Politics of Design in French Colonial Urbanism* (University of Chicago Press).

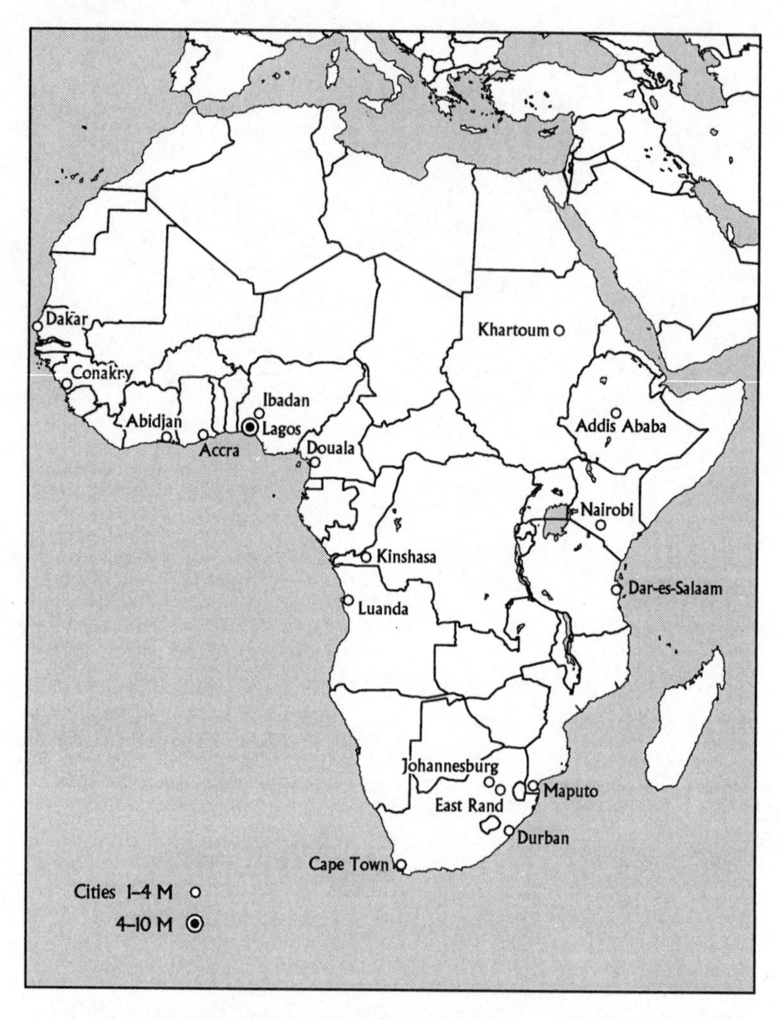

FIG. 7.1. Urban Agglomerations in Africa South of the Sahara with more than 1 Million Inhabitants in 1990

7

Urbanization in Africa South of the Sahara: New Identities in Conflict

JOSEF GUGLER

> I thought I should go to the capital of Kenya to look for work. Why? Because when money is borrowed from foreign lands, it goes to build Nairobi and the other big towns. As far as we peasants are concerned, all our labour goes to fatten Nairobi and the big towns.
>
> (Ngũgĩ wa Thiong'o, *Devil on the Cross*)

AFRICA South of the Sahara is the cradle of humanity, but its people have been separated from the rest of the world by vast expanses of water and, since about 3000 BCE, by the Sahara, the world's largest desert. Within the large subcontinent a relatively small population is widely dispersed, establishing quite distinct cultural traditions. Some African people have traded with Arabs for more than a thousand years. With the foreigners came Muslim clerics, and today perhaps half of Africa's population professes Islam. More than five hundred years ago Europeans appeared on the shores of Africa. For a long time they remained on the margins of the region, but by the end of the nineteenth century they had imposed colonial rule on virtually the entire subcontinent. The first Christian missionaries arrived with traders in the fifteenth century, but it was only with colonial rule that Christianity rapidly spread throughout the region—except where it encountered Islam.

The impact of colonialism varied as the British, the French, the Portuguese, the Belgians, the Italians, and the Spanish (and, until the

'Africa South of the Sahara', while cumbersome, is preferable to the racist 'Black Africa' and the Eurocentric 'sub-Saharan Africa'. Throughout this chapter, 'Africa' will stand as a shorthand for Africa South of the Sahara. I wish to thank, without implicating, William G. Flanagan and Anthony M. O'Connor for helpful comments on an earlier version.

First World War, the Germans) pursued quite distinct policies. Perhaps even more important was the divide between those colonies where European settlers influenced policies to their advantage and the other colonies where colonial policy focused on encouraging the export of raw materials: agricultural products, timber, minerals.

Most of Africa became independent more than a generation ago. At the latest count, there are forty-three countries in Africa South of the Sahara, not counting outlying islands such as Madagascar, the Comoro Islands, the Seychelles, Mauritius, and Cape Verde. Since independence they have moved along distinct trajectories. Several have been ravaged by civil war, some wars lasting more than a decade. Most African countries have experienced military coups, usually more than once. And their foreign relations, especially their relations with the former colonial power, and their position during the Cold War, varied, and often shifted over time. In part because of different colonial and post-colonial histories, in part because resource endowments vary greatly across the subcontinent, there are major differences in income levels and living conditions across African countries as well (Table 7.1).

Africa South of the Sahara is thus characterized by great economic, political, and cultural diversity. Any discussion of the subcontinent calls for an emphasis on distinctions. Where I venture generalizations, they will be subject to major exceptions.

African Cities through more than Two Millennia

Parts of Africa South of the Sahara have a long urban history. The magnificent granite stelae of Aksum, the voluminous chronicles of Timbuktu, the beautiful bronze sculptures of Ife bear witness to the splendour of urban civilizations that emerged at various sites across the continent centuries ago (Figure 7.2).[1]

The earliest cities of Africa were situated along the middle Nile in present-day Sudan. Meroë, the capital of the kingdom known as Cush or Nubia, was mentioned as early as the fifth century BCE by Herodotus, but its own accounts, rendered in hieroglyphs or in a distinctive cursive alphabet, wait for a full understanding of the Meroitic language. Meroë's antecedents lay in Napata and Kerma, the latter dating back to before 1500 BCE. These cities eventually disappeared, but Qasr Ibrim, whose occupation dates back to perhaps as early as 1500 BCE, was not

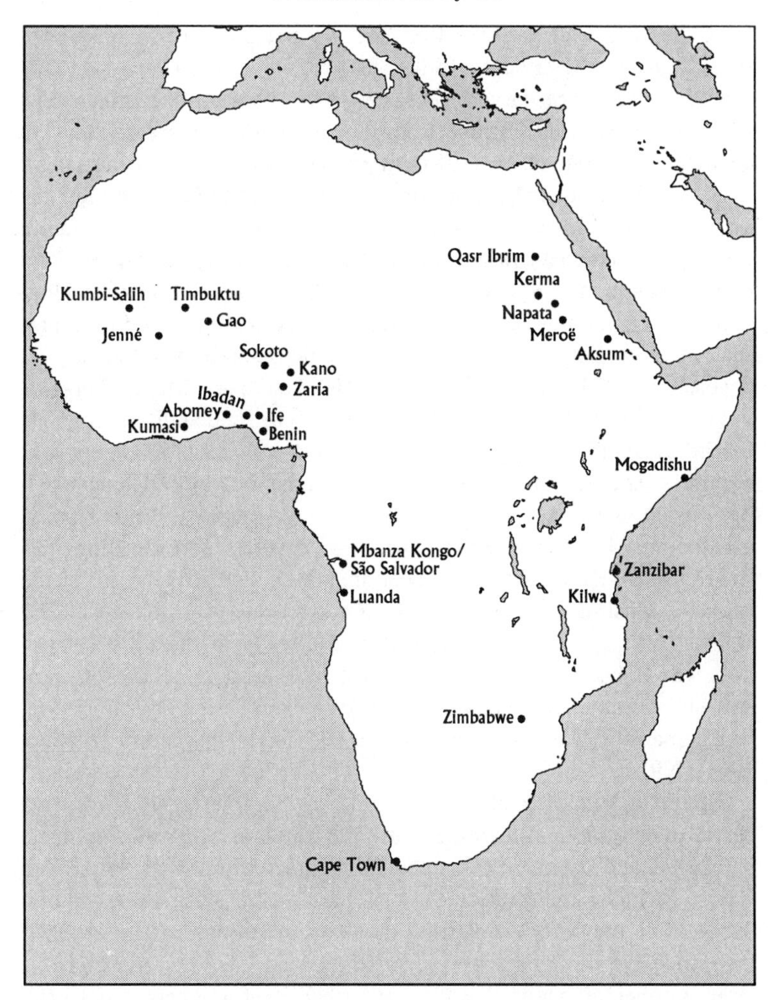

FIG. 7.2. African Urban Centres, 1560 BCE–CE 1850

finally abandoned until the early nineteenth century CE. It was one of several major urban centres that boasted episcopal cathedrals under the Christian kingdoms of Nubia that existed from the sixth to about the fourteenth century CE.

Aksum arose as the capital of the Aksumite state in the highlands of Northern Ethiopia in the first century CE. Astride trade routes leading from the Nile Valley to the Red Sea, it carried on a rich trade with the Roman world through Adulis, its port on the Red Sea. Aksum's unique stelae, large carved monoliths, have been lasting monuments to rulers who issued coinage from the late third to the early seventh century and repeatedly waged military campaigns on the Arab peninsula. Aksum produced a fair body of written material, and, after the conversion to Christianity of the 'king of kings' Ezana in the fourth century, the bible and various other literary works were translated from Greek, Arabic, and Syriac into Ge'ez. The mediaeval Ethiopian kingdoms, successively based further south, connect Aksum with the present: the last 'king of kings', emperor Haile Selassie of Ethiopia, was deposed in 1974 (Munro-Hay, 1991: 71–88, 172–6, 202–6, 244–8).

In the Western Sudan a number of cities flourished at different times. In the eleventh century the capital of the kingdom of Ghana was described as consisting of two towns: one inhabited by Muslims boasted a dozen mosques, while the royal town was located six miles away, and the area between the two towns was covered with dwellings (al-Bakrī, 1068). Kumbi-Salih has been identified as the likely site of the capital's Muslim section, its population estimated at 15 000 to 20 000 (Mauny, 1961: 481–2). Jenné, occupied already in the third century BCE, was to become a major city in the Mali empire, perhaps its capital, in the fourteenth century. Gao, the capital of the Songhai empire, had perhaps 75 000 inhabitants in the late sixteenth century (Mauny, 1961: 499).

Timbuktu had a smaller population, but became the distinguished centre of religion and learning under the rulers of Songhai who made it their second capital. Scholarship, in contact with North Africa and Egypt, included many disciplines of the Islamic sciences: grammar, exegesis of the Koran, doctrinal theology, traditions of the Prophet, jurisprudence, the philosophy of jurisprudence, rhetoric, and logic. A large number of commentaries, transcribed by foreigners and locals, were studied with scholars who specialized in different disciplines. Ahmad Baba is reported to have lost as many as 1600 volumes during the Moroccan invasion (Saad, 1983: 74–81). Sons of Timbuktu, al-Sa'dī (1655) and Ibn al-Mukhtār (1655), wrote the two great chronicles, the Ta'rīkh al-Sūdān and the Ta'rīkh al-Fattāsh, which constitute principal sources for the history of the Western Sudan.

On the East African coast close to two hundred settlement sites with stone ruins have been found. Every site has at least one stone-built mosque. The origins of some of these settlements date back to the closing centuries of the first millennium. By the beginning of the sixteenth century Kilwa had 12 000 inhabitants according to a Portuguese estimate (Freeman-Grenville, 1962: 66). Coins were minted in Kilwa, Zanzibar, and Mogadishu. A distinctive culture developed along the East African coast that was at least in part urban, mercantile, literate, and Islamic. Swahili, a north-eastern Bantu language rich in loan-words from Arabic and several other languages, became the mother tongue of many coastal peoples and a lingua franca in a vast hinterland that by the nineteenth century reached as far as present-day Zaïre and Zambia.

The world was created at Ife, in present-day south-western Nigeria, according to Yoruba traditions. Occupations of the site probably commenced during the late first millennium CE. Related Benin was occupied by about the thirteenth century. The world-famous sculptures in terracotta and bronze date from the fifteenth century at the latest, perhaps as early as the twelfth century in Ife, from the fifteenth to the seventeenth century in Benin. By the nineteenth century most Yoruba, including the agriculturalists, lived in towns where artisans and long-distance trade flourished (Mabogunje, 1968). Each town had its own ruler, but alliances were established, especially during the protracted wars that ravaged the region in the nineteenth century and led to the establishment of Ibadan, the largest metropolis in tropical Africa until the 1950s when it was home to more than half a million people.

Other sites have revealed even less about their past glory. Igbo-Ukwu, near the lower Niger, yielded bronzes, dated to the end of the first millenium CE, that rival those of Ife and Benin in sophistication of technique and style. But the single burial site where they were found, however remarkable, does not provide sufficient evidence to assume an urban culture. The imposing ruins of Zimbabwe, dated to the period from 1250 to 1450 CE, may have housed as many as 18 000 people about whom we know nothing. The sites of other urban centres remain to be discovered if they are not lost forever.

The proper assessment of the forces that brought forth these various urban civilizations remains a matter for debate. Connah (1987: 231) has put forward the 'productive land hypothesis': the crucial common factor underlying the emergence of African élites, and hence of social complexity, was the access to and control of land that was more productive than usual. Connah (1987: 228 and *passim*) emphasizes that all these early urban civilizations had a strong subsistence base with a potential for producing a storable, transportable surplus. All the people concerned were agriculturalists, growing a range of food plants suited to the local environments, and kept livestock. In addition, there are signs that in some areas there was an intensification of agriculture. Thus, on the middle Nile more efficient irrigation was achieved by introducing the ox-driven water-wheel; in the Inland Niger Delta of Mali recessional cultivation of rice was developed to a sophisticated level; in the Ethiopian Highlands slopes were terraced and water-storage dams constructed. In each case food production was enhanced by the application of a varied and sometimes sophisticated metallurgically based technology.

Connah (1987: 230) distinguishes between local trading networks and external long-distance trade. He acknowledges that local trading networks also played an important role in the emergence of African élites who controlled important resources such as copper, iron, and salt. Long-distance trade beyond the subcontinent, in contrast, he sees merely as an intensifier or a catalyst of changes that had already commenced. The distinction appears to be born from a desire to offer 'a purely indigenous explanation for the origin of social complexity in some parts of tropical Africa' (Connah, 1987: 231). The distinction, however, is arbitrary. Copper, iron, and salt were traded over long distances. While they were going to be used in the main by people elsewhere in Africa, millions of slaves, and much gold and ivory, eventually reached outside Africa South of the Sahara, beyond the Sahara and the seas.[2] But the eventual destination of trade goods made no difference to local rulers who controlled such valuable resources, ensured the safety of the trade, and exacted their due. Nor did it make a difference to the traders who connected one centre of trade with the next. Nor did it affect the clergy, the functionaries, and the artisans supported by the rulers and wealthy traders. Rather, long-distance trade beyond the confines of the subcontinent mattered because of the volume of demand it created for certain commodities, and because of the resources it provided: weapons vital to the preservation and extension of political power and luxury goods to reward followers.

A local subsistence base was a prerequisite for urban agglomerations at a time when bulk could be transported only over short distances and long-distance trade remained limited to precious goods. But it was long-distance trade that brought the resources to reward warriors, courtiers, and artisans. Urban agglomerations emerged at the crossroads of commerce. The more powerful ones came to control the access to many trade goods or indeed their very production. Crafts flourished in these cities and some might be exported, export crops might be raised on plantations nearby, elephants hunted for ivory not so far away, but the deposits of gold, copper, iron or salt were usually at quite some distance, and 'foreign' people rather than subjects were the victims of slave raids. Crop production would require fertile land, but invariably the critical resource was labour. Most long-distance trade was heavily dependent on slave labour. State organization allowed the capture and control over slaves to be employed in production—mining gold, copper, iron, or salt, harvesting kola nuts or cloves, growing cotton and weaving cloth; to transport such goods over long distances; to be recruited into slave-raiding armies; or to be traded themselves to the salt mines in the Sahara, across the Sahara, across the Indian Ocean and the Atlantic. Cities were the centres of states that typically derived resources from both: control over long-distance trade as well as production for that trade.

The significance of long-distance trade for these early cities is attested to by the effect of a decline or interruption in the trade. With the development of trans-Saharan camel caravans during the first millennium CE, the middle Nile Valley began to lose its significance as a major world trade route. The rise of maritime trade around Africa's coasts in the sixteenth and seventeenth centuries finally destroyed the middle Nile Valley trade route and eclipsed the trans-Saharan trade in the Western Sudan as well. Virtually all the cities in the middle Nile Valley disappeared, and most of the cities of the Western Sudan declined. After the rise of Islam in the seventh century, Christian Ethiopia became increasingly isolated. While the Ethiopian state survived, Aksum all but vanished. Zimbabwe was abandoned when the gold trade declined, probably because of falling world prices and the depletion of the more easily worked deposits on the Zimbabwe Plateau.

The urban history of Africa followed the vagaries of long-distance trade until the twentieth century. Cities were located where trade arrived across a major obstacle: the Sahara or the seas. The Sahara was the greater obstacle, but the trans-Saharan trade survived the development of the transatlantic trade, and the quantity of foods passing annually across the desert appears to have been greater in the eighteenth and the nineteenth centuries than in the pre-European era (Austen, 1990). To the East of Kumbi-Salih, Jenné, Gao, and Timbuktu, the Hausa cities of present-day Northern Nigeria became important termini of the trade. Foremost amongst them were Sokoto, with an estimated population of 120 000 in the early nineteenth century, Zaria, and Kano.

The arrival of European traders on the West African coast led to the establishment of cities which rose as the maritime trade increasingly outpaced the trans-Saharan trade. The role of the coastal states took on a new dimension with the dramatic expansion of the slave trade in the seventeenth century. Ample contemporary documentation has survived for the Kongo kingdom and its capital Mbanza, renamed São Salvador when the king converted to Christianity in 1491 (Hilton, 1985), and to a lesser extent Ashanti (Wilks, 1975), and it reveals increasingly efficient communication, elaborate organization, and complex political manœuvring in internal and external affairs, especially with regard to the encroaching European powers.

For the most part the Europeans remained limited to the coast where they established their forts. They traded by leave of local rulers, whose position was strengthened over time through the acquisition of firearms. Their capitals brought together the court resplendent in the wealth of the lucrative trade and large standing armies, most famous Kumasi, the Ashanti capital, and Abomey, the capital of Dahomey.

African cities were located on the periphery of larger trading systems. Some rulers ventured successfully into the centre of those systems. Nubians conquered Egypt and ruled as pharaohs during the latter part of the eighth and the earlier part of the seventh century BCE. Aksumite inscriptions speak of King Aphilas's conquest of parts of South Arabia in the third century CE and King Ezana's destruction of Meroë in the fourth century CE. But more commonly peripheral cities and states are the subject of attack from the centre. The Egyptians conquered Kerma in the fifteenth century BCE. The Egyptians returned in the sixth century BCE to sack Napata, the Romans destroyed it again in the first century BCE. The Almoravids came from the North to sack the capital of Ghana in 1076 or 1077. A Moroccan army defeated Songhai at Tindibi in 1591, the last Sudanic empire collapsed, and its cities withered. When Europeans followed centuries of trade along the African coasts with conquest in the nineteenth century, the weakness of the African periphery became obvious. The intruders might be defeated once, or even twice as in the case of Ashanti, but the various powers at the centre were inviolable and their attacks continued until they had established control over the entire region but for two exceptions. Liberia had its independence guaranteed by the United States. And Ethiopia, having defeated the Italians at Aduwa in 1896, managed to remain independent except for the Italian occupation from 1936 to 1941.

Urbanization in the Colonial and Post-Colonial Eras

That foreign powers should divide up an entire continent among themselves is quite extraordinary, but the immediate impact was limited. Luanda had been founded by the Portuguese in 1575 and become the administrative centre of Angola in 1627. Cape Town had been established by the Dutch in 1652 as a supply station on the Dutch East India Company's sea-route to the East. To settle freed slaves the British founded Freetown in 1792, the Americans Monrovia in 1822, and the French Libreville in 1848. Dakar grew up around a French fort built in 1857. Lagos, an old Yoruba town, was annexed by the British in 1861. In 1881, Kinshasa was renamed Léopoldville after the king of the Belgians. But at the end of the nineteenth century these beachheads of European expansion were still small towns. Most Africans remained involved in subsistence farming complemented by some local trade, and only about 5 per cent of the population lived in urban centres.[3]

Eventually Europeans opened up the vast hinterlands for trade with the far-distant colonial metropolis. Crops were developed for export. Major opportunities for mining were discovered. Kimberley was established when rich diamond fields were found in 1867; Johannesburg was born from the discovery of gold in 1886; copper-mining spawned Elisabethville, since renamed Lubumbashi, and Kitwe. Elsewhere mining operations that had gone on for centuries were intensified, e.g. the exploitation of the Ashanti gold-fields that had constituted a key resource for the Ashanti state. Colonial rule made trade secure, and modern means of communication increased the trading potential of the interior manifold.

Railway construction, some of it prompted by strategic considerations, dramatically improved communication to coastal ports. The 'iron snake' facilitated the import of equipment for mining operations and made the growing of crops and mining of minerals for export to Europe commercially attractive even in the remote interior. The railway from Cape Town and Port Elizabeth reached Kimberley in 1885. The line from Beira connected with it to reach the Zambian Copperbelt in 1906, then crossed the border to reach Lumumbashi in 1910: it ferried copper, cobalt, lead, and zinc over more than a thousand miles to the sea. In 1901 the Uganda Railway connected Mombasa with Lake Victoria at what is now Kisumu, 580 miles to the East, and a cotton boom ensued across the lake in Southern Uganda. In 1914 the rail from Dar es Salaam covered 782 miles to Lake Tanganyika at Kigoma. In 1916 seven hundred miles of rail connected Lagos with Kano to carry groundnuts and tin. In 1924 the rail from Dakar reached Bamako 775 miles to the East to evacuate groundnuts (Hollingsworth, 1979). Along these and lesser railways, towns were established that served both trade and the colonial administration.

The basic pattern of overseas trade continued from the pre-colonial era through the period of colonial rule. There was little industrialization as the colonial state and commercial interests were content to have their colonies import manufactures from the metropolis. Only South Africa began to produce consumer goods for its own market and neighbouring colonies in the 1920s. Africa was firmly connected to the economies of the various imperial powers as a supplier of raw materials. In this 'drainage economy' port cities played a key role. Peripheral to the colony, but at the juncture of collection of raw materials and trans-shipment, they took pride of place in the urban landscape. As a rule, colonial capitals were established in port cities, reflecting the colonial economy's central concern with the extraction of raw materials to be traded for goods manufactured in the metropolitan centre. African cities remained centres of commerce and administration, apart from a few mining centres. By 1960 only about 12 per cent of Africa's population lived in urban centres.

The towns newly constructed by the colonial regimes and European settlers presented a distinct contrast to the long-established centres of Ethiopia, the Western Sudan, the West African coast, and the Swahili coast. They differed in layout, in architecture, and in construction material. At the same time, just as pre-colonial towns differed from one region to another, so there were distinct differences in the creations of the Portuguese, Dutch, British, French, Germans, Belgians, Italians, and Spanish. There were major differences as well depending on whether a small number of colonial officials had come for a limited sojourn or a settler community was planning to stay.

O'Connor (1983: 28–41) distinguishes six types of African cities:

- indigenous cities such as those of the middle Nile, Ethiopia, and the West African coast which tend to be well integrated with their immediate hinterland;
- cities influenced by Islam such as those of the Western Sudan and the Swahili coast with their distinctive architecture;
- colonial cities characterized by sharp contrasts between functional

and residential zones, the latter segregated initially by race, then increasingly by class;

- settler cities which were intended to be European towns and pushed racial segregation to extremes—they were subject to rapid transformation as colonies became independent and majority rule was achieved in South Africa;
- dual cities constituted of physically separate parts which represent different city types, e.g. an Islamic city and a colonial city—while they are interdependent, each has a full range of urban functions and develops in its own way; and
- hybrid cities which combine indigenous and alien elements and integrate them to a large extent, rather than juxtaposing them as in the dual city.

Some of these differences are noticeable to this day, but there has been a measure of convergence as skyscrapers have come to mark the skyline of cities and shanty-towns have spread.

Africa South of the Sahara continues to be one of the least urbanized regions in the world,[4] but over the last three decades Africa has been transformed by urban growth faster than in any other region (Chapter 1).[5] This veritable urban explosion has been fuelled by large-scale rural–urban migration and the world's highest rate of natural population growth (Table 7.1).[6] Improved health conditions brought about not only a decline in mortality but also an increase in fertility.[7] These trends have been reversed though in some countries with the decline in living conditions brought on by the severe economic crisis in the 1980s and the onslaught of the AIDS epidemic.

Rural–urban migration accelerated with the advent of independence. The Sudan became independent in 1956, Ghana followed in 1957, Guinea in 1958, nearly all remaining French colonies, Nigeria, and the Belgian Congo in 1960. By the end of 1966 most of Africa was independent, the major exceptions the Portuguese colonies, Zimbabwe, and Namibia, where armed struggles were to bring independence in 1974–5, 1986, and 1990 respectively. With the approach of independence two developments emerged that encouraged rural–urban migration. Industrialization now came to be actively promoted. Governments provided incentives for the establishment of import-substituting and export-processing industries as well as the development of mineral resources. And they made major investments in public utilities, especially in the construction of hydro-electric projects. The industrialization drive soon

TABLE 7.1 Demographic, Economic, and Human Development Indicators for Larger Countries in Africa South of the Sahara, about 1990

Country	Population (m) 1992	Annual population growth-rates (%)		Total fertility[a] rate (%) 1992	Urban population (% of total population)		Urban population annual growth-rates (%)		Per cent of urban population in largest city 1990	Urban sex ratio (males per 1000 females) 1970–90s[b]	GNP per capita (US$) 1992	GNP per capita annual-growth (%)		Real GDP[c] per capita (US$) 1992	Infant mortality (per 1000 live births) 1992
		1960–92	1992–2000		1960	1992	1960–92	1992–2000				1965–80	1980–92		
Angola	9.9	2.3	3.6	7.2	10	30	5.7	6.0	63	—	—	0.6	—	751	124
Benin	4.9	2.5	3.0	7.1	9	30	6.3	4.7	—	928	410	−0.3	−0.7	1630	86
Burkina Faso	9.5	2.4	2.6	6.5	5	22	7.4	10.1	—	1045	310	1.7	1.0	810	130
Burundi	5.8	2.2	2.9	6.8	2	7	6.1	6.8	—	1266	210	2.4	1.3	720	102
Cameroon	12.2	2.6	2.8	5.7	14	42	6.3	4.9	22	1077	830	2.4	−1.5	2390	63
Chad	5.8	2.0	2.8	5.9	7	21	5.7	4.0	—	1062	220	−1.9	3.4	760	122
Côte d'Ivoire	12.9	3.9	3.4	7.4	19	42	6.4	4.9	45	1070	680	2.8	−4.7	1710	92
Ethiopia	50.3	2.5	3.0	7.0	6	13	4.7	5.1	31	867	110	0.4	−1.9	330	119
Ghana	16.0	2.7	3.0	6.0	23	35	4.0	4.5	28	949	460	−0.8	−0.1	2110	81
Guinea	6.1	2.1	3.0	7.0	10	28	5.4	5.7	76	970	490	1.3	—	592	134
Kenya	25.4	3.5	3.1	6.3	7	25	7.6	6.2	27	1216	330	3.1	0.2	1400	69
Malawi	10.2	3.4	2.3	7.2	4	12	6.8	5.1	—	1103	230	3.2	−0.1	820	143

Mali	9.8	2.6	3.1	7.1	11	25	5.2	5.7	—	982	310	2.1	−2.7	550	159
Mozambique	14.7	2.1	3.2	6.5	4	30	9.1	7.4	41	1097	70	0.6	−3.6	380	148
Niger	8.3	3.2	3.4	7.4	6	16	6.5	5.9	—	—	290	−2.5	−4.3	820	124
Nigeria	102.1	2.8	2.9	6.5	14	37	5.9	5.1	23	1149	330	4.2	−0.4	1560	84
Rwanda	7.4	3.1	2.6	6.6	2	6	6.0	4.6	—	1216	250	1.6	−0.6	710	110
Senegal	7.7	2.8	2.6	6.1	32	41	3.6	4.0	55	973	780	−0.5	0.1	1750	68
Somalia	8.9	2.7	2.5	7.0	17	25	3.8	4.0	37	1008	—	−0.1	—	1001	122
South Africa	38.8	2.5	2.2	4.1	47	50	2.7	3.0	13	1032	2830	3.2	0.1	3799	53
Sudan	25.9	2.7	2.7	5.7	10	23	5.3	4.7	35	1133	—	0.8	—	1620	78
Tanzania	27.2	3.1	2.9	5.9	5	22	8.2	6.0	27	1075	100	0.8	—	620	85
Uganda	19.3	3.4	3.1	7.3	5	12	6.2	5.7	38	1191	180	−2.2	—	860	115
Zaïre	39.9	3.0	3.1	6.7	22	29	3.8	4.3	33	992	—	−1.3	−1.8	523	93
Zambia	8.7	3.2	2.7	6.0	17	42	6.2	3.4	29	1027	370	−1.2	—	1230	104
Zimbabwe	10.5	3.2	2.3	5.0	13	30	6.0	4.6	30	1140	580	1.7	−0.9	1970	67
AFRICA SOUTH OF THE SAHARA[d]	510	2.8	2.9	6.3	15	30	5.0	4.5	—	—	559	1.4	−1.8	1346	97

[a] The total fertility rate is the average number of children born to a woman during her lifetime.
[b] Urban sex ratios are for earlier dates for Guinea (1955), Nigeria (1963), and Uganda (1969).
[c] Real GDP is based on conversion in terms of purchasing power parity.
[d] The values for Africa South of the Sahara are appropriately weighted, except for total population.

Sources: United Nations Development Programme (1995), tables 2, 4, 15, (corrected by UNDP), 16, 20; urban sex ratios from Gugler and Ludwar-Ene (1995) and United Nations (1995), table 6.

faltered because of the limits of the interior market, restricted access to the markets of other African countries, and shortages of foreign exchange that jeopardized the supply of raw materials and spare parts even to the established industries. A greater and more sustained impulse to urban growth came from the rapidly expanding bureaucracy of the newly independent states.[8] Both the industrialization drive and the expansion of the government apparatus were carried out at the expense of the rural sector. The opportunity structures were characterized by an urban bias (Bates, 1981; Lofchie, 1996)[9] that provided strong incentives for rural–urban migration.[10]

While the urban areas experienced large-scale in-migration, rural populations continued to grow as well: the masses of migrants are huge in the urban sector, but they are outnumbered by natural population growth in the much larger rural sector. In consequence, population pressure has become severe in some rural areas.

Eventually conditions deteriorated for most urban dwellers throughout virtually the entire region. First of all, there was the rapid growth in urban populations. In some countries large numbers of the new arrivals had not chosen the city over the village but were refugees from civil war or drought. Second, Africa South of the Sahara is not only the poorest region in the world, but it suffered sustained declines in per capita income (Table 7.1).[11] Third, as a consequence of the severe economic crisis, structural adjustment programmes were instituted in various countries that hit urban workers hard: large-scale lay-offs, especially in the public sector; abrupt abolition of subsidies that entailed large increases in the price of food, transport, and housing; readjustments of the foreign exchange rate that, from one day to the next, multiplied the price of imported food, fuel, and medical drugs. This was, of course, precisely one intent of these programs: to do away with urban bias. They succeeded all too well at the urban end. Whether they sufficiently increased the attraction of farming and rural manufacture relative to urban opportunities to substantially reduce rural–urban migration remains a matter of debate (Jamal and Weeks, 1993).

Urban households dealt with their changed circumstances as best they could. Workers took on second and third jobs, and additional household members entered the labour-force. In consequence the 'informal sector' greatly expanded just about everywhere. Cultivation and the keeping of livestock in and around cities and towns became more common.[12] Urban residents returned to their rural area of origin to farm, to establish a workshop, or to trade.

Urban services declined drastically as extremely poor countries, getting even poorer, proved unable to deal with rapid urban growth. Stren and White (1989) and their collaborators describe a veritable urban management crisis. Local and national governments are unable to finance critical urban services, administrative and skilled technical personnel are inadequate to the task of operating urban services and to maintaining infrastructure, and local communities are ineffective in the local administrative and political decision-making process (White, 1989).

'Informalization' correctly conveys that public control over working conditions, the use of urban space for trade and shelter, housing standards, and the quality of urban transport has by and large broken down.[13] But it is also a euphemism that obscures that large parts of the urban population are reduced to extremely low incomes and abysmal living conditions.

Still, until the 1980s at least, governments throughout Africa privileged the urban sector. And within that sector resources were disproportionately directed towards capital cities. As a consequence, capital cities were the preferred destinations for migrants and grew at an even faster pace than the urban population in general.[14] Most African capitals were primate cities already in colonial days; today, the pattern is even more pronounced.[15]

The Changing Parameters of Rural–Urban Migration

In colonial days there was considerable variation in the context of rural–urban migration. The policies pursued by the various colonial powers varied, and they changed over time. Economic conditions changed with the Great Depression, the impact of the Second World War, and the boom in raw materials brought about by the Korean War. Most important, where Europeans had settled in substantial numbers, they invariably opposed the settlement of Africans in town. The *apartheid* policies of South Africa were condemned across the world, but settler interests had a strong impact on colonial policy in Zimbabwe, Malawi, and Kenya as well. In each of these countries major efforts were directed at making the urban residence of Africans temporary. A variety of policies were pursued in different countries at various times: wages were kept low, recruitment was limited to men, only bachelor accommodations were provided, workers were recruited on limited-time contracts, migration was controlled outright.

South Africa presented the extreme case of temporary migration imposed by law on a large part of the urban labour-force for over a century. The origins of pass controls can be traced back to the period of slavery in the Cape in the eighteenth century. After the abolition of slavery in the early nineteenth century, pass controls were adapted to support various forms of indentured and child labour. When the diamond- and gold-mines were established in the late nineteenth century, pass controls were refashioned to institutionalize a system of temporary migrant labour. They were used to promote organized recruitment, enforce contracts of limited duration, and prevent permanent settlement of workers in the mining districts. This 'cheap labour power system' benefited the mines and other sectors employing migrant labour: the costs of the reproduction of labour power were borne by the rural sector (Hindson, 1987: 15–27).

Agricultural production in the South African 'reserves' declined sharply, not only in per capita, but even in absolute terms, in the late 1950s. By 1970 agricultural production as a proportion of subsistence requirements was negligible (Hindson, 1987: 55, 72). The 'cheap labour power system' had collapsed: the cost of reproduction of the urban labour-force could no longer be displaced on to the rural sector. Pass controls continued, however, to serve the interests of *apartheid*: not only was urban surplus labour resettled in the 'bantustans', but a large proportion of the urban labour-force continued to be recruited on temporary contracts while their dependants were forced to live in the 'bantustans'. Thus the mining companies were prohibited by law from providing family accommodation for more than 3 per cent of their African work-force until 1986. Many men were recruited on short-term contracts not only for the mines, but in various other sectors of the economy. Turok and Maxey (1985: 252–3) estimate that substantially more than half the Africans in registered employment were migrant workers in the 1970s. Racial oppression had given rise to a paradox: the continent's most industrialized country—where large numbers of Africans have worked in mines, factories, and urban services for several generations—had the highest proportion of short-term recruits in its labour-force.[16]

Short-term migration was common outside the settler colonies as well. It was a function of the recruitment of men at low wages. Where employees were only provided with bachelor accommodations, both aspects—the cheap labour policy and the exclusive recruitment of men—were brought into sharp relief. Most strikingly, in environments thus characterized by circular migration, some employers did establish a more stable labour-force by providing conditions that encouraged workers to bring their wives and children. The Union Minière du Haut Katanga changed its labour policy in the copper-mines in what is now Shaba Province, Zaïre, as early as 1927. A measure of compulsion was involved: workers had to bring their wives, but the region soon boasted a stable labour-force. By 1957 the average length of service of African employees in the mines was eleven years. On the Copperbelt, in what is now Zambia, a policy to establish a permanent labour-force was initiated in 1940. Employers and colonial government provided permanent accommodations for married employees, adequate schooling in the urban areas, and a pension scheme. The average length of employment of African workers increased from four and a half years in 1956 to seven in 1964. Cheap labour policies were modified in different colonies, by various employers, at different times. By the time independence arrived, urban wages had increased substantially everywhere. The urban labour-market was completely transformed as a consequence.

Sabot (1979) describes and analyses the transformation of the urban labour-market in Tanzania. The colonial labour-market was characterized by a persistent shortage of labour. Government and private employers paid low wages, arguing that at higher wages migrants would work for shorter periods. Sabot could find no evidence in the historical record to support this proposition. Moreover, even if higher wages were to induce the individual migrant to work for a shorter period, they could also be expected to attract a greater number of migrants. It appears than that a small number of large estate owners and the government took advantage of their oligopolistic position to administer wages at a level below that which would attract a sufficient number of workers. The shortfall was made up in part through forced labour until the 1920s and again during the Second World War, and through unscrupulous methods of recruiting labour in distant regions. After the Second World War urban wages increased steadily while rural incomes stagnated. When independence was granted in 1961, the rise in wages accelerated. Growth of the rural labour-force, and the widening gap between income from farming and from urban wage employment, increased the size of the migrant stream, and labour surplus replaced labour scarcity. At the same time a labour-force that had been characterized as 'uncommitted' was stabilized: turnover and mobility rates became low by international standards.

Sabot explains the dramatic rise in wages, and the transformation of the problem of labour scarcity into a problem of urban unemployment, in terms of the convergent interests of employers, trade unions, and government. As Tanzania became independent, a burst of import-substituting industrialization occurred. These relatively capital-intensive factories had to invest in training the industrially disciplined, semi-skilled workers they required. In order to secure this investment, the labour-force had to be stabilized, and employers raised wages to that end. The trade unions were identified with the nationalist movement, gaining in strength with the emergence of educated leaders and the stabilization of the labour-force. They had only begun to develop in the 1950s, but more than half the labour-force was unionized by 1965. The unions brought pressure to bear on wages through both political action and collective bargaining. The government felt that aspirations raised by the advent of independence had to be met. It legislated a minimum wage in 1957, and increased the wages of low-level government employees.

Migration Strategies

Migrating is not a solitary affair, even when individuals move alone. The days when village elders disapproved of young men 'running away' and left them no alternative but to abscond at night (Banton, 1957: 48–59) are long past. Rural folk have come to appreciate the opportunities the city offers. The urban experience often took on positive connotations. Thus, in the 1950s, young men in many parts of the Sudanic belt were expected to spend one or several spells of seasonal migration in Ghana (Rouch, 1956). Cultural norms exalted the challenge to the young to prove themselves, the experience to be gained. Such norms became so generally established that individuals were swept along even when they did not share the economic rationale for going to the city.

Today, villagers throughout Africa actively promote the outmigration of young adults. Rural communities have developed migration strategies which are informed by the experience of migrants who have kept in touch, who return to the village on visits or to stay, and by villagers who have visited kin and friends in the city. These strategies are modified over time as experience dictates. Potential migrants are thus presented with quite well-defined options.

The decision to migrate in turn is rarely an individual one, rather it is usually a family decision. Much rural–urban migration of individuals is part of a family strategy to ensure the viability of the rural household (van Velsen, 1960). And migrants typically receive considerable assistance in their move. Frequently a wide range of relatives, in the countryside as well as in the city, is drawn on to help pay for an education for the future migrant, provide a home for children who are sent to town to go to school, offer the newly arrived migrants shelter and food for a while, assist in the search for employment, take care of parents and assist wife and children who stay behind. The kinship group thus acts as an agent of urbanization (Flanagan, 1977).[17]

The rural–urban migration of single men was fostered in the settler colonies. Urban sex ratios were extraordinarily high under such circumstances: Several years after independence, in 1969, Kenya reported 1386 men for 1000 women in the urban areas; Zimbabwe, still ruled as Rhodesia by a white minority regime, 1412 men for 1000 women in the same year; and Namibia, controlled by South Africa, 1268 men for 1000 women in 1960 (Gugler and Ludwar-Ene, 1995). However, contrary to generalizations commonly made, high urban sex ratios are not the rule throughout Africa. Several countries have consistently reported women outnumbering men, sometimes by a substantial margin (Table 7.1).[18]

Research on rural–urban migration in Africa has focused on men—women were assumed to come as their dependants. As the high urban sex ratios that characterized the settler colonies declined sharply, the assumption was commonly made that more balanced sex ratios reflect an increased tendency for women to join their husbands in town. However, substantial numbers of men continue to leave their wives and children in the rural home, while a large and growing number of women come to town on their own.

Three categories of women, in particular, move on their own. In many countries, substantial numbers of young, unmarried women come to town to work as domestic help. In some settings their remittance of part of their wages plays a significant role in the survival strategies of rural households. Elsewhere they come at an early age as foster children of urban families, typically relatives, with the promise of better educational, career, and marriage prospects in the urban setting.[19]

With the expansion of secondary education outside the major cities, increasing numbers of educated young women move to the cities in search of employment. Many postpone marriage or prefer to remain unmarried, while improving their standard of living with the contributions of male friends.[20]

Abandoned, separated, divorced, and widowed women constitute a third category of women who frequently move from the countryside to the city on their own. Most African societies are patrilocal and patrilineal. Women leave their community of origin upon marriage, and they are integrated into their husband's community through their husband and their sons. And patrilineal control over property tends to leave women dependent on their husbands and their sons to access rural resources, in particular land and cattle. The woman who has lost her husband thus typically finds herself in rather precarious circumstances. Unless a well-established son is prepared to support her in the rural setting, the city offers a promising alternative for many such women. A move to the city is particularly attractive for those older women who can stay there with a married son. They may resume child-rearing, this time of their grandchildren.

In many cases rural–urban migration is not a 'once for all' move. Return migration to the rural area is common. Many of the young women who have come to the city to work as domestics eventually marry in the village. Most of the men who have left their wives and children in the rural home plan to retire there. And when families migrate to the city, theirs is not necessarily a permanent move either.

The strategy of family separation may be characterized as 'one family—two households' (Weisner, 1972). It is a function of rural and urban conditions. Living costs in the city are high, urban earning opportunities narrowly circumscribed for most women. Wife and children who remain on a family farm can grow their own food, perhaps raise cash crops as well. A measure of communal control over land obtains in much of Africa South of the Sahara and precludes selling one's landholding. A wife who comes to the city abandons a source of food and cash to join a husband on low wages.

If the Industrial Revolution established the distinction of workplace and home, the separation of men from their wives and their children has been drastically magnified for many migrants in Africa. The frequency of visits varies a great deal with employment conditions and distance. As transport has become faster and cheaper, monthly and even weekly commuting has become common. Still, the separation strains the migrants' relationship with wife, children, kin, and village community.[21]

Marital relationships are often profoundly affected by the single migrant's involvements in the urban setting. Two threats have always been important but have become even more severe in recent years. The sufferings caused by the transmission of sexually transmitted diseases have become excruciating with the spread of the AIDS epidemic. And the neglect or even outright abandonment of rural households threatens the very survival of growing numbers of such households that have become heavily dependent on urban–rural transfers.

Many families settle down in the city for the long term, perhaps for a working life, all the while planning to return to the rural community eventually. Strong ties to the kinship group, perhaps to the village community at large, can make an eventual return attractive. In the city most migrants, even when they manage to support a family, enjoy little economic security. And uncertain political times add to a sense of insecurity. Unemployment and underemployment are widespread, but there is no unemployment compensation. Few qualify for support in disablement. And retirement benefits are meagre for the tiny minority who are entitled. Many migrants rely instead on the security the village provides. They maintain their position in the rural community, and even during an extended urban career remain assured of access to land on their return.

Partaking of the city as well as the village, may be characterized as 'life in a dual system'. I found such a pattern well established in south-eastern Nigeria in 1961—and enduring when I returned there in 1987 (Gugler, 1971, 1996). Urban dwellers were fully committed to their urban career, all the while maintaining strong ties with the (husband's) community of origin. They visited there, welcomed visitors (more or less enthusiastically), let themselves be persuaded to contribute to development efforts, built a house, planned to retire, and wanted to be buried 'at home'.

Independent women—abandoned, separated, divorced, widowed, never married—are less likely than men to retire to the countryside. If the village offers the ultimate security for many urban dwellers, this security is problematic for a single woman. She cannot expect much of a welcome on her return to the village—whether she try for the village of a former husband or her village of origin.

The Urban Labour-Market

For several decades now, an excess of labour has characterized urban labour-markets throughout Africa South of the Sahara. Initially it was unskilled workers who faced severe competition for scarce jobs, but increasingly secondary school leavers and even university graduates have had to confront unemployment and underemployment.[22] When urban unemployment first appeared, rural–urban migrants could be seen to try their luck at an urban job lottery. Since much labour migration had been short-term until then, the new arrivals faced little competition from entrenched urban workers and their descendants. Moreover, there was typically a significant expansion of urban employment around the time of independence. And the system of recruiting labour with limited skills approximated a random process. Since minimum wages were high, relative to rural incomes, even an extended job search was a promising strategy (Gugler, 1969).[23]

The urban job lottery pattern corresponded to a transitional phase. As a rule, workers cling to their jobs, few new jobs are created, and recruitment is anything but random. Access to the labour-market is largely a function of three criteria: education and training, patronage, and gender. Differential access in turn shapes the composition of the migration stream.

Conventional education adds little to earnings in agriculture, but it makes a major difference in the urban labour-market. In the years leading up to independence and for some years after, Europeans vacated administrative positions in government and commerce. Initially a completed secondary education ensured easy access to such positions and excellent career prospects in many countries. Today, the career prospects of even a university graduate are problematic in all but a few African countries. Still, the highly educated quite clearly have better earnings prospects in the urban than the rural sector.

Labour-markets are not only stratified by education, they are also segmented by patronage. Where employers are besieged by an army of suitably qualified applicants, they are likely to rely on personal recommendations. After all, skills and knowledge are not as important for many positions as other qualities: dependability, potential for training, persistence, and initiative. A relative, a friend, a powerful politician or an employee vouches for the applicant, exerts pressure on him/her to perform, and may return the favour. Such sponsorship favours relatives and 'home people'. As it weaves networks of exchange, it becomes a key factor delineating ethnic identities.

Thirdly, labour-markets are segregated by gender. Urban earning opportunities remain severely circumscribed for African women. They are handicapped threefold: because of their reproductive burden—African women bear more children than women anywhere else, because they carry nearly the entire responsibility for domestic tasks, and because of a substantial gender gap in education. Thus women hold only a small fraction of the jobs in manufacturing: 2 per cent in Chad, 22 per cent in The Gambia, 12 per cent in Kenya, 5 per cent in Niger, 10 per cent in Tanzania, and 5 per cent in Zimbabwe; only in Botswana and Swaziland, countries where men prefer to find work across the border in South Africa, do women hold about a third of the manufacturing jobs (International Labour Office, 1992, 1994). In some African countries, few women are found even in those occupations that are considered a female preserve elsewhere: most domestics, secretaries, and teachers are men (Hansen, 1989). Finally, the stereotype about women traders omnipresent in West African cities holds only for part of the region—and, on closer inspection, a distinct gender hierarchy appears in that sector as well (Afonja, 1990).

Nelson's (1996) detailed study of the economic activities of women in a shanty-town in Nairobi shows women to be much more restricted than men in their choice of economic activity. Few of the local business establishments are run by women. Illegal beer-brewing and prostitution are the principal sources of income for women. They are at a disadvantage because they are less well-educated than men, have fewer skills of commercial value, and support and care for children. The last point is demonstrated by the fact that a disproportionate number of successful women entrepreneurs are childless, that other women begin to expand and consolidate their business only in their late forties, when most or all of their children have grown up and perhaps contribute to household income. The handicaps experienced by women may be seen as structural constraints, but they are a function of the cultural context: the education and training thought appropriate for girls; the occupations considered suited for women; the emphasis on women as mothers who bear children, many children, and have the primary responsibility for raising them.

Social Integration and Change

Many urban dwellers remain firmly rooted in a rural-based descent group.[24] Even if they are isolated in the urban setting, they can feel secure in the knowledge that they continue to be members of the community they came from. Wives and children who had to be left behind, parents, kin, and friends continue to define a rural place as home. Regular visits, whether at short or longer intervals, confirm their sense of continued integration.

Most migrants move to a city where they can expect to be received by relatives or friends. They will be offered shelter and food for a while, they will be introduced to the urban environment, and efforts will be made to find them an opportunity to earn a living. This pattern of initial urban association encourages people from the same area to form residential clusters. Allocation of housing by public authorities or employers may inhibit such clustering. But even when residentially dispersed, people of common origin frequently maintain close ties and think of each other as 'home people'.

Links with home people in the city and ties to the common home tend to reinforce each other: each enhances communication and social control in the alternate context. Marriages are arranged and establish additional ties with affines in the village or close by. Frequently the urban–rural connection takes on normative character. Whether to remain involved in the rural community is no longer a matter of individual affection and calculus, rather such involvement is expected among members of the social network of home people. An ideology of loyalty to home is established, and it tends to hold sway even over those few who succeed in establishing themselves securely in the urban setting. Their patronage will go first and foremost to home people.

Urban communities of home people have established formal organizations in many African cities. Some such 'ethnic' associations go to considerable lengths to have all from home join in.[25] They meet regularly, elicit intensive participation, and serve a wide range of explicit and implicit purposes. Members relax in one another's company, they evolve common responses to the urban milieu, provide assistance in personal crises, settle their disputes within the association, and are often deeply involved in the affairs of the home community. Some make major contributions towards improving rural conditions (Gugler and Flanagan, 1978: 81–8).

The connection with the rural community is more precarious for women than for men, as we have seen. And in the urban setting 'home people' are usually defined in patrilineal terms as well. A married woman is connected through her husband with the social network of his home people. If there is an ethnic association, it may have a women's branch, but that branch is subordinate and composed of women who have their husbands' rather than their own origin in common. Many women create a community afresh in the city in a church gathering (Ludwar-Ene, 1991). In the place of kinship ideology they put ritual ideology.[26]

The city, unlike the typical rural setting, gives a measure of choice: how closely to associate with various relatives, whether to discontinue relationships with some of them altogether. Epstein (1981) describes the pattern in the 1950s on the Copperbelt in what has since become Zambia. The corporate kin group of the village is transformed into a social network of urban kin, a network maintained and developed selectively. At the same time the range of recognized kin may be extended. A minimum level of support may be expected from all people sharing a common origin.

Links with home people in the city and ties to the common home are a key variable affecting the pattern of association among migrants. Two other variables come into play. Mayer (1971: 283–93) emphasizes the interaction between cultural traits and the structure of social relationships. Migrants who have a traditionalist outlook will tend towards encapsulation in a group of like-minded home people who uphold shared rules of behaviour. Where Mayer's analysis represents cultural background as the prime determinant of patterns of association in the urban setting, Banton (1973) focuses attention on a third variable: opposition among social groupings in town. He suggests that the social density that characterizes the village is encouraged among urban groups both by the degree of discontinuity between the rural and the urban system and by the extent and strength of structural opposition in the urban system.

The commitment many migrants have to their community of origin may be taken to suggest that they remain peasants at heart, that they do not become urbanites. But most migrants have some familiarity with urban conventions of behaviour and ways of thinking. They have learned about urban conditions in school and from visiting or returned migrants. Some have been in town before to sell rural products, to make purchases or obtain services, as guests of kin or friends.

As soon as migrants arrive in town, they have to adopt behaviour that will allow them to pursue their goals effectively. The point was made forcefully by Gluckman (1960: 57) when he dismissed an earlier perspective that presented African urban workers as 'tribesmen' in his classic dictum: 'An African townsman is a townsman, an African miner is a miner'. That is, the migrant's behaviour is defined by the immediate context. Yet such a model of 'situational change' captures only one aspect of the migrant's adaptation. There is also a drawn-out process: migrants continue to modify their behaviour as they gain urban experience, as they undergo 'biographic change'. [27]

The mastery of several languages is an important element in the adaptation to African cities where discourse is invariably multilingual. A two-tier pattern of language use obtains in all but a few African countries. Popular discourse is dominated by the language(s) of one or two groups that impose themselves by their numbers and/or status. Many migrants thus need to learn an African language quite unrelated to their own to get along with their neighbours, to trade, to find employment. But government, much of the educational system, and most intellectuals are committed to the language of the former colonizer. In consequence, the uneducated are severely handicapped in their dealings with the bureaucracy. The migrants' switch in language illustrates the propositions of both the situational and the biographic change model: as soon as migrants arrive in the city they will need to employ what little they know of the lingua franca; as they stay on, their language skills will develop.

Learning new behaviour, adopting new norms, some individuals grow away from their rural ways. But most do not forget how things were done at home. They continue to behave in urban or rural ways as the situation demands. Those who maintain the rural connection are 'peasants' as well as 'townsmen'—depending on the arena in which they operate.

New Identities in Conflict

The great majority of adults in African cities are first-generation migrants from rural areas. Their identity of origin is reinforced by ties to what many consider their home community and by the assistance received from and given to people from home. Larger ethnic networks are established as people reach beyond the limited pool of kin and home people and draw on schoolmates and affines. These ethnic networks tend to be delineated by the regional recruitment of schools, by patterns of endogamy, and by common language.

Ethnic groups were delineated and recognized by colonial governments, their languages demarcated by mission publications in the vernacular. Common language facilitates communication.[28] It is also the medium of ideology. In Nigeria, the Ibo Union was established in Lagos in 1936. A Pan-Yoruba cultural society was founded in London in 1945. Both played a principal role in enlisting popular support for two major Nigerian parties after the Second World War, parties which were soon identified with the Igbo (the current spelling) and Yoruba people respectively. A third party came to represent Northern interests. These divisions shaped civilian politics until 1966, underlay the civil war that visited widespread suffering from 1967 until 1970, and continue to be a salient feature of Nigerian politics. In the past neither the Igbo nor the Yoruba actually constituted a political unit. Still, these new groups follow lines of cultural affinity, and in particular of shared language. The various dialects of the Igbo are clearly distinct from the languages of their neighbours. At the same time, there is such variation within Igbo, that some dialects are barely intelligible to speakers of others. Significantly, major efforts were made to standardize Igbo speech and writing. In addition, certain forms of traditional culture, such as music and dance, were reaffirmed, and a new pride in Igbo identity cultivated.

Most urban dwellers identify themselves and others in ethnic terms. To refer to this pattern as 'tribalism' is problematic. The term 'tribe' acquired pejorative connotations in the colonial era, and the concept of 'tribalism' is quite misleading. Ethnic identities in the urban setting bear a rather tenuous relationship to traditional societies and their culture. The past provides the raw materials, but ethnic identities are fashioned in the confrontations of the urban arena. As Schatzberg (1988: 9) puts it, ethnicity 'is a protean, contextual, and intermittent phenomenon'.

Urban dwellers draw on a whole series of identities of origin: their kinship group, their home village, the village group to which it belongs, their region, the speakers of their language, their nation, their race. Some identities represent traditional bonds and shared culture, others are new. Which identity is salient depends on the situation. Banton (1965: 145) observed that a Temne migrant to Freetown, Sierra Leone, was involved in a series of oppositions in colonial days: African versus European, tribesman versus Creole, Temne versus, say, Mende. The three categories—African, tribesman, Temne—were in a straight hierarchy, each one was relevant only when a higher-order opposition did not enter. Such a series of identities of origin may be seen as a nested hierarchy (Leeds, 1973). A set of concentric circles provides a graphic model of the more narrowly or more largely defined ethnic groupings (Gugler, 1975).

Members of an ethnic group frequently monopolize major political and/or economic opportunities. Two processes are at work. First, an ethnic group builds up an important lead. Thus in many countries

Christian missions focused their work on particular regions, giving those people a head start in Western education, and hence in careers requiring such education. When cash crops were introduced, particular regions were favoured, say country suitable for growing cocoa. And the military officers who took power in most countries at one time or another, frequently came from ethnic groups, the 'martial tribes' as colonial administrators were wont to characterize them, preferentially or even exclusively recruited into colonial armies.

Second, once some members of an ethnic group are in a privileged position, they come to control access to various opportunities. When their patronage goes to kinsmen, fellow villagers, or other 'brothers', an entire group can be seen to enjoy privilege, while outsiders are excluded. The perceived coincidence of ethnic difference and economic opposition may prompt a reaffirmation of cultural distinctiveness and culminate in an ethnic renaissance. Cohen (1969) describes such a process among the Hausa controlling the trade in cattle and kola nuts in Ibadan, Nigeria.

In some cities conflicts arise between indigenous peoples and migrants. In Nigeria this has occurred in Kano (Paden, 1973), Ibadan (Sklar, 1963: 289–320), and Lagos (Baker, 1974), urban centres established in pre-colonial times. However, such conflicts are the exception. Because of the recent origin of most cities and their explosive growth, the urban-born are usually a small minority. More commonly, people from a city's hinterland and migrants from more distant lands are juxtaposed in conflict.

Violent ethnic conflict can erupt quite suddenly as has been demonstrated most recently in South Africa. The premier political party, the African National Congress, had pursued a multi-ethnic and indeed multi-racial policy from its very establishment in 1912. It always had a fair succession of leaders from among the Zulu as well as the Xhosa and had managed to overcome ethnic consciousness to a large extent, at least as far as the political activists were concerned. The ANC initially encouraged Mangosuthu Buthelezi, a former member of the ANC Youth League, to assume the leadership of the Kwazulu 'homeland' the South African regime had established. The *Inkatha* movement he established was envisaged as an internal wing of the ANC under the protective umbrella of the 'homeland'. Buthelezi, however, pursued an independent line. While he consistently maintained an anti-*apartheid* stance, his rejection of the armed struggle and international sanctions drew *Inkatha* closer to the South African regime and away from the

ANC exiles abroad. The acknowledgement by the South African regime in 1989 that the majority could no longer be excluded from the legitimate political process, gave urgency to the conflict over the respective roles of the ANC and *Inkatha* in post-*apartheid* South Africa. The conflict has escalated in cities where competition over jobs and housing has heightened—and where confrontations between more- and less-skilled workers, between long-time urban residents, squatters, and migrants in single-men hostels, are readily cast in ethnic terms: the majority of Zulu are migrants. At the same time a Zulu cultural revival has taken place. Hostel dwellers and unemployed migrants, in particular, find solace from their material and symbolic deprivation by taking pride in their King and the reconstructed memory of Zulu resistance against colonial conquest (Adam and Moodley, 1996).

The notion of an African working class has elicited a good deal of research and argument.[29] The obstacles to the emergence of a strong, self-conscious working class have been several. The proportion of workers in large-scale organizations, other than the public bureaucracies, is small in all but a very few countries. Accordingly, the few efforts to establish parties based on the trade unions have remained inconsequential. Also, most workers are semi-proletarians: they remain involved in the rural economy. Finally, and relatedly, their organization is subject to ethnic cleavages.

Some workers have considerable leverage because they control key resources. Strikes by railroad or dock workers strangle economies heavily geared towards exports and imports; strikes by workers producing a major export commodity, such as petroleum, copper, or gold, put the financing of crucial imports in jeopardy; strikes by workers in electricity or public transport paralyse the cities. The skilled workers in these sectors cannot be readily replaced, and governments have little choice but to respond to their demands. Most of the time such leverage has been used to exact relatively generous wages and benefits, only rarely to effect broader political changes.

Organized workers have played a major role in popular protest movements. Trade unions were at the forefront of independence movements in Africa. But after independence authoritarian regimes everywhere managed to subject them to close control. The economic crisis of the 1980s brought rapid inflation and a precipitous decline in real incomes. Civil servants, in particular, in many countries saw the purchasing power of their salaries shrink to the point where their monthly cheque would at most buy a couple of days' food for their family. If they held on to their job it was because of benefits such as subsidized housing—and the opportunities their position might give them for graft.

Since the 1980s trade unions in a number of countries have come to the forefront of popular protest movements once more. In South Africa, they played a major role at several stages of the struggle against the minority regime (Friedman, 1987). Elsewhere they were involved in movements that toppled authoritarian regimes or at least forced them to make more or less substantial concessions towards the demands for human rights and democratization.

Elements of the élite tend to pursue class interest more consistently than workers. As independence approached in Nigeria, Sklar (1983: 480–505) found a new and rising class engaged in class action and characterized by a growing sense of class consciousness. Since then politicians, bureaucrats, and commercial interests have come together in just about every country to form a political class: they use the state machinery to gain privileged access to public resources (Diamond, 1987).

The powerful and wealthy dispense patronage to a large clientele. In a detailed study of Mushin, a recently established suburb of Lagos, Barnes (1986) shows that most political participation is based on patron–client relationships. A clientele that can be mobilized for various forms of action—voting, influencing, disturbing—is the basis for power, and this power in turn is used to strengthen and expand client support.

Patrons recruit much of their following on the basis of common origin.[30] If the largesse they dispense mutes class antagonism, the nature of their clienteles fosters a perception of the conflict over the allocation of resources in terms of the competition of different ethnic groups over government jobs, commercial opportunities, housing.[31]

Schatzberg (1988: 9–11) characterizes class as contextually fluid: the identity, composition, and boundaries of social class often vary according to the contexts of the moment. He argues that social dynamics in Africa South of the Sahara best be viewed as a triple-stranded helix of state, class, and ethnicity. The state and ethnicity would appear to be the dominant strands. Access to economic resources is largely predicated on political power, and political alignments are typically established along ethnic lines.

The various ethnic élites were not prepared to share the spoils for long in most African countries. Those in power came to be perceived as representatives of an ethnic minority and to lose legitimacy. The process was reinforced by two developments. More than a decade of economic decline throughout much of Africa severely curtailed the resources governments could allocate to their supporters. And with the end of the Cold War foreign support for some of the regimes considered unrepresentative was withdrawn. Suddenly the pace of political change accelerated.

The democratic institutions bequeathed to the new nations at independence had soon become farce or been discarded altogether. Government changes were usually effected by military coups. By the 1980s the majority of African countries were ruled by military regimes, and most civilian regimes were similarly authoritarian. Popular protests were virtually always effectively suppressed (Wiseman, 1986). And even when a government fell, the successor regime usually turned out to be similarly oppressive and corrupt. But 1990 ushered in a new era. In Benin, a velvet revolution took place: a national conference turned into a civilian coup that stripped the dictator of effective power, established a transitional government, and prepared the way for multiparty elections. In South Africa, Nelson Mandela was released from prison, the opposition parties were 'unbanned', and negotiations towards enfranchising the majority began. Pressures for democratic change increased dramatically across the continent. In one country after another long-suffering people—some of whom had witnessed the revolutions in Eastern Europe on television screens—took to the streets, and trade unions that had been effectively muzzled sprang into action. A number of regimes were forced to change their ways if they did not collapse altogether.[32]

These mass movements took place in cities, their impact was most strongly felt in capital cities. African politics no doubt will continue to be shaped in the urban arena. There is little prospect that the interests of the majority of Africans, the rural masses, will be represented effectively.[33]

A large gap—cultural, economic, and political—continues to separate African cities from much of the countryside. But many urban dwellers bridge this gap: they remain deeply involved in what they see as their rural home. This connection reinforces the ties they have established with home people in the urban setting. The social networks reaching beyond this core tend to be delineated in terms of region of origin, language, and culture. They constitute new ethnic identities that divide the urban populace throughout the continent—and pose a daunting obstacle to the efforts at democratization.

Notes

1. The following account of early urban civilizations in Africa South of the Sahara is heavily indebted to Connah (1987). For a wide-ranging discussion of pre-colonial cities in the region, based on a comprehensive bibliography of French as well as Anglo-Saxon sources, see Coquery-Vidrovitch (1993). Chandler (1987) provides population estimates and key information on the world's cities from a vast array of sources. One or more African cities appear in most of his compilations of the world's seventy-five largest cities at various dates since 1800 BCE.
2. West Africa was the leading supplier of gold to the international economy between the eleventh and the seventeenth century. It is thought to have accounted for almost two-thirds of world production in the later Middle Ages (Hopkins, 1973: 82).
3. For comprehensive annotated bibliographies of urbanization in the region, see O'Connor (1981) and Stren (1986).
4. The low level of urbanization fails to convey the full impact of urbanization. Many rural residents have experienced urban life: return migration to rural areas continues to be significant as we shall see. And, of course, villagers are connected with the city through contacts with migrants, through market transactions, through government, and through radio and, increasingly, television.
5. For a wide-ranging review of the literature on post-colonial trends in urbanization in the region, see Mabogunje (1990), for a review of African urban research since the late 1980s, Stren (1992). Stren (1994) presents the findings of a large international project surveying past urban research and proposing an agenda for future research in Africa.
6. The migrant share of urban growth between 1975 and 1990 has been estimated at 64 per cent for Kenya, 75 per cent for Senegal, and 85 per cent for Tanzania (Findley, 1993).
7. The eradication of smallpox, the expansion of vaccination programmes, the increasing availability of potable water, and rising levels of education have been probably the key factors drastically lowering mortality, at least until recently.
8. Little data on the expansion of the bureaucracy is readily available. Abernethy (1988: 189) suggests that employment in central and local governments in the region grew from 1.9 to 6.5 million between 1960 and 1980, and that the public sector, including parastatals, accounted for half of those in non-agricultural wage employment by 1980. Robinson (1990: 9–11), however, produces data for only 14 countries, and most of these statistics include the health and education sectors. Expansion of these sectors may be considered highly desirable, the more so as the population is growing rapidly. And in terms of urban growth the question arises what proportion of government services is based in rural areas.

Bradshaw and Noonan (1996) found Africa South of the Sahara, as well as Latin America, to be highly urbanized relative to GNP in a cross-national analysis that controlled for the effects of investment dependency and pressures from the International Monetary Fund on the level of urbanization in less developed countries.

9. Smith and London (1990) report urban bias much more pronounced in Africa South of the Sahara than in any other major region of the world. Their measures include the ratio of the output of the average non-agricultural worker to the output of the average worker in agriculture in 1965 and a Gini index of sectoral inequality based on data on product-per-worker in the agricultural, service, and manufacturing sectors in 1970.
10. The migration model implicit in this discussion may be characterized as rent-seeking. Becker and Morrison (1993) describe such a rent-seeking as well as a neoclassical and a demographic model and evaluate their empirical relevance.
11. In spite of the dismal economic performance, life expectancy increased from 40 to 51 years, infant mortality declined from 165 to 97 per thousand, and adult literacy increased from 27 to 54 per cent between 1960 and 1992 (United Nations Development Programme, 1995: 163).
12. 'Urban agriculture' has begun to attract the interest of policy-makers and researchers, e.g. Memon and Lee-Smith (1993).
13. The breakdown of such urban controls may be seen as the demise of the last remnants of migration control: it removes obstacles to the insertion of rural–urban migrants and facilitates their escape from rural deprivation.
14. For an account of capital city bias in West Africa, see Gugler and Flanagan (1978: 40–3).
15. In about two-thirds of the countries that are not land-locked the capital city continues the colonial inheritance of being also the major port. Over the last two decades Abuja in Nigeria, Dodoma in Tanzania, and Yamoussoukro in Côte d'Ivoire were established as new capital sites to deal with the issues of both primacy and peripheral location, but they have remained rather empty show-cases.
16. The South African police state succeeded in keeping urban population growth extraordinarily low (Table 7.1) and, by and large, in enforcing urban planning. The abandonment of various controls has unleashed large-scale migration to the major cities and the rapid spread of squatter settlements.
17. Stark and Lucas (1988), drawing on data from Botswana, present migration as an intertemporal proposition generating streams of various benefits to both migrants and their families. This approach was further pursued by Hoddinott (1994) with data from Western Kenya.
18. For the urban sex ratios of additional African countries, historical data, and a more detailed discussion of the role of gender in rural–urban migration, see Gugler and Ludwar-Ene (1995).

19. In terms of educational opportunities, fostering has lost much of its promise: schools have become more widely available in rural areas, the quality of urban education has declined in all but a few élite schools, and the career opportunities for the educated have sharply diminished.
20. Behrman and Wolfe (1984) emphasize that the rural–urban migration of women is affected not only by the labour-market. They propose a model which—like the Harris and Todaro job search model—includes both the probability of finding a companion and the earnings of such a companion.
21. Brown reports from Botswana (1983) that with large-scale male migration to South Africa marriage patterns have changed dramatically. Men delay marriage, and most women have borne several children before marriage and, frequently, even before an engagement has been formalized. Most of the unmarried fathers offer little or no support to the mother of their children.
22. The informal sector turned out less of a panacea for the urban employment problem than once assumed (International Labour Office, 1972). On the problematic nature of the very concept of an 'informal sector', see Gilbert and Gugler (1992: 94–100). For an annotated bibliography of the urban informal sector in Africa, see International Labour Office (1991).
23. Harris and Todaro (1968, 1970) incorporated the probability of securing urban employment along with the rural–urban income differential into an econometric model of rural–urban migration and job search. They defined the probability of obtaining urban employment as the proportion of the urban labour-force actually employed. However, the assumptions underlying this definition were problematic even in the early stages of urban unemployment in the region (Gugler, 1976).
24. Ekeh (1990) argues that kinship systems were elaborated and strengthened in Africa during the many centuries in which the slave trade ravaged much of the continent. Kinship systems provided a measure of protection in stateless societies as well as in predatory states. The only exceptions were well-established states: Ethiopia which did not participate in the slave trade and Benin which never became involved in a major way. In these states kinship institutions remained weak.
25. 'Ethnic associations' suggests organizations more broadly based than most of these are. The terms current in francophone countries, 'associations d'originaires' and 'associations de ressortissants', more accurately convey the basis of their membership.
26. The distinction is made by Cohen (1969: 208–9) who suggests that political ideology tends to take the form of a predominantly kinship ideology in segmentary political systems, while it tends to take the form of a predominantly ritual ideology in centralized societies. I rather focus on the relative position of men and women in patrilocal kin groups.
27. 'Situational change' and 'biographic change' refer to changes in the behaviour of individuals and may be distinguished from the changes a

society undergoes over time, or 'historical change'. For a discussion of this conceptual distinction, an account of its intellectual history, and an application to West Africa, see Gugler and Flanagan (1978: 73–104).

28. In addition to language, manifold clues allow the urban dweller to gauge the ethnic identity of a stranger: patterns of speech, even when a lingua franca is used; dress and ornaments; cultural modifications of physical appearance such as hair-style, beard, and facial scars; food and drink preferences; behaviour ranging from minor physical mannerisms to dance. Furthermore, particular occupations as well as residential areas are known to be the more or less exclusive preserve of one or another ethnic group. Ease in categorizing encourages the rapid establishment of interaction with group members and the maintenance of distance toward outsiders. Easily discernible characteristics guide participants in communal riots in selecting their victims.
29. For discussions of the literature on the African working class, see Freund (1988) and Copans (1987).
30. Achebe's novel *Things Fall Apart* presents the classic portrayal of the politics, patronage, and corruption of post-independence Nigeria.
31. Ngũgĩ's novel *Petals of Blood*, situated in Kenya, provides a fine exposition of the élite strategy of obscuring the emerging division of class by fostering resentment between ethnic groups.
32. For a review of the literature on military regimes and the democratization experience of the early 1990s, see Luckham (1994) who presents a summary of various types of 'democratic transitions' in a large number of African countries. Riley and Parfitt (1994) provide an account of recent political changes that focuses on Zambia, Côte d'Ivoire, and Zaïre.
33. I have argued previously that contemporary revolutions are largely urban in character (Gugler, 1988). However, in Uganda in 1986, in Somalia in 1991, and in Liberia in 1992 government forces were disorganized to such an extent that guerrilla armies were able to constitute themselves in rural areas. It is quite unlikely though that the rural elements in these revolutionary movements will bring forth policies that move beyond solemn declarations to effectively address rural needs on a basis of equality in relation to the demands forcefully articulated in the urban arena.

References

Abernethy, David (1988), 'Bureaucratic Growth and Economic Stagnation in Sub-Saharan Africa', in Stephen K. Commins (ed.), *Africa's Development Challenges and the World Bank: Hard Questions, Costly Choices* (Boulder, Colo.: Lynne Rienner), 179–214.

Achebe, Chinua (1958), *Things Fall Apart*, African Writers Series 1 (London: Heinemann).

Adam, Heribert and Moodley, Kogila (1996), '"Tribalism" and Political Violence in South Africa', in Josef Gugler (ed.), *Cities in Asia, Africa, and Latin America: Multiple Perspectives* (Oxford University Press).

Afonja, Simi (1990), 'Changing Patterns of Gender Stratification in West Africa', in Irene Tinker (ed.), *Persistent Inequalities: Women and World Development* (New York: Oxford University Press), 198–209.

Austen, Ralph A. (1990), 'Marginalization, Stagnation, and Growth: The Trans-Saharan Caravan Trade in the Era of European Expansion, 1500–1900', in James D. Tracy (ed.), *The Rise of Merchant Empires: Long-Distance Trade in the Early Modern World, 1350–1750* (Cambridge University Press).

Baker, Pauline, H. (1974), *Urbanization and Political Change: The Politics of Lagos, 1917–1967* (Berkeley: University of California Press).

al-Bakrī (1068) *Kitāb al-masālik wa'l-mamālik*, French trans. of parts dealing with the Western Sudan in Joseph M. Cuoq (1975), *Recueil des sources arabes concernant l'Afrique Occidentale du VIIIe au XVIe siècle (Bilād al-Sūdān)* (Sources d'Histoire Médiévale, Paris: Centre National de la Recherche Scientifique), 80–109.

Banton, Michael (1957), *West African City: A Study of Tribal Life in Freetown* (London: Oxford University Press).

—— (1965), 'Social Alignment and Identity in a West African City', in Hilda Kuper (ed.), *Urbanization and Migration in West Africa* (University of California Press), 131–47.

—— (1973), 'Urbanization and Role Analysis', in Aidan Southall (ed.), *Urban Anthropology: Cross-cultural Studies of Urbanization* (Oxford University Press), 43–70.

Barnes, Sandra T. (1986), *Patrons and Clients: Creating a Political Community in Metropolitan Lagos*, International African Library (Manchester University Press, and Bloomington, Ind.: Indiana University Press).

Bates, Robert H. (1981), *Markets and States in Tropical Africa: The Political Basis of Agricultural Policies* (Berkeley: University of California Press).

Becker, Charles M. and Morrison, Andrew R. (1993), 'Observational Equivalence in the Modeling of African Labor Markets and Urbanization', *World Development*, 21: 535–54.

Behrman, Jere R. and Wolfe, Barbara L. (1984), 'Micro Determinants of Female Migration in a Developing Country: Labor Market, Demographic Marriage Market and Economic Marriage Market Incentives', *Research in Population Economics*, 5: 137–66.

Bradshaw, York and Rita Noonan (1996), 'Urbanization, Economic Growth, and Women's Labor Force Participation: A Theoretical and Empirical Reassessment', in Gugler (ed.), *Cities in Asia, Africa, and Latin America*.

Brown, Barbara B. (1983), 'The Impact of Male Labour Migration on Women in Botswana', *African Affairs*, 82: 367–88.

Chandler, Tertius (1987), *Four Thousand Years of Urban Growth: An Historical Census* (Lewiston, NY: St. David's University Press).
Cohen, Abner (1969), *Custom and Politics in Urban Africa: A Study of Hausa Migrants in Yoruba Towns* (London: Routledge & Kegan Paul, and Berkeley: University of California Press).
Connah, Graham (1987), *African Civilizations: Precolonial Cities and States in Tropical Africa: An Archaeological Perspective* (Cambridge University Press).
Copans, Jean (1987), 'A la recherche d'une classe ouvrière', in M. Agier, J. Copans, and A. Morice (eds.), *Classes ouvrières d'Afrique noire* (Paris: Karthala and ORSTOM), 23–43.
Coquery-Vidrovitch, Catherine (1993), *Histoire des villes d'Afrique noire: des origines à la colonisation* (Paris: Albin Michel).
Diamond, Larry (1987), 'Class Formation in the Swollen African State', *Journal of Modern African Studies*, 25: 567–96.
Ekeh, Peter (1990), 'Social Anthropology and Two Contrasting Uses of Tribalism in Africa', *Comparative Studies in Society and History*, 32: 660–700.
Epstein, A. L. (1981), *Urbanization and Kinship: The Domestic Domain on the Copperbelt of Zambia, 1950–1956* (London: Academic Press).
Findley, Sally E. (1993), 'The Third World City: Development Policy and Issues', in John D. Kasarda and Allan M. Parnell (eds.), *Third World Cities: Problems, Policies, and Prospects* (Newbury Park, Calif.: Sage), 1–31.
Flanagan, William G. (1977), 'The Extended Family as an Agent in Urbanization: A Survey of Men and Women Working in Dar es Salaam, Tanzania', Ph.D. diss., University of Connecticut.
Freeman-Grenville, G. S. P. (1962), *The East African Coast: Select Documents from the First to the Earlier Nineteenth Century* (Oxford: Clarendon Press).
Freund, Bill (1988), *The African Worker* (Cambridge University Press).
Friedman, Steve (1987), *Building Tomorrow Today: African Workers in Trade Unions, 1970–1984* (Johannesburg: Ravan Press).
Gilbert, Alan and Gugler, Josef (1992), *Cities, Poverty and Development: Urbanization in the Third World*, 2nd edn. (Oxford University Press).
Gluckman, Max (1960), 'Tribalism in Modern British Central Africa', *Cahiers d'Études Africaines*, 1: 55–70.
Gugler, Josef (1969), 'On the Theory of Rural–Urban Migration: The Case of Subsaharan Africa', in J. A. Jackson (ed.), *Migration*, Sociological Studies 2 (London: Cambridge University Press), 134–55.
—— (1971), 'Life in a Dual System: Eastern Nigerians in Town, 1961', *Cahiers d'Etudes Africaines*, 11: 400–21.
—— (1975), 'Particularism in Subsaharan Africa: "Tribalism" in Town', *Canadian Review of Sociology and Anthropology*, 12: 303–15.
—— (1976), 'Migrating to Urban Centers of Unemployment in Tropical Africa', in Anthony H. Richmond and Daniel Kubat (eds.), *Internal Migration: The New and the Third World* (London: Sage), 184–204.

—— (1988), 'The Urban Character of Contemporary Revolutions', in Josef Gugler (ed.), *The Urbanization of the Third World* (Oxford University Press), 399–412; repr. rev. from (1982) *Studies in Contemporary International Development*, 17/2: 60–73.

—— (1996), 'Life in a Dual System Revisited: Urban–Rural Ties in Enugu, Nigeria, 1961–1987', in Gugler (ed.), *Cities in Asia, Africa, and Latin America*, repr. rev. from (1991) *World Development*, 19: 399–409.

Gugler, Josef and Flanagan, William G. (1978), *Urbanization and Social Change in West Africa*, Urbanization in Developing Countries (Cambridge University Press).

—— and Ludwar-Ene, Gudrun (1995), 'Gender and Migration in Africa South of the Sahara', in Jonathan Baker and Tade Akin Aina (eds.), *The Migration Experience in Africa* (Uppsala: Nordiska Afrikainstitutet).

Hansen, Karen Tranberg (1989), *Distant Companions: Servants and Employers in Zambia, 1900–1985*, Anthropology of Contemporary Issues (Ithaca, NY: Cornell University Press).

Harris, John R. and Todaro, Michael P. (1968), 'Urban Unemployment in East Africa: An Economic Analysis of Policy Alternatives', *East African Economic Review*, 4: 17–36.

—— —— (1970), 'Migration, Unemployment and Development: A Two—Sector Analysis', *American Economic Review*, 60: 126–42.

Hilton, Anne (1985), *The Kingdom of Kongo* (Oxford: Clarendon Press).

Hindson, Doug (1987), *Pass Controls and the Urban African Proletariat in South Africa* (Johannesburg: Ravan Press).

Hoddinott, John (1994), 'A Model of Migration and Remittances Applied to Western Kenya', *Oxford Economic Papers*, 46: 459–75.

Hollingsworth, Brian (1979), *Railways of the World* (New York: Gallery Books).

Hopkins, Anthony G. (1973), *An Economic History of West Africa* (London: Longman, and New York: Columbia University Press).

Ibn al-Mukhtār (1665) *Ta'rīkh al-Fattāsh*. French trans. Octave Victor Houdas and Maurice Delafosse (1913) *Rarikh el-Fettach ou chronique du chercheur pour servir à l'histoire des villes, des armées et des principaux personnages du Tekrour*, Documents Arabes Relatifs à l'Histoire du Sudan (Paris: Ernest Leroux).

International Labour Office (1972), *Employment, Incomes and Equality: A Strategy for Increasing Productive Employment in Kenya* (Geneva: ILO).

—— (1991), *The Urban Informal Sector in Africa in Retrospect and Prospect: An Annotated Bibliography*, International Labour Bibliography 10 (Geneva: ILO).

—— (1992), *1992 Yearbook of Labour Statistics* (Geneva: ILO).

—— (1994), *1994 Yearbook of Labour Statistics* (Geneva: ILO).

Jamal, Vali, and Weeks, John (1993), *Africa Misunderstood or Whatever Happened to the Rural–Urban Gap?*, Macmillan Series of ILO Studies (Basingstoke: Macmillan).

Leeds, Anthony (1973), 'Locality Power in Relation to Supralocal Power Institutions', in Aidan W. Southall (ed.), *Urban Anthropology: Cross-Cultural Studies of Urbanization* (New York: Oxford University Press), 15–41.

Lofchie, Michael F. (1996), 'The Rise and Demise of Urban-Biased Development Policies in Africa', in Gugler (ed.), *Cities in Asia, Africa, and Latin America.*

Luckham, Robin (1994), 'The Military, Militarization and Democratization in Africa: A Survey of Literature and Issues', *African Studies Review*, 37/2: 13–75.

Ludwar-Ene, Gudrun (1991), 'Spiritual Church Participation as a Survival Strategy among Urban Migrant Women in Southern Nigeria', in Gudrun Ludwar-Ene (ed.), *New Religious Movements and Society in Nigeria*, Bayreuth African Studies 17 (Bayreuth: Eckhard Breitinger), 53–67.

Mabogunje, Akin L. (1968), *Urbanization in Nigeria* (London: University of London Press).

—— (1990), 'Urban Planning and the Post-Colonial State in Africa: A Research Overview', *African Studies Review*, 33/2: 121–203.

Mauny, Raymond A. (1961), *Tableau géographique de l'Ouest africain au Moyen Age d'après les sources écrites, la tradition et l'archéologie*, Mémoires de l'Institut Français d'Afrique Noire 61 (Dakar: IFAN).

Mayer, Philip (1971 [1961]), *Townsmen or Tribesmen: Conservatism and the Process of Urbanization in a South African City,* Xhosa in Town 2, 2nd edn. (Cape Town: Oxford University Press).

Memon, Pyar Ali and Diana Lee-Smith (1993), 'Urban Agriculture in Kenya', *Canadian Journal of African Studies*, 1: 25–42.

Munro-Hay, Stuart (1991), *Aksum: An African Civilization of Late Antiquity* (Edinburgh University Press).

Nelson, Nici (1996), 'How Women and Men Got by and Still Get by (Only Not so Well): The Gender Division of Labour in a Nairobi Shanty-Town', in Gugler (ed.), *Cities in Asia, Africa, and Latin America*, repr. rev. from Ray Bromley and Chris Gerry (eds.) (1979) *Casual Work and Poverty in Third World Cities* (Chichester: John Wiley & Sons), 283–302.

Ngũgĩ wa Thiong'o (1977), *Petals of Blood,* African Writers Series 188 (London: Heinemann).

—— (1980), Caitaani *Mũtharaba-inĩ* (Nairobi: Heinemann); English trans. by the author (1982), *Devil on the Cross*, African Writers Series (London: Heinemann).

O'Connor, Anthony M. (1981), *Urbanization in Tropical Africa: An Annotated Bibliography* (Boston, Mass.: G. K. Hall).

—— (1983), *The African City* (New York: Africana Publishing Company).

Paden, John N. (1973), *Religion and Political Culture in Kano* (Berkeley: University of California Press).

Riley, Stephen P. and Parfitt, Trevor W. (1994), 'Economic Adjustment and Democratization in Africa', in John Walton and David Seddon (eds.), *Free*

Markets and Food Riots: The Politics of Global Adjustment (Oxford: Blackwell), 135–70.

Robinson, Derek (1990), *Civil Service Pay in Africa* (Geneva: ILO).

Rouch, Jean (1956), 'Migrations au Ghana (Gold Coast): Enquête 1953–1955', *Journal de la Société des Africanistes*, 26: 33–196.

Saad, Elias N. (1983), *Social History of Timbuktu: The Role of Muslim Scholars and Notables 1400–1900* (Cambridge University Press).

Sabot, R. H. (1979), *Economic Development and Urban Migration: Tanzania 1900–1971* (Oxford: Clarendon Press).

al-Sa'dī (1655), *Ta'rīkh al-Sūdān*, French trans. by Octave Victor Houdas (1900) *Ta'rikh es-Soudan*, Documents Arabes Relatifs à l'Histoire du Sudan 1 (Paris: Ernest Leroux).

Schatzberg, Michael G. (1988), *The Dialectics of Oppression in Zaïre* (Bloomington, Ind.: Indiana University Press).

Sklar, Richard L. (1983 [1963]), *Nigerian Political Parties: Power in an Emergent African Nation*, paperback ed. (New York: NOK Publishers International).

Smith, David A. and London, Bruce (1990), 'Convergence in World Urbanization? A Quantitative Assessment', *Urban Affairs Quarterly*, 25: 574–90.

Stark, Oded, and Lucas, Robert E. B. (1988), 'Migration, Remittances, and the Family', *Economic Development and Cultural Change*, 36: 465–81.

Stren, Richard E. (1992), 'African Urban Research Since the Late 1980s: Responses to Poverty and Urban Growth', *Urban Studies*, 29: 533–55.

—— (1994) (ed.), *Urban Research in the Developing World*, ii. *Africa* (Toronto: Centre for Urban and Community Studies, University of Toronto).

—— with Letemendia, Claire (1986), *Coping with Rapid Urban Growth in Africa: An Annotated Bibliography in English and French on Policy and Management of Urban Affairs in the 1980s*, Bibliography Series 12 (Montréal: Centre for Developing Area Studies, McGill University).

—— and White, Rodney R. (1989) (eds.), *African Cities in Crisis: Managing Rapid Urban Growth* (Boulder, Colo.: Westview Press).

Turok, Ben, and Maxey, Kees (1985 [1976]), 'Southern Africa in Crisis', in Peter C. W. Gutkind and Immanuel Wallerstein (eds.), *The Political Economy of Contemporary Africa*, 2nd ed. (Beverly Hills, Calif.: Sage), 243–78.

United Nations (1994), *1992 Demographic Yearbook* (New York: United Nations).

United Nations Development Programme (1995), *Human Development Report 1995* (New York: United Nations).

Van Velsen, Jaap (1960), 'Labor Migration as a Positive Factor in the Continuity of Tonga Tribal Society', *Economic Development and Cultural Change*, 8: 265–78.

Weisner, Thomas Steven (1972), 'One Family, Two Households: Rural–Urban Ties in Kenya', Ph.D. diss., Harvard University.

White, Rodney R. (1989), 'Conclusion', in Richard E. Stren and Rodney R. White (eds.), *African Cities in Crisis: Managing Rapid Urban Growth* (Boulder, Colo.: Westview Press), 305–12.

Wilks, Ivor (1975), *Asante in the Nineteenth Century: The Structure and Evolution of a Political Order* (London: Cambridge University Press).

Wiseman, John (1986), 'Urban Riots in West Africa, 1977–85', *Journal of Modern African Studies*, 24: 509–18.

FIG. 8.1. Urban Agglomerations in Latin America with more than 1 Million Inhabitants in 1990

8

Urban Development and Social Inequality in Latin America

ORLANDINA DE OLIVEIRA AND BRYAN ROBERTS

In this chapter, we analyse the patterns of urbanization in Latin America, concentrating on changes in urban social structure, and especially the changes in occupational structure. We document three component processes of urbanization: the rate and timing of urbanization in six major countries of the region; the different stages of industrialization; and the growing importance of the service sectors in the economy, both traditional services and modern ones linked to the growth of government bureaucracy and to contemporary business practices (administrative, financial, and technical).

In the developed world, these processes resulted in a convergence of their social structures: the expansion of the middle classes, the consolidation of an industrial working class, and improvements in the general welfare of the population. In the case of Latin America, urbanization, industrialization, and the changes in social structure have resulted in a greater heterogeneity in patterns of stratification, with marked regional contrasts within countries, and also divergences between countries. The dependence of the region on foreign technology and on external finance, combined with Latin America's role in the world economy as a supplier of primary commodities has resulted in an uneven development both between countries and between regions of the same country. We will emphasize these differences, paying attention to the specific situation of each country.

In terms of social stratification, there has been an uneasy relation between urban growth, economic development, and social equity. The cities multiplied and concentrated economic resources. Urban industrial growth stimulated an increase in levels of education, the proletarianization of the labour-force, and the expansion of non-manual sectors. It led to changes in customs, and, in general, to attitudes favourable to economic growth and to social change. On the other hand, this same urban growth did little to diminish the existing polarization of the social structure, either in terms of income or labour conditions, as shown by the persistence of non-waged forms of labour (self-employed workers, unpaid family workers) and highly skewed income distributions. The cities of Latin America at the end of the twentieth century, just as those at the beginning of the century, are scenes of social inequality. Opulence coexists with the poverty of large sectors of the population.

The socio-economic and political changes in the region drastically modified urban social stratification. Some social actors such as, for instance, independent professionals and craftsmen, became less important. Others became stronger such as the working class in industry and the services. New actors came on stage, such as the salaried middle class dependent on the state and on private enterprise. With these changes, the bases for the formation of collective identities altered, as did the social roots of urban politics.

To capture the heterogeneity of these changes in Latin America, we use data mainly from six countries—Argentina, Brazil, Chile, Colombia, Mexico, and Peru—which, in 1990, made up 76 per cent of the total population of Latin America and the Caribbean (ECLAC, 1993: table 103). We emphasize the contrast in the patterns of their development and urbanization, looking at the difference between their middle and working classes, showing how women's roles in the labour-market have changed, and seeing the implications of these processes both for social mobility and social inequality.

As background to our analysis in Latin America and as a benchmark to assess the contemporary situation, we begin by periodizing the changes from 1930 to 1990 in three stages. Remember, however, that, in reality, the years from 1930 to 1990 were ones of relatively continuous change. Moreover, the three stages overlap. The countries of Latin America differ in the timing of the stages and in the extent to which they were affected by the dominant trends of the period. The first stage, which began in the 1930s and ended at the beginning of the 1960s was that of the expansion and consolidation of industrial centres. In this period there was a strong industrialization based on import substitution of basic consumer goods—textiles, food and drink, and shoes. There were high levels of urban growth and an intense rural–urban migration. The industrial activities that expanded made an intensive use of labour. The second period began at the end of the 1950s and ended in the early 1970s. It was characterized by a marked internationalization of the urban economies, a new stage of industrialization, the growth of commercial agriculture based on modern inputs, and the stagnation of the peasant sector. The phase of import substitution of basic consumer goods began to exhaust itself, and investments concentrated in intermediate and capital goods industries. The technology needed by these industries was expensive and often could be obtained only through association with foreign companies. This new stage of industrialization, compared with former ones, was more capital- than labour-intensive, leading to linkages with multinational companies. This shift in emphasis was in part produced by changes in the organization of the world economy, in which multinational companies integrated hitherto fragmented national markets and, in developing countries, sought new markets for their products.

Finally, the third stage can be dated from the sharp rise in oil prices in the early 1970s, its consequences for world financial markets, and the growth of Latin America's external debt. Through the 1970s and 1980s, the Latin American economies became more highly dependent than ever before on external financing as a result of factors such as capital flight and the high monetary costs of attempting to develop rapidly industrial and urban infrastructure with imported technology and capital goods. Seeking to meet the requirements of international creditors and wanting new investments, Latin American governments enacted austerity measures. Markets were liberalized, reducing tariff protection, privatizing state-owned companies, and easing restrictions on foreign investment (Canak, 1989). The resultant economic restructuring altered the political as well as the economic balance in Latin America. Direct state control over the economy diminishes. Private sector growth is encouraged in both manufacturing and the services. Export-oriented enterprise flourishes, while enterprise dependent solely on the internal market stagnates. The corporate institutions that hindered the development of a free market, such as trade unions or social property sectors, are weakened by government action throughout the region, from Argentina and Chile in the south to Mexico in the north.

The most dramatic signs of the impact of economic restructuring in Latin America came in the 1980s. These were years of economic crisis for Latin America that resulted in an abrupt halt to the modernization of certain sectors of the urban economies. The crisis had negative consequences for per capita incomes and employment. Together with high rates of inflation and the inadequacy of social security, these contributed to a marked decline in the standards of living of the region's population. The crisis manifested itself as an urban crisis. The withdrawal of urban subsidies of various kinds together with a general reduction in government expenditures led to a deterioration in urban infrastructure and services and to increasing environmental problems. Also, social problems became more evident. Many of the cities of Latin America were unquiet places in the 1980s with high rates of urban violence, frequent protests against price rises, and outbursts of looting. Social and economic exclusion was visible in the resurgence of street begging in its many forms and the proliferation of street sellers of all kinds. Apart from these immediate consequences, the crisis marked the relative failure of the social mobility that was apparent in the 1960s and 1970s, based on improvements in consumption resulting from economic growth and on high rates of occupational mobility from agricultural to non-agricultural jobs and from manual jobs to non-manual ones (Durston, 1986; CEPAL, 1989).

Changes in the Pattern of Urbanization, 1930–1990

At the beginning of the 1930s, Latin America was still a predominantly rural region in terms of where its population lived and in terms of economic activity. The important cities of the period depended, with few exceptions, on their links to the agricultural sector. Some of them, such as Buenos Aires, São Paulo, or Medellín, prospered through the commercial and transportation activities associated with the new agricultural exports, such as coffee, cereals, and meat, that developed at the beginning of the century. Others were mainly regional administrative and commercial centres in which landowners had their main residences. The economic activities of these centres chiefly served the rural economy and population.

National differences in urban systems were also important since countries differed considerably in area, population, and level of economic development. These differences combined with the nature of their external trade, had already produced sharp contrasts between, for example, Buenos Aires or São Paulo bustling with European migrants and the wealth brought by the export economy, Lima or Mexico City whose period of rapid population growth was only just beginning and was to be based on internal migration, and the much smaller capitals of the Central American countries, all of which were below 100 000 in population by 1930.

Many of the important urban centres of the 1930s and 1940s succeeded in attracting new economic activities and consolidating their position. Others, in contrast, stagnated. Some towns, previously insignificant, were able to compete with established centres. In this section, we will not deal with the specifics of this process, but consider some general traits of Latin American urbanization, notably the changes in the rate and pattern of urban growth through time.

The 1930s saw the beginnings of fundamental changes in the spatial distribution of population in the region. In these years, Latin America was still linked to the world economy, though insecurely, through the export of raw materials and the import of manufactured goods. The world recession and the Second World War initiated a gradual change in this situation by stimulating import-substitution industrialization. Combining with the modernization of agriculture, this industrialization gave rise to a rapid urbanization based on the rural–urban migration that began, on a large scale, in the 1940s.

Urban growth in Latin America has been more rapid than that of the advanced industrial world in its comparable period of growth. For example, England in the nineteenth century experienced an urban 'explosion'. Yet in no decade of that century did the rate of urban growth exceed 2.5 per cent (Lawton, 1978: table 3.2). In Latin America, the rates of urban growth have been substantially higher throughout our period. There have been dramatic changes in the levels of urbanization and in urban structure in the four decades, 1940–80 (Table 8.1).

In 1940, only 37.4 per cent of the population of the six countries we are considering lived in urban areas. Many of these were little more than villages that served as administrative centres of a rural area. In contrast, the figure had risen to 69.5 per cent by 1980. In 1940, the urban structure was highly polarized: small and medium-size towns had 20 per cent of the total population and the metropolitan centres contained 13.1 per cent, whilst intermediate cities had only 4.3 per cent of the total population. By 1980, the distribution of the urban population had diversified: the relative weight of intermediate cities increased, reaching 19 per cent of the total population, though metropolitan concentration continued, and the urban system of Latin America remained highly unbalanced.[1] The overall rates of increase of the urban population remained high throughout our period and reached their highest levels (4.6 per cent) in the 1950s. As we will see in the next section, a large part of this urban growth was due to migration from rural areas, and from smaller urban places to larger ones.

TABLE 8.1 Distribution of Population and its Growth in Six Latin American Countries, 1940–1980 (percentages)

Category	Distribution of population (1940)	Distribution of population (1980)	Annual rate of growth (1940–80)
Rural	62.6	30.5	0.08
Urban[a]	37.4	69.5	4.1
towns[b]	20	23.3	3
intermediate cities[c]	4.3	19	6.3
metropolises[d]	13.1	27.2	4.4
TOTAL POPULATION (m)	(95.7)	(268.3)	2.6

[a] 'Urban' follows the census definition of each country, usually administrative centres and places of more than 2000 population.
[b] Towns are urban places of less than 100 000.
[c] Intermediate cities are between 100 000 and metropolis.
[d] Metropolis is defined as those cities with more than 2 million inhabitants in 1985.

Sources: Estimated from the population censuses of Argentina, Brazil, Chile, Colombia, Mexico, and Peru.

The Significance of Migration

At the apogee of urbanization in Latin America rural–urban migration became the salient issue in public policy discussion and in social science research. This research concentrated on issues such as the differences between migrants and natives, migrant assimilation to urban life, and the contribution of migrants to the urban labour-markets. The main features of this migration are summarized in various texts, but can be outlined as follows (Roberts, 1979; Butterworth and Chance, 1981). In general, migration flowed from villages to the large cities, bypassing intermediate size places. Migrants were concentrated in the economically active ages, particularly between 18 and 25. Female migrants tended to outnumber male migrants, partly as a result of the opportunities for domestic service in the large cities. Most migrants came initially alone, but were followed by other family members, establishing bridgehead settlements in particular urban neighbourhoods.

Push and pull factors were important in these migration flows. Small-scale agriculture throughout Latin America was stagnant in the period. Declining plot size and the semi-subsistence nature of peasant production meant that the growing rural population could not be maintained through farming alone. The poverty of most rural areas, and their lack of infrastructure, meant that other economic activities, such as trade, transport, and craft industry, could not provide an adequate supplementary income to retain population.

The diversification of rural economic activities was, however, a marked feature of the ways in which rural households attempted to adapt to population pressure and inadequate land resources. In most parts of Latin America, this resulted in extensive social networks linking farm to town and city through family-based trading and transport enterprises or through labour migration. These activities helped sustain the household economy in the village, but also channelled further migration to the city. The operation of these types of network is described in Smith (1989) for Peru, by Arizpe (1982) for Mexico, and is reviewed in Butterworth and Chance (1981: ch. 5).

The expansion of commercial agriculture also contributed to population expulsion, displacing peasant farming, either directly as in the case of ranching or indirectly through price competition. Income opportunities grew rapidly in the cities, partly as a result of the industrialization of the years from 1940 and partly because of the 'urban bias' of economic development in Latin America that concentrated infrastructural investment in the large cities. The growth of these cities demanded not only skilled labour, but also unskilled and semi-skilled labour in construction and in the services.

The impact of migration was different between countries. United Nations (1980: table 11) estimates of the contribution of migration and reclassification to urban growth show that migration was most important to urban growth in Argentina from 1947 to 1960 (50.8 per cent). The next most important contribution of migration to urban growth was in Brazil from 1950 to 1960 (49.6 per cent) and from 1960 to 1970 (44.9 per cent), Peru from 1961 to 1972 (41.6 per cent), Chile from 1952 to 1960 (36.6 per cent) and from 1960 to 1970 (37.4 per cent), Colombia from 1951 to 1964 (36.6 per cent), and Mexico from 1960 to 1970 (31.7 per cent). The contribution of migration to urban growth was greatest throughout most of the period in Brazil, followed by Chile and Colombia, least important in Mexico, in Peru, except in the 1960s, and in Argentina except in the 1950s. As for metropolitan growth, rural–urban migration was a relatively insignificant factor only in Argentina, accounting for about 20 per cent of the growth of Buenos Aires between 1940 and 1980. In Mexico, migration, though more important to metropolitan growth than in Argentina, was never the principal component, mainly because of the high rates of natural increase of the population. In contrast, migration, not natural increase, contributed half or more to metropolitan growth in Brazil and Peru in two of the periods between 1940 and 1970, and in Colombia throughout this period.

International migration also became significant in the region and, again, its importance varied among countries. In Mexico, the permanent or temporary migration of people from rural areas to the United States probably helped diminish the flow of migrants to the Mexican cities (Massey *et al.*, 1987; Cornelius, 1991; Garcia y Griego, 1990). In Argentina, the role of international migration has been a different one, increasing the contribution of rural migration to urban growth. In Buenos Aires, a large part of the foreign migrants reported in 1970 were from Bolivia and Paraguay, probably from the rural areas of those countries. In contrast, internal migration to Buenos Aires in the same year principally came from urban areas (Marshall, 1978: 88; Lattes, 1984; Balán, 1985).

From the 1970s, rural–urban migration became a less important issue in urban development, while inter-urban, intra-urban, and international movements became more significant themes. This tendency was clearest in those countries which had high levels of urbanization and in which the rural population has declined in absolute numbers, such as in Argentina. Bearing these country differences in mind, we will use Mexican migration patterns to illustrate some general tendencies.

In the 1970s, despite a change in the direction of migration towards the north of Mexico, the metropolitan area of Mexico continued to show a net gain from in- and out-migration flows. The gain through migration was, however, concentrated in the municipalities of the state of Mexico which are part of the city's metropolitan area. From 1970, the Federal District, the old core of Mexico City, lost both male and female population (Cantú and González, 1990; Corona, 1988). Out-migration from the Federal District to other urban areas and to the United States increased in the 1970s and, even more sharply, in the 1980s. Migration to the United States, mainly illegal, increased in the 1980s due to Mexico's economic crisis. The origins and skills of these migrants diversified to include more people of urban origin and with higher educational levels than had previously been the case (Bean *et al.*, 1990; Durand and Massey, 1992).

Gender has an atypical impact on migration patterns in Latin America. In comparison with other developing regions, Latin America has the most feminized net urban in-migration. For every three rural–urban male migrants there were four female migrants. Argentina is the only Latin American country which had a balanced sex ratio in net urban in-migration. The age pattern of female net urban in-migration resembles that of males, except that the peak age for female migration is approximately five years less (Singelmann, 1993). The highest rates of net urban in-migration for females occur at the ages 15–19 years and 20–24 years. Although net urban in-migration in most Latin American countries became more male-oriented during the 1970s, the region continued to have the lowest sex ratio by the end of the 1970s (Singelmann, 1993).

In Mexico, women predominated in internal migrations from 1930 onwards (Brambila, 1985; Corona, 1989; Goldani, 1977). Female migrants are 10 per cent more numerous than male migrants in interstate migration from 1940 to 1980. Figures from the 1970s show that, at all ages, there are more women migrants to Mexico City than men. This predominance is most marked in the ages 10–19. Women tend to migrate at earlier ages than do men, especially from those places nearest the capital. Female migration is as important to the new, externally oriented economic centres in the North, Centre, and Coast, as to the major metropolitan areas (Chant, 1991; Fernandez-Kelly, 1983). The tourist centre of Puerto Vallarta has proved especially attractive to women migrants who head their own households because of the numerous income opportunities available to them in trade and services (Chant, 1991).

The reasons for female migration are varied, including search for job opportunities, to study or to accompany other family members, whether husbands, parents, or children. Some of this migration is a direct result of changes in patterns of consumption and production. Industrialized foodstuffs and other manufactures have displaced home production, thus freeing some female labour from the domestic sphere. Changes in family norms of residence, marriage, and inheritance have reduced the restrictions on single females leaving their families. Because of the authority structure of the rural family and the subordinate place of women within this, it is often the young daughter that migrates, while the mother concentrates on looking after the young children, domestic chores, and agricultural tasks (Arizpe, 1978, 1985).

The rural women who migrate to the large cities are principally employed in activities that demand little education and few skills. In 1970, more than 40 per cent of the women who were economically active in Mexico City were migrants, mainly from rural areas though some came from other Mexican cities. Those from the rural areas were generally from impoverished peasant families, while those from urban areas were usually daughters of professionals, clerical workers, or factory workers. Most women who migrate alone from the countryside to Mexico City work first as domestic servants (Leff, 1976). The case of Mazahua women migrants to Mexico city—the so-called 'Marías'—illustrates another pattern of migration. Those Mazahua women who come from a village in which land still provides a basic subsistence, come as seasonal migrants, with husbands and children, and sell fruit in the street. Their male companions work as casual labourers in construction or as unskilled service workers (Arizpe, 1978). Some of these families end by remaining in the city.

Because of their social background and their high levels of education, urban-born women have a greater range of possibilities when they migrate to other towns and cities. The majority work in white-collar occupations in commerce, public administration, educational, financial, and health services. These activities grew rapidly in the large Mexican cities, as elsewhere in Latin America, as a result of the expansion of government employment and the modernization of economic and social infrastructure. The migration of urban-origin female migrants to the northern border cities has been particularly marked in the 1980s, seeking work in the expanding in-bond industries (Carrillo Huerta, 1990).

The heterogeneity of female migration in Mexico is also present in other metropolises of Latin America (Recchini de Lattes and Mychaszula, 1993). The stereotype of the young female rural migrant who comes alone to the city to work as a domestic servant is not as applicable as it was in the past. As Raczynski (1983) shows for Chile, this is due to the diversification of urban employment, the importance of inter-urban migration, and the expansion of education for women.

The Diversification of the Urban System

The growth of the smaller urban places (the towns in table 8.1) was slower than that of the larger places throughout the period, suggesting that much rural–urban migration bypassed the smaller urban places to move directly to the bigger cities. The analysis of the growth of the metropolises by decade, not included in Table 8.1, shows that their growth was rapid between 1940 and 1970, but there was a marked decline in growth rates in the 1970–80 decade. By the end of the 1980s, there was a reduction in urban primacy in these countries (Portes, 1989).[2]

Intermediate cities had the highest rates of increase throughout the period, growing much faster than both the small cities and the large metropolitan centres. The high rates of growth of intermediate cities was due in part, however, to increases in the number of such cities as smaller towns grew and passed into the intermediate category. The United Nations (1980: table 21) estimates that 15 per cent of the growth of cities with over 250 000 population was due to graduation from a smaller-size class between 1970 and 1975. The rate of increase of individual intermediate cities was, at times, lower than that of the metropolitan centres. The growth of intermediate cities became associated with greater urban specialization accompanying the recent phases of industrialization. In the 1970s, the increasing complexity of the industrial structure, with the production of intermediate and capital goods, resulted in the location of new plants outside the large cities. For example, the big steel plants of Brazil and Mexico were located in secondary cities. The automobile industry and heavy engineering were also located outside the metropolitan centres: Córdoba in Argentina, Puebla, Toluca, Saltillo, and Hermosillo in Mexico, and to the intermediate cities outside the city of São Paulo.

Two further tendencies reinforced this dispersion of industrial activities. First, the separation of administrative functions from those of production occurred in Latin America just as in the advanced industrial world. Administrative offices were located in the large cities where modern services were found, whilst manufacturing facilities were placed where land was cheap, infrastructure adequate, and subsidies provided. The increased emphasis in the 1980s on export manufacturing in certain countries contributed further to the dispersion since plants producing for export had little incentive to locate near the major internal markets. Multinational companies in particular sought out low-cost areas for their plants. The automobile and autoparts industry of Mexico began to transfer its operations to the northern border region to be closer to the US market, becoming closely integrated with US automobile production (Carrillo, 1989). In Brazil, Manaus, in the heart of the Amazonian region, attracted large-scale assembly industries for electronics as a result of the establishment of a duty-free area (the Zona Franca) in 1967 (Despres, 1991: 29–37). The electronic firms imported 74 per cent of their capital goods and 73 per cent of their assemblage components from the industrial south of Brazil. They exported the bulk of their products, either directly or through the many tourists who came to buy at the low prices of the zone (ibid. 48). This pattern of industrialization has led to the rapid growth of Manaus with a regional concentration of population in the city that is much higher than at the time of the rubber boom at the turn of the century which, though it led to Manaus's early splendour, with opera house and trolley lines, was associated with a more dispersed pattern of Amazonian settlement (ibid. 23–46).

The major cities of the six countries that we analyse differed in their trajectory during the period. There were 'successes' in terms of cities that, industrializing early on the basis of protected markets, soon approximated the economic and social organization of the advanced industrial countries. Some of these cities, such as São Paulo, were to retain their importance in face of the lowering of tariffs on industrial products and increasing economic integration on a world scale. In the case of other cities, such as Buenos Aires, early success did not guarantee a smooth transition to becoming leading regional and international industrial or service centres. There were also the relative 'failures', provincial centres and some national capitals, such as Lima, which neither industrialized substantially in the early phases nor captured, at later phases, specialized economic functions demanded by the new economic order.

Comparing the six countries shows significant differences between them in the levels of urbanization and in the rates of urban growth. The most important contrasts are between those countries which, starting with high levels of urbanization, have had relatively low rates of urban growth in the last four decades and those which beginning with a low level of urbanization have subsequently had high rates of urban growth. Amongst the first group are Argentina and Chile, which in the 1940s were the most urbanized, with 61 per cent and 52 per cent of their population living in urban areas. The rates of urban growth of these two countries between 1940 and 1980 have been the lowest of the six. Chile experienced the more rapid rate of urban growth in the 1950s, had rates above 3 per cent in the 1960s and lower rates in the 1970s, reaching Argentina's level of urbanization by 1980, with more than 80 per cent of its population in urban areas. The four countries with the highest rates of urban growth between 1940 and 1980 had between 30 and 35 per cent of their population living in urban areas around 1940. By 1980, this proportion had increased to around 65 per cent. The period of most rapid urban growth varied, however, from country to country: Mexico experienced it in the 1940s; Colombia and Brazil in the 1950s, and Peru in the 1960s (Oliveira and Roberts, 1994: app. I).

Though these differences in timing were important, equally strong were the contrasts between the urban systems of each country. The contribution to urban growth of the metropolitan centres, of intermediate cities, and of towns (urban places of less than 100 000 population) showed marked contrasts from one country to another. Geography and population size were factors accounting for why Brazil and Mexico were the only countries to have more than one city of more than two and a half million population by 1985. The proliferation of large cities was, however, a common feature in most countries of Latin America. In Argentina, intermediate cities grew, as a class, faster than Buenos Aires in the four decades. Buenos Aires was the only metropolis in the region that lost in relative importance between 1940 and 1980. The predominance of Lima continued in Peru, but this must be set against the increasing significance of intermediate cities. In the 1940s, Peru had no intermediate city, while by 1980, eight cities had more than 100 000 population. The growth of intermediate cities, such as Arequipa, Trujillo, Chiclayo, Chimbote, and Huancayo, was marked and occurred at the expense of smaller urban places. Colombia, where intermediate cities already had a substantial presence in 1940, showed the strongest decline in the weight of the towns. There, intermediate cities, such as Medellín, Cali, and Barranquilla, concentrated most of the urban population by 1980 despite the rapid growth of Bogotá in the last decades. In Mexico intermediate cities also grew rapidly, as we noted above, proliferating in these years: there were seven in 1940 and fifty-two in 1980 containing 30 per cent of the urban population.

Chile and Brazil had the most polarized urban systems by 1980, but for different reasons. The sparse population of most regions of Chile was an insufficient base for the growth of intermediate cities that could counter population concentration in the central valley, in Santiago, and its port of Valparaiso. In Brazil, intermediate cities grew rapidly, but this growth was not sufficient to offset early population concentration in towns and the strong growth of its six metropolitan centres—São Paulo, Rio, Belo Horizonte, Pôrto Alegre, Salvador, and Recife. Preliminary analyses of the 1991 Brazilian Census show, however, that the metropolitan regions grew much less rapidly from 1980 to 1991 than in the previous decade. The fastest growing size class was municipal districts with a population of between 100 000 and 500 000 (IBGE, 1992).

These contrasts were based on differences in agrarian systems and their consequences for rural–urban migration. Colombia and Peru—the countries in which the relative decline of the town was most marked—had, in the 1940s, a predominantly peasant agrarian structure. The majority even of the urban population in these countries lived in small towns closely linked to agriculture and to craft production. The stagnation of the peasant economies of these countries, together with the weakness of the local economy, resulted in large-scale migrations to the major regional urban centres, without needing the pull of urban economic development. The continuing polarization of Brazil's urban system was due to marked regional differences: in the north-east, the decline of agricultural production (both plantation and peasant) stimulated migration to the cities of the south and concentrated the regional population in local centres that reached metropolitan size, such as Recife, Salvador, and Fortaleza. In contrast, in the centre-south and south of the country—Minas Gerais, São Paulo, Paraná, and Rio Grande do Sul—an economically dynamic agriculture and small-scale commerce was integrated through a local transport and trading network. This pattern of regional development, that included continuing industrialization, sustained an urban system based on towns, intermediate-size cities, and metropolises.

The growth and modernization of regional economies appeared more clearly in the case of Argentina, in face of the declining importance of Buenos Aires. It was also apparent in Chile which, though highly centralized in Santiago, contained a number of dynamic local economies in which the agricultural labour-force lived, at times, in towns, as in the case of the valley of Putaendo (Rodriguez, 1987). Mexico showed a combination of all these tendencies: zones in which the peasant population was expelled to Mexico City, to other regional centres or to the United States; zones in which commercial agriculture combined with new industries such as tourism, automobile production, and microtechnology to sustain the growth of cities linked to the export economy.

To provide a detailed overview of these trends, Table 8.2 shows the patterns of growth of Mexican urban areas for each decade from 1940 to 1990. The classification into types is meant to illustrate some of the major factors affecting the development of Mexico's spatial economy.[3]

The first type, the primate city, is Mexico City, the nodal centre of economic and demographic concentration in Mexico since before the Conquest. Then there are the regional metropolises, Guadalajara and Monterrey, representing the only cities that have, over time, stimulated a relatively autonomous regional development. The other types represent the major economic and ecological trends affecting the shape and growth of the Mexican urban system in the period, 1940–90. The fringe of Mexico City results from the decentralization of economic activity and the dispersal of population to cities neighbouring the capital. This occurred as a result both of government policy and the

TABLE 8.2 Growth of Mexican Cities, 1940–1990[a] (percentages)

City type	Annual rate of growth				
	1940–50	1950–60	1960–70	1970–80	1980–90
The primate city	5.2	4.9	5.1	4.3	1
The regional metropolises	5.1	6.4	5.6	4.3	2.3
The fringe of Mexico City[b]	3.2	2.7	4.8	4.4	3.3
The old intermediate cities[c]	4	2.3	2.7	4.2	2.7
The oil cities	2.7	4.4	5.1	4.1	2.6
The northern border cities[d]	9.9	7.9	4.9	3.2	3.5
Interior cities of the outward-looking economy[e]	3.7	4.7	5.1	4.5	3.6
All other cities	4	4.3	4.9	4.2	2.7
Total cities	4.6	4.8	5	4.2	2.1
TOTAL POPULATION	2.7	3	3.2	3.3	1.9

[a] These are the 52 cities that had over 100 000 population in 1980, based on central city and its conurbation.

[b] The cities on the fringe of Mexico City are Cuautla, Cuernavaca, Pachuca, Puebla, Toluca, and Querétaro.

[c] The old intermediate cities are the three cities that had over 100 000 population in 1940 and do not appear in another category of the Table: Torreón, San Luis Potosí, and Mérida.

[d] The border cities are the six border cities with 100 000 or more population in 1980: Ciudad Juárez, Matamoros, Mexicali, Nuevo Laredo, Reynosa, and Tijuana.

[e] The interior cities of the outward-looking economy are Acapulco, Aguascalientes, Celaya, Chihuahua, Hermosillo, Irapuato, León, and Saltillo.

Sources: Garza and DDF (1987: table 4.2) for city size between 1940 and 1980; INEGI (1990), for the 1990 figures, estimating conurbation size following definitions in M. Negrete Salas and Héctor Salazar Sánchez (1986: table A-1).

diseconomies of scale associated with population concentration in the megalopolis. The old intermediate cities are the historically important regional centres in Mexico, though various of these have been placed in other categories, as is the case of Puebla, classed as part of the fringe of Mexico City. The oil cities are the cities that developed as part of the growth of the oil economy, providing refining capacity, support services and distribution. They are mainly located on the Gulf coast. The northern border cities grew as the result of the increasing importance of the border with the United States, first for trade, then tourism, and finally assembly industries. The interior cities of the outward-looking economy are those that we judged to have received substantial urban investments in the 1970s and 1980s, relative to their size, aimed at taking advantage of the export market in manufacturing, agro-industry, and tourism.

Rapid population concentration in Mexico City is marked from 1940 to 1970, but it is rivalled by the growth of the two other major metropolitan areas. In 1940, the combined populations of Guadalajara and Monterrey were only a fifth of Mexico City's two million population. The two regional metropolises grew rapidly in the following decades, absorbing population from their extensive hinterlands and playing an important role in the development of import substitution in Mexico: Guadalajara through basic goods for the regional and national markets; Monterrey through intermediate and capital goods industries. From 1970, Mexico City's growth begins to decline. This is marked in the 1980–90 decade, though the extent of the decline may be exaggerated by the lack of census comparability (Corona, 1991).[4] Note, however, that the cities on the fringe of the capital increase their growth rate from 1960 on. In the two following decades, they are amongst the fastest growing cities. The cities that grow relatively slowly throughout the period are the 'old' intermediate cities, provincial centres whose economies were relatively stagnant when compared to the metropolitan areas and the cities based on new economic opportunities. With the nationalization of oil in the 1930s and the discovery of important new deposits, Mexico began an intensive period of developing port and refinery facilities that stimulated urban development, especially in the 1960s and the 1970s. The rapid growth of the northern border cities from 1940 onwards is explained partly by the fact that in 1940 they were very small places (for example, Tijuana with 17 000 population). More important, their location became economically strategic as Mexico consolidated its customs frontier with the United States through the development of import substitution, the resultant international trade flows, and, subsequently, through in-bond industries (*maquiladora* plants). The interior cities of the outward-looking economy have had a rate of growth that has been higher than most Mexican cities throughout the period. In the 1980s they became the fastest growing cities. These cities vary in location and economic significance, including northern cities, cities of the centre, traditional industrial cities, centres of commercial agriculture, of mining, and tourist centres. Their previous economic dynamism in combination with favourable location, assisted, often, by the initiative of local political élites, enabled them to seize the opportunities offered by the opening of the Mexican economy.

As a consequence of these various trends, there are interesting changes in the pattern of urban growth in the last decades. Most marked is the sharp decline in the growth of the big metropolitan areas, particularly Mexico City, though the other metropolises barely exceed the natural increase of the population. The rapidly growing cities, in which migration makes an important contribution to urban growth are of three types: the cities on the fringe of Mexico City, the northern border cities, and, above all, the interior cities of the outward-looking economy. The Mexican case illustrates the ways in which changes in the global economic context diversify the patterns of urban growth. The primacy of the system declines. Intermediate cities grow faster than the major metropolitan areas, acquiring their own economic niches in the national and international economy.

This shift from the concentration of population and economic activities in a few urban places towards a more diversified, and specialized urban system took place throughout the region. It did not occur to the same extent in each country, nor follow the same pattern, producing contrasts between urban systems. In the next section, we outline some of the major contrasts within the overall pattern described above.

The Transformation of Urban Space

The trends in urban spatial organization are the final components of the context of Latin American urbanization that we will briefly outline (Hardoy, 1975; Gilbert and Ward, 1985). Urban land use patterns in Latin America between 1930 and 1990 were rarely neatly ordered by such market factors as urban land rent gradients tied to the costs and benefits of central location (Yujnovsky, 1976). The residential and economic land use patterns within Latin American cities became, if anything, more mixed than they had been at the beginning of the period. Spatial organization, and the changes therein, provided only a loose framework channelling social and economic interaction.

In 1930, the normal pattern for towns and cities in Hispanic America was to be organized around a central square near to, or around, which were found the major government offices, the principal religious edifices, the mansions of the élite, and the major commercial establishments (Hardoy, 1975). Distance from this centre meant, on the whole, declining social importance, with respectable urban craftspeople inhabiting the next ring of buildings which provided both shelter and a place of business. On the outskirts of the city were found the poorest urban inhabitants who worked as day labourers, street sellers, or by offering a variety of personal services. The proximity to the countryside meant that the urban outskirts merged economically as well as spatially with the rural world, with inhabitants cultivating kitchen gardens or working as paid labourers in agriculture.

This account was truer for the older and less dynamic cities than for those that were industrializing in the 1920s and 1930s. Already, the élites of cities such as São Paulo, Buenos Aires, and Mexico City, had begun to move out from the centre to neighbourhoods that were free of the noise and pollution of the busy centres. The 'frontier' cities of the 1930s were already spatially heterogeneous with industry, business, and housing sharing space. Rich and poor lived in close proximity.

Economic and population growth was to foster heterogeneity. Because of low wages and the precarious nature of economic development, housing was rarely purpose-built for the working classes. Even in the few cities where such housing appeared—Buenos Aires, São Paulo, Monterrey—this housing covered only a fraction of these classes. The working classes found what housing they could—through the subdivision of the abandoned mansions of the rich as in some of the *vecindades* of Mexico City or through the intensive occupation of other central spaces (Ward, 1990: 42–62). Increasingly, they sought out alternative forms of cheap shelter, such as self-construction after invading land or semi-legal purchases from property speculators (Roberts, 1973*a*). This land occupation was not spatially ordered since, though most unoccupied space was on the urban periphery, its availability depended on political factors such as whether it was public or private, the strength of popular organization, and the speculative intentions of its owners—what has been described as the 'logic of disorder' (Kowarick, 1977, 1979; Rolnik, 1989; Valladares, 1989). Moreover, some of the unoccupied spaces were in the centre of the city, as in the case of the hillsides of Rio de Janeiro or the ravines of Guatemala City. Despite self-construction, renting continued to be the major means of access that the poor had to shelter, with its incidence probably increasing towards the end of our period as even squatter settlements became a 'normal' part of the city and the original owners rented out space as a means of supplementing their incomes (Gilbert and Varley, 1990; Gilbert and Ward, 1985).

The flight of the middle and upper classes from the centre of cities was tempered by poor communications and inadequate infrastructure in prospective suburban areas. Also, the proximity of squatter settlements to most middle-class suburban subdivisions diminished their social exclusivity. Further complications were created by the economic heterogeneity of the cities. The persisting importance of informal economic activities, to be discussed later, meant that small-scale industrial and commercial establishments were to be found throughout the city, in middle-class as well as in working-class neighbourhoods, as entrepreneurs saved the cost of renting space by using part of their family residence to conduct their business.

By 1990, conflicting tendencies were apparent in urban spatial organization with varied consequences for social segregation. Portes (1989: 22) points to a 'qualitative leap' in class polarization in Santiago when he contrasts its patterns of spatial polarization with those of Bogota and Montevideo. In Santiago, polarization was a result of the land market and urban administrative policies of the military government. In contrast, though patterns of residential segregation according to occupation and income had clearly emerged by the 1970s in Montevideo and Bogotá, they were partially reversed in the 1980s as economic crisis led most classes to seek affordable shelter irrespective of location. In other cities, such as Rio de Janeiro and São Paulo, the occupation of space had become more socially and economically mixed as high-rise middle-class housing was built in poor areas. Poor squatters were expelled to make way for new middle-class subdivisions, while the poor sought whatever niche they could find within established residential neighbourhoods (Rolnik, 1989; Valladares, 1989). Though substantial progress had been made up to 1980 in the provision of basic urban services, such as water, electricity, and sewage disposal, a substantial part of the urban population in most of the major Latin American cities still remained without adequate access (Edel and Hellman, 1989; Ward, 1990: 138–77). Much of the progress was due to popular movements, and their pressure on urban and national governments.

A starting point for understanding the diverse patterns of urban social segregation is the fact that the urban land market is an integrated one, to which even the supposedly 'marginal' squatter settlements belong (Gilbert and Ward, 1985). It is this integration that underlies spatial heterogeneity. Over time, land and housing will be bought and sold in squatter settlements, or housing rented, at prices that reflect, though imperfectly, factors such as distance from centres, space, facilities, as well as the social 'cachet' of an area. This land market is only likely to segregate spatially low-and middle-income groups when absolute prices of segments of the residential land market are sharply differentiated (Gilbert and Ward, 1985; Ward *et al.*, 1993). Piecemeal urban development creates, however, a mosaic of tenure and infrastructural conditions even within a relatively small urban area.

The heterogeneity of most urban areas and the relatively low incomes earned by most urban households mean that middle-income households are likely to make trade-offs of space or convenient location against the low status and poor services of an irregular settlement. The money saved can, in any event, be used to purchase services, such as private education for the children, that may be seen as more crucial to mobility aspirations. This phenomenon of the middle-income raiding of low-income neighbourhoods has been reported throughout Latin America.

The argument can be illustrated by a comparison over time of spatial segregation, incomes, and land prices in three Mexican cities, Toluca, Querétaro, and Puebla (Ward *et al.*, 1993). In all three cities, the trend in land prices has been cyclical, with current prices similar to those twenty years previously. The factors that have most impact on urban land prices in the three cities are the macro-economic conditions, notably the expansion or contraction of the economy and wage trends. Land and housing affordability was a problem even for middle-income households as a result of the declining real incomes of the 1980s. And for the poor, the problem became severe.

Only in Querétaro, however, is there substantial spatial segregation between middle- and low-income classes. This is explained by the greater availability of blocks of low-priced and poorly serviced land in Querétaro due to the inability of *ejidatarios* to resist encroachment and to set their own terms of sale. The availability of low-priced land means that the differential with middle-income land prices is marked, entailing that while the poor have access to land, they have little possibility of mobility into middle-income subdivisions. Conversely, middle-income households are less likely in Querétaro than in the other two cities to trade off a larger plot for a lower status settlement since these settlements are more homogeneously poor than in Puebla and Toluca.

Urban Class Structure and Social Inequality in Latin America

Alongside the economic changes acting to modernize the urban occupational structure of Latin America—more capital-intensive systems of production and marketing and the associated growth of administrative, financial, technical, and social services—were two special trends that need to be noted. The growth of state employment and of female paid work, while not unique to Latin America, had important consequences for class structure and social inequality.

One result of the internal and external pressures to modernize the Latin American economies was a substantial increase in state employment in administration, in state-financed development agencies, in public enterprises, and in social services, such as health and education. In Argentina in 1980, public employment represented 34 per cent of registered urban wage employment, 29 per cent in Brazil, 21 per cent in Colombia in 1982, and 49 per cent in Peru in 1981 (Echeverría, 1985; Saldanha, Mais, and Camargo, 1988). The increase in public employment was particularly significant in the non-manual strata in all countries of the region. In some countries, public employment became the major source of non-manual employment. By 1981, the public sector employed 57 per cent of non-manual workers in Peru, and 52 per cent of non-manual workers in Argentina (Censo de Población de Peru, 1981, Resultados Definitivos, vol. A, pt. ii. table 28; Censo Nacional de Población de Argentina, 1980, series D: Población Total de País, table A. 10, p. 59).

The increase in state employment had both direct and indirect consequences for the status of non-manual as well as manual urban employment. Latin American governments became active agents in stabilizing and formalizing urban labour-markets, creating clear-cut categories of worker with different entitlements and contracts. Social security benefits, such as health care and pension rights, were extended first to state employees, and mainly to white-collar employees. Key manual worker groups, such as railroad workers or workers in the energy sectors, who were often state employees, received such benefits next (Mesa Lago, 1978, 1983, 1986). In the 1950s and 1960s, such benefits were extended to many sectors of the urban working class, especially those working in large-scale formal enterprises. Many factors were involved in these increases in workers' rights such as labour-union organization, the mobilization by populist governments of certain sectors of workers as a source of support, and the pressures of international agencies such as the International Labour Office.

There was also a growing regulation of the labour-market, resulting in labour codes that gave some security of employment, established minimum wages, and set up health and safety requirements. These labour codes were often not enforced, but they did create a distinction between formal and informal employment that was to become an increasingly significant feature of urban class structure in the 1970s and 1980s. Moreover, social security benefits and employment protection made non-manual employees—administrators, professionals, technical staff, and office workers—a distinct urban class, a 'new' middle class that contrasted with the 'old' middle class of independent professionals and small-scale entrepreneurs.

In the 1960s and 1970s, state non-manual employment was not only amongst the fastest growing sectors of employment, it was also amongst the most secure and better paid, if non-wage benefits were taken into account. Access to government loans, specially built housing, and government-owned subsidized stores were amongst the attractions of state employment throughout Latin America. They made possible a distinctive style of life, based on comfortable single-family housing often located in purpose-built housing estates. It included cars, well-equipped houses, domestic servants, and, increasingly, private education for the children, who, in increasing numbers, were going on to university. Whereas for the generation of the parents, secondary education was sufficient for middle-class status, by the 1970s university education was becoming the norm for their children.

By the end of the 1970s, state employment was ceasing to be a dynamic factor in changing the urban class structure. Between 1980 and 1988, the growth of state employment slowed in most countries of the region, and real incomes in state employment declined. Though, as Blanco Sánchez (1990) notes, public employment grew during the economic crisis in most Latin American countries. Despite this growth, the labour-market for clerical and semi-professional workers, both in state employment and outside it, is likely to have become more difficult in this period. And it is likely that substantially higher levels of education were required to obtain even modest positions. Escobar (1992) shows that relatively few people with high educational qualifications were able to enter professional and managerial positions in 1990 in Guadalajara, Mexico, as compared to 1982 before the crisis. García and Oliveira (1992) point to the much higher qualifications required of women entering the white-collar labour-market in 1987 as compared to 1982.

Declining real incomes are likely to have brought the interests and preoccupations of this lower middle class closer to those of manual workers in the 1980s. This class depended on the state for social services since they did not have the money to pay for private education, health care, or child-minding. Furthermore, the problems of daily urban life (light, water, transport, sanitation, pollution) increasingly became shared by the different social sectors, resulting frequently in a common opposition to a state that failed to provide basic services or take remedial action. An increasing social heterogeneity of residential areas was reported for various cities, resulting from middle and working classes invading each other's spaces to find cheap accommodation (Portes, 1989; Rolnik, 1989; Ward *et al.*, 1993).

Another result of economic diversification and increasing urbanization is the rapid growth of female participation in Latin American labour-markets in the 1970s and 1980s. The increase in female participation rates was due to changes in the supply of labour and to modifications in the social and spatial division of labour. Increase in educational levels delayed the age of entry into the labour-force, but also increased female participation. Highly educated women became more likely to seek work outside the household. Low levels of fertility, already present in Argentina and Chile and being rapidly reached in Brazil and Mexico, also encouraged the growing labour force participation of women.

The rapid urbanization of Latin America was perhaps the most important factor affecting both the demand and supply of female labour. Female participation rates were highest in the large metropolitan areas. These areas grew markedly, as we noted above, through natural increase and rural–urban migration. The occupational structures of these metropolitan areas were particularly open to female employment: in domestic services, in other personal services, in commerce, and in the expanding ranks of office workers. Female employment became part of the increasing polarization of the urban occupational structures. Women have had more job opportunities than in the first half of the twentieth century in 'middle'-class occupations such as teachers or skilled secretaries, but they also entered informal employment, in increasing numbers, as personal service workers or as domestic outworkers.

The increase in the numbers of women in certain types of work had little effect on the gender division of labour. Women remained segregated in the labour-market, despite the changes in occupational structures. Opportunities for women were restricted not as a result of direct competition in the labour-market, but by factors such as the possibility of combining domestic with extra-domestic work and by social norms which fixed which occupations were accepted as suitable for women (Jelin, 1978; De Barbieri, 1984*a*; 1984*b*; Humphrey, 1987).

We need to emphasize that the expansion of the service sector in

Latin America resulted not only in an increase in informal employment (low-paid workers in personal services, street sellers, etc.) but also created middle-class occupations in the public and private sector, a significant part of which were held by women. Peru was an exception in that the expansion of female employment in the services occurred basically through personal services.

Women's employment in manufacturing industry did not increase to the same extent as it did in the services. It also differed between countries. In Argentina, Chile, Brazil, and Peru, women's employment in industry declined absolutely between 1960 and 1980, mainly due to the decline in craft industry (CEPAL, 1986). Female manufacturing employment increased in Mexico from 1950 to 1970 due to the expansion of assembly operations by multinational companies, the persistence of industries (such as the garment industry) that traditionally were heavy users of female labour, and the spread of domestic outwork (Oliveira and García, 1988). Brazil showed a slight expansion of female manufacturing workers and craft workers between 1970 and 1980, due probably to the expansion of industrial activities that made heavy use of female labour.

In the 1980s, the economic crisis made it even more necessary for the urban poor to use various monetary and non-monetary resources to make ends meet. A single salary became increasingly inadequate to maintain a family in face of the decline in real wages. Even the low salaries of the young and women became necessary to sustain the household, along with increased domestic work. This has been the major factor increasing female labour-force participation in the poorest households, but has also contributed to increase employment among women with high levels of education and married to men with high-status jobs—the married female category that has the highest levels of participation in the labour-force (De Barbieri and Oliveira, 1987; Gonzalez de la Rocha, 1987; García and Oliveira, 1992). Also, mutual-help networks increased amongst relatives and friends (Ramos, 1984; Raczynski and Serrano, 1984; De Barbieri and Oliveira, 1987; González de la Rocha and Escobar Latapi, 1988).

In the metropolises of Latin America, women's paid work had become an essential part of the domestic budget by 1980. There are no detailed studies of the changes in the female labour-market from 1980 to 1990 in the six countries, but data from Mexico and Brazil can be used to illustrate the trends. In Mexico there was a marked expansion in female employment during the 1980s, with an increase of 6.5 per cent per year in the participation rate of economically active women between 1979 and 1987. The equivalent rate of increase between 1970 and 1979 was 3.5 per cent annually (Oliveira and García, 1990; Pedrero, 1990). The trend in Brazil was similar with an increase of 7.6 per cent in female participation between 1980 and 1985, as against one of 4.6 per cent between 1970 and 1980 (Bruschini, 1989).

Economic recession in the 1980s led in Mexico to the mobilization of a potential supply of labour mainly made up by adult women (35 to 49 years) of low levels of education, married and with young children. In contrast, young, single women (20 to 34 years), with middling or high levels of education showed a relative decrease in their relative participation in the labour-market. This contrast was likely to have been produced by the contraction in non-manual employment opportunities and the increase in informal employment (Oliveira and García, 1990). The Brazilian data indicate a similar tendency in terms of educational levels and age of the female labour force. Women with low levels of education increased their participation rates by 56.3 per cent between 1980 and 1985, whilst women with five or more years of study showed more modest increases. Women between 30 and 49 years had higher increases in participation in the same period than younger women (Bruschini, 1989).

The changes in the characteristics of women entering the labour-market occurred in conjunction with transformations in the form of their insertion. In Mexico, the percentage of non-manual workers (professionals, technicians, and clerical workers) in the female economically active population decreased significantly. Only the most qualified workers succeeded in obtaining the few non-manual jobs that were created. Data for the years 1982 and 1987 show that women with low levels of education showed a clear drop in their participation in manual wage work, but those with middling levels of education increased their presence (Oliveira and García, 1990). Both these trends showed greater credentialism and stricter requirements for contracting labour in periods of recession.

Domestic servants became a significantly smaller proportion of the female economically active population, as did manufacturing workers. Only manual wage workers in the services increased their share of female wage work. The female self-employed increased their share of employment, especially those with low levels of education, living in common-law unions, and with young children. The increases in self-employment occurred not only in the tertiary sector—the sector with the most female employment—but in manufacturing (García and Oliveira, 1992). This expansion of self-employment was not only due to survival strategies on the part of poor families, but to the restructuring of manufacturing activity through the use of subcontracting to workshops and to domestic workers (Escobar, 1986; Tokman, 1987; Roldán and Benería, 1987; Roberts, 1989*a*, 1991; Marshall, 1987; Arias, 1988; García, 1988; Portes, Castells, and Benton, 1989).

The Nature of Urban Stratification

Economic, demographic, and social changes resulted by 1980 in a pattern of urban stratification that in certain crucial respects has become increasingly similar throughout Latin America (Table 8.3). The most salient divisions within this class structure can be identified in terms of several criteria: the degree of control over the means of production, control over labour power, and type of remuneration (Portes, 1985). These criteria approximate the major divisions of collective interest that are likely to be found in Latin American cities: between those who depend on the profits of enterprise and those who are employed by them, between those who manage the labour of others and those who don't, between those on stable incomes and those who are casually employed. A further division needs to be added, that between manual and non-manual work, differentiating those whose jobs, and life-style, depend on schooling and those who do not.

This class division includes firstly, a proportionately small dominant class, based mainly on ownership of large-scale enterprises concentrated in the service and manufacturing sectors. Below that class had emerged a clearly defined bureaucratic-technical stratum with high levels of education and employed in managerial and administrative positions in both public and private sectors. Both classes probably have been more concentrated in metropolitan centres and large urban areas. Portes (1985: table 2) estimates that the dominant class and the bureaucratic technical class make up about 8 per cent of the economically active population. These classes correspond to the occupational categories of higher non-manual workers used in Table 8.3 which, by 1980, made up approximately 16 per cent of the urban population.

The basis for making class distinctions among the rest of the urban population is complicated by the importance to life chances of whether the economic activity and employment is formally regulated by the state or not. As noted above, the informal/formal distinction became

TABLE 8.3 Urban Occupational Stratification in Six Latin American Countries, 1940–1980 (percentages)

	1940	1950	1960	1970	1980
Non-agricultural population					
Higher non-manual strata	6.6	9.4	10.1	12.7	15.9
Employers, independent professionals	4.4	5.2	1.9	2.6	2.4
Managers, employed professionals, and technical personnel	2.2	4.2	8.2	10.1	13.5
Lower non-manual strata	15.2	16.0	16.9	18.5	19.0
Office workers	8.4	10.0	11.1	11.7	13.2
Sales clerks	6.8	6.0	5.8	6.8	5.8
Small-scale entrepreneurs	0.8	2.5	2.6	2.5	2.5
Commerce	0.8	2.3	1.5	1.2	1.3
Other (manufacturing, services)	0.0	.2	1.1	1.3	1.2
Self-employed	28.5	19.8	20.5	17.4	18.6
Commerce	9.5	7.1	7.5	6.6	5.8
Other	19.0	12.7	13.0	10.8	12.8
Wage workers	35.9	41.3	40.4	39.4	36.4
Transport	6.1	3.8	4.5	3.7	2.7
Construction	5.4	7.0	7.1	7.8	7.1
Industry	20.1	19.2	19.1	16.3	16.5
Services	4.3	11.3	9.7	11.6	10.1
Domestic servants	13.0	11.0	9.5	9.5	7.6
TOTAL	100.0	100.0	100.0	100.0	100.0
Agriculture (% of the active population)	61.6	52.5	46.7	39.5	30.6

Sources: Calculated from the population censuses of Argentina, Brazil, Chile, Colombia, Mexico, and Peru. The years are approximate: 1940 includes interpolated figures from the 1914 and 1947 Argentine Censuses; the figures for 1980 are not available for Colombia.

an important factor in the stratification of the urban populations of Latin America. Both informal workers and informal employers had a different set of interests and different levels of income than their formal counterparts. The partial extension of social security coverage to the Latin American population created two classes of wage-workers—those who received a range of benefits, including contractual security, and those who did not. These benefits constituted a premium for workers who were further benefited through higher wages resulting from trade union negotiations and collective contracts. We label wage workers covered by contract and social security the formal working class and those not covered, including most self-employed and family workers, the informal working class (Portes, 1985).

Likewise, the partial extension of state regulation of economic activity and the uneven development of urban economies created two classes of entrepreneurs. One group—mainly large-scale operators—were increasingly enmeshed in sophisticated credit, marketing, and supply networks that necessitated their having legal status that made it more difficult to avoid fiscal and social security obligations. A second group of mainly small-scale entrepreneurs worked so close to the margins of profitability, often in markets that experienced sharp fluctuations, that savings on overheads, such as their fiscal and social security obligations, became an important part of their survival strategies. The distinction between small- and large-scale entrepreneurs should be seen as a continuum rather than a sharp break (Tokman, 1991). Most enterprises observed one or more of their legal obligations, though only a minority observed all of them. However, the small size of the enterprise, the low level of technology, and precarious market position compared to larger and better endowed enterprises gave the group of small-scale entrepreneurs, whom we label the informal petty bourgeoisie, an especial class position.[5] Both the informal petty bourgeoisie and the informal working class are likely to be concentrated in the smaller cities and towns. For example, Briones (1991) shows that the smaller urban places of El Salvador have higher proportions of both the informal bourgeoisie and the informal proletariat. The formal working class represents a more important share of the urban economically active population in the large cities (Pérez Sainz and Menjívar, 1991; Telles, 1988: table 2.4; Roberts, 1991; SPP, 1979).

Portes (1985: table 2) estimates the informal petty bourgeoisie in 1970 as 10 per cent, the formal proletariat or working class as 22 per cent, and the informal proletariat (which includes the peasant population) as 60 per cent. Telles' study of occupations (1988: table 2.4), earnings, and stratification in the six major metropolitan areas of Brazil in 1980 indicates that the formal working class averages just under a third of the economically active population of these cities, and the informal proletariat about the same proportion.

A further stratum of lower non-manual workers needs to be added to this description of the urban class structure. This lower middle class included semi-professional occupations, such as teachers, nurses or other health workers, secretaries, bank clerks, and sales clerks. Evidence from Mexico suggests that in the 1960s, this stratum of the urban population earned more than a skilled industrial worker, but that the differential had decreased or been eliminated by the 1970s (Reyes Heroles, 1983: table 11.21). In Brazil, by 1980, the mean occupational income and prestige scores of semi-professional occupations such as teachers, nurses, secretaries, and bookkeepers were higher than those of most manual workers in industry, construction, and the personal services (Telles, 1988, app. B, 173–81).

Though the differences were small between the incomes of this lower middle-class group and that of the skilled, formal working class, four factors make it important to include this group as one of the six urban strata. Average levels of education of this class were higher than for the manual working class, with formal education being essential not only for carrying out the job but for reaching the better-paid positions. Conditions of work were generally better, with office workers, teachers, and health workers enjoying more social security protection than did the manual working class as a whole. Finally, this was the class most dependent on state employment. It represented most social mobility opportunities for women, e.g. female office workers earned significantly higher salaries than did females in the manual wage-earning categories of domestic and other personal services. By the 1970s this class was likely to have amounted to about 21 per cent of the total urban population.

From the 1970s, the demand for workers in urban areas was negatively affected by technological changes that saved on labour and by the decline in the regional economy. The result was the persistence of unpaid family workers and the self-employed. By 1990, these categories of urban employment were increasingly important throughout the region. When added to changes in the pattern of migration and a substantial increase in female participation rates, the increasing differentiation of the structure of urban employment created a more heterogeneous and polarized urban social structure than before.

The decline in real wages had important implications for the class structure, especially since it occurred in the context of a cut-back in state expenditures and employment. In the 1980s, the Latin American economies increasingly adopted free market policies aimed at stimulating the private sector and reducing state intervention in the economy. Dramatic consequences were probably felt by the urban middle and working classes, especially the groups that we have labelled lower middle class, formal, and informal working classes. The incomes of intermediate and lower-level state employees, including teachers and health personnel, appear to have dropped sharply in these years, so that public-sector workers in Uruguay, for example, earned 56 per cent of their 1975 wage by 1985 (ILO, 1989: 85). In Mexico, the decile of household incomes that mainly included the lower middle class became differentiated from the one above. In terms of income and sources of income, the lower middle class decile became more similar to those below (Cortés and Rubalcava, 1991).

There was a sharp decline in incomes at the top of the income distribution relative to the period before the crisis, but here there were signs of differentiation between the entrepreneurial section and high-level administrators and professionals. The income from profits rose during the years of crisis, while salaries dropped substantially. However, for the upper urban classes incomes were still substantially above those of other classes. Though consumption may have diminished in these years, income remained adequate for a comfortable life-style. Escobar and Roberts (1991) indicate some of the monetary and non-monetary benefits that these upper strata continued to receive—company cars, productivity-linked bonuses, school fees, free travel.

The bottom end of the urban class structure appears to have suffered also from the crisis relative to the previous period, despite beginning with very low levels of income. The formal working class saw reductions in their incomes that were not offset by the increasing importance of non-wage benefits and by other sources of income, including remittances from abroad. Furthermore, the formal working class lost importance within the Latin American working classes in the 1980s, as the informal working class grew substantially both in numbers and as a proportion of the urban labour-force. Initial PREALC (1988: table 1) estimates suggest that in the region as a whole urban informal employment grew rapidly, particularly in the mid-1980s.[6] Using PREALC calculations, Portes and Schauffler (1993: table 3) estimate that by 1989 informal employment made up 31 per cent of total urban employment in the region, compared with 30 per cent in 1980.

Data from Mexico and Central America provide a more detailed picture of these trends. In the three major metropolitan areas of Mexico, there was an increase in employment in small-scale enterprises, in self-employment, and in unpaid family employment. This increase was particularly marked in the repair services and in commerce (INEGI, 1988, 1977; Escobar, 1988; Gonzalez de la Rocha, 1988*a*; Roberts,

1991). Informal employment (including domestic service) was estimated at 33 per cent of the urban labour-force in 1987. The self-employed and workers in small enterprises appear to have suffered a drop in real incomes. The informal sector had by 1989 become synonymous for many of its workers with bare subsistence. Only the owners of small-scale enterprises and informal workers with skills in demand earned significantly more than the minimum wage, but their enterprises were, in general, poorly equipped and showed little sign of capital accumulation.

Surveys showed that the urban labour-markets of most Central American countries were highly informalized by 1989 in terms of the high numbers of the self-employed, unpaid family workers, and workers and owners of firms with less than five workers (Pérez Sainz and Menjívar, 1991). The percentages informally employed by these criteria were 33 per cent in Guatemala City, 30 per cent in Tegucigalpa, 28 per cent in San Salvador, and 23 per cent in San José. In Managua, suffering the effects of the economic blockade and the war in the countryside, and with a weak industrial base, the informally employed were 48 per cent of the urban labour-force, not counting domestic servants (Chamorro, Chávez, and Membreño, 1991). In Managua, the informal sector was already large by the time of the 1979 revolution, having expanded rapidly after the earthquake of 1972. From 1979 onwards, the informal sector was alternatively encouraged and discouraged by the Sandinista regime, though by 1989, it was viewed as providing essential services within the war-torn economy. Costa Rica, whose economy had been less affected than most Latin American economies by the recession of the 1980s, was the only one of the Central American countries not to have experienced increasing informalization in the 1980s.

Informal workers in the Central American cities were disproportionately drawn from the younger and older age groups, from migrants, from women, and from those with low levels of education. Informal employment was mainly in commerce, though about a quarter of the informally employed were in the industrial sector. In all the cities, the informal sector was socially and economically diverse with large differences in income between the owners of small-scale enterprise, their employees, and the self-employed. Poverty was concentrated in the informal sector, though in all the cities, including San José, a substantial minority of formal workers earned incomes that placed them below the poverty line. Case studies of samples of the self-employed and small-scale enterprises in these cities indicated that informality for the self-employed was basically a household survival strategy in face of unemployment and declining real wages. Only the small-scale entrepreneurs earned a wage significantly above the minimum, but even this sector showed little economic dynamism.

Because informal employment provided relatively easy access to incomes that could supplement household incomes, it facilitated a household strategy of placing more members on the labour-market as a means of offsetting the declines in real wages (Gonzalez de la Rocha, 1988*a*; Selby *et al.*, 1990; Oliveira, 1989*a*). Households containing members of the informal working class may, as a consequence, have experienced a smaller reduction in overall income than other working-class families. Evidence from Mexico suggests that this was the case, with non-wage sources of income also having become more important—remittances, self-provisioning, renting-out accommodation (Cortés and Rubalcava, 1991). The result of these various tendencies was a continuing polarization of incomes in Mexico despite a slight decline in income inequality. The poorest survived at the margins of subsistence through strategies such as using child labour, mothers of young children seeking paid work, or cutting food consumption, that were likely to perpetuate their disadvantages.

Studies from Santiago show that, in those cases where the male head of household was unemployed, sharp reductions in expenditure inevitably occurred in face of the difficulty of either men or women finding some form of paid work, leading to a worsening of nutritional levels, the sale of household goods, and exclusion from public utilities such as water, light, and gas (Raczynski and Serrano, 1984). Studies in Mexico also found a worsening of nutritional levels, the abandoning of children's education, and a decline in general welfare (Tapia Curiel, 1984; Arizpe, 1990; Benaría, 1991).

One of the most important aspects of urban life by 1990 was the pressure on households and on the residential community as they became the essential means of survival, particularly for low-income families and for those who had recently migrated to the city. Complementing low incomes and sharing housing, whether amongst the *allegados* (the non-nuclear members of a household) of Santiago, the poor of Lima, or migrants to the city of Mexico, was crucial for the survival of the poor. Mutual help amongst neighbours and collective strategies of survival—such as communal kitchens—were equally important (Tironi, 1987; Valdes and Weinstein, 1993).

Both household and community survival strategies generated tensions which led to household break-up and community fragmentation (de la Peña *et al.*, 1990; Gonzalez de la Rocha *et al.*, 1990). The pressure on family relationships was considerable, particularly for women. Women, as mothers and housewives, carried a double responsibility: they had to look after the house and care for other household members, whilst seeking income sources through domestic outwork or employment outside the home. Male heads, though unable to maintain the household on their low salaries, often remained resistant to their wives working outside the home. Males were reluctant to contribute their entire income to the family budget, increasing the potential for family conflicts. Fathers expected both their sons and daughters to contribute to the family pot, while these, in turn, wished to use their earning for individual needs. The gender and generational conflicts that arose within households were a marked characteristic of urban life in Latin America by 1990.

Equally important, the neighbourhood also contained many potential sources of conflict since the actions that benefited individual household interests, such as use of collective services without paying for them, weakened community spirit. As the economic crisis continued, so the possibilities of obtaining help from family and friends diminished, whilst state-provided services also deteriorated. One consequence was unorganized forms of protest, such as the proliferation of gangs of youths, collective looting of shops and of shopping malls. Likewise, there was an increase in violence, robbery, alcoholism, drug consumption, and prostitution (De Barbieri and Oliveira, 1987).

The changes in the labour-market tended to weaken the job as the central factor structuring daily life, redefining the bases for social stratification. The intensification of subcontracting by large-scale enterprises to domestic outworkers or workshops, the increase in unemployment, the casualization of much of the labour-force, and the increase in labour-force turnover produced greater instability in employment. These processes worked against the consolidation of the urban social classes in Latin America. An individual's occupation became a less useful indicator than in the past of social and class position. This was especially true for the working class, for whom employment was decreasingly linked to particular skills and to a stable work career and for whom the individual salary was usually insufficient to maintain a household.

The importance of occupation in defining life chances and social position was replaced by that of position within the household structure, by stage in household cycle—heads of nuclear or extended households, with children or not, with or without spouse, or living alone—and by access to community and family help and to information networks. Stable occupational careers (characterized by remaining in the same enterprise, obtaining skills and promotion, benefiting from seniority and social security) became rare. The increase in inter-urban mobility was one indication of frequent changes of job. Residential mobility also implied that neighbourhoods became less stable than in previous decades, as did the neighbourhood basis for social solidarity.

The Nature of Urban Inequality

The transformation of the occupational structure of Latin America had contradictory consequences for income distribution and for the share of the different urban classes in that distribution. In general, in the period up to the mid-1970s, there was a rise in real incomes for all strata of the urban populations (Iglesias, 1983). This general trend was interrupted by economic cycles and by political conjunctures, such as the 1964–78 decline in real wages in São Paulo, Brazil, resulting from the policies of the military government to contain inflation. This drop in real wages is likely to have had a negative effect on infant mortality in São Paulo (Wood and Carvalho, 1988: 115–18). During this same period, income concentration increased in Latin America with the top 10 per cent of households by income receiving a greater share in 1975 than they had in 1960 (Portes, 1985: table 3).

The impact of the crisis on labour-markets varied between the Latin American countries. For example, the unemployment rate in Chile increased dramatically from 8.3 per cent in 1974 to 18.6 per cent ten years later. In Peru, the increase was less marked in the same period. In Argentina, Brazil, Colombia, and Mexico change was minimal (Iglesias, 1985). Despite such differences, the increase in the supply of labour during the crisis had a generally negative effect on urban wages. PREALC (1983) data show that, in the 1970s, there were clear signs that the long secular trend of rising real urban wages was coming to an end. In Mexico, the over-expansion of government expenditures and increases in inflation resulted in adjustment policies which reduced government expenditures and wages from 1979 onwards. Urban minimum salaries were the most affected, declining by 8 per cent between 1978 and 1981. In contrast to Mexico, Brazil had a high dependence on petrol imports. Brazil increased its external debt as a means of offsetting the negative shift in the terms of trade. From 1978, however, the continuing worsening of the terms of trade resulted in a marked drop in the rate of increase of the basic Brazilian real minimum wage, and in the minimum-wage categories for industry and construction.

In Argentina and Chile, economic policy in the 1970s was directed towards controlling salaries as a means of economic stabilization and of improving international competitiveness. Tight control over wages was accompanied by a breaking-up of the trade union structure to lessen organized opposition to fiscal and tariff policies that removed protection from local industry and led to a rise in unemployment. In Argentina, the reduction in real salaries from 1975 to 1978 was of the order of 50, 55 and 56 per cent for the urban minimum wage, the industrial minimum wage, and the construction minimum respectively. In Chile, the sharp decreases took place between 1970 and 1975 when urban minimum salaries dropped by 41 per cent, the industrial minimum by 42 per cent, and construction by 18 per cent. In Peru, real wages also dropped markedly between 1975 and 1978.

In the 1980s, the decline in real wages was more general and consistent throughout the region. Between 1980 and 1987, the real minimum wage declined by 14 per cent in Latin America, though with some recuperation between 1985–7 (ILO, 1989: table 1.10). The decline appears to have been most severe in public-sector wages which declined by 17 per cent, and least severe in manufacturing which declined by 10 per cent. This overall trend concealed important variations by country and city. The declines were most severe in Peru and Mexico, while in Colombia real wages appear to have increased. Of the six countries, Colombia was the only one to carry out an economic policy that, at one and the same time, permitted the expansion of the volume of exports, the neutralization of the effects of the drop in international prices, and an increase in real wages from the 1970s up to 1987.

In terms of levels of poverty, the Chilean data indicate that the proportion of homes below the poverty line (with incomes insufficient to meet minimal nutritional levels) increased from 12 per cent in 1979 to 23 per cent in 1984. In Peru, the equivalent figures indicate an increase from 8 to 21 per cent between 1970 and 1982 (PREALC, 1987). In Mexico, as a result of the drop in public-sector income, there were reductions in the subsidies for basic foodstuffs and in the budget of the social services. Families with incomes below two minimum salaries experienced a decline in consumption as measured by the basic food-basket. Even so, expenditures on food represented 52 per cent of the minimum salary in 1986, whereas in 1982 they only represented 34 per cent (Lustig and Ros, 1986).

The national figures on income distribution when related to class differences enable us to explore this urban inequality (Cortés and Rubalcava, 1991). The social strata that have been described above are likely to be differentiated by sources of income and by the amounts they receive. The highest income levels are likely to correspond to the dominant and bureaucratic-technical class. The upper middle-income levels are likely to have been made up of aspiring members of those classes, just beginning their careers in government or in the private sector, some small-scale entrepreneurs in industry and the services, and, in some countries, skilled workers of key industries, such, for example, as the petroleum or, in this period, the car industry.

The intermediate levels of income in this period probably included the lower middle class—teachers, ancillary personnel in the health or welfare services, bank clerks, and office workers in private and public sectors. The intermediate level also included the less-successful members of the informal bourgeoisie, the more successful among the self-employed, and skilled workers in industry, transport, and communications. The lower middle levels of income included the bottom end of the lower middle class—sales clerks and non-specialized office workers—skilled workers in basic goods industries, in construction, and in the services and the self-employed in industry and certain of the services. The bottom income levels included semi- and unskilled formally employed workers in industry, construction, and the services, together with most self-employed workers and informally employed workers. The self-employed and skilled informally employed workers were, in this period, likely to be in the upper end of these strata, especially the craftworkers, while those in personal services, were at the bottom end.

Studies from Mexico, Brazil, and some other countries of Latin America suggest that by the mid-1970s, urban income distribution, while demonstrating sharp inequalities, also evidenced the consolidation of these various strata (Portes, 1985: table 3; Iglesias, 1983; Cortés and Rubalcava, 1991; Wood and Carvalho, 1988: table 3.5; Escobar and Roberts, 1991). The bureaucratic-dominant group—who are likely to be synonymous with the richest 10 per cent of households in income—had clearly benefited from the rise in real salaries and in profits. In the period from 1960 to 1975, they probably increased their share of household income slightly (Portes, 1985: table 3). However, the informal bourgeoisie, the lower middle class, and the formal working class had also increased their real wages. Brazil may have been an exception since Wood and Carvalho's (1988) figures for Brazil show a decline in the share of income of the intermediate and lower middle urban strata.

These gains reflected a series of factors in the pattern of economic development since the 1960s: the dominant groups and the informal bourgeoisie benefited from the general dynamism of the economies of the region and the entrepreneurial opportunities they generated; the lower middle class from the growth of state employment and the benefits given to state employees; and the formal working class from the power of organized labour to extract wage concessions. Up to 1980, all these groups, with the exception of the informal bourgeoisie, were also increasing their share of the population as waged and salaried work increased as a proportion of the urban labour-force. While the unskilled and informal workers appear to have decreased their share of total income, thus creating a certain polarization in the class structure, this was offset by a rise in real incomes even among these strata.

The discussion of class structure and income inequality is further complicated by the need to take account of the household in determining patterns of stratification. By the mid-1970s, most urban households in Latin America had more than one member who was economically active. Female participation rates had risen sharply, as we have seen. It was increasingly common for wives, and not just adult children, to work for a wage or help with the family business.

The distribution of extra wage-earners was not, however, even across the class strata, both for economic and demographic reasons. Overall, households in the highest strata had, by the 1970s, smaller families and fewer members in the labour-market. The lowest urban strata had larger families and more members in the labour-market since, as data from Mexico show, children in these strata were more likely than children in the higher strata to enter the labour-market rather than stay on at school (Selby *et al.*, 1990). Members of very poor households were more likely than members of other households to be economically active even controlling for the number of available workers in the household (García, Muñoz, and Oliveira, 1981: 217–18).

In the higher strata, the income of the head of household was usually sufficient to maintain the family at its expected level of subsistence.

Other members worked to pursue a professional career and/or to increase the level of consumption which, at times, resulted in higher rates of participation in the labour-force among married females. The Mexican Urban Employment Survey shows that married women from households whose male head had a professional, managerial, or technical occupation were more likely to work for a wage than women whose husbands had working-class occupations, and at all stages of the household cycle (Roberts, 1993). In Mexico, working-class wives are burdened with considerable domestic chores, and it is their sons and daughters that provide the supplementary income. Women with higher levels of education can obtain much higher incomes than their working-class counterparts and can earn enough to substitute their own domestic labour with paid domestic services (García, Muñoz, and Oliveira, 1982: table V.1; García and Oliveira, 1992). In the bottom deciles of income, the salaries of the head of household were often insufficient to make ends meet or were close to the margin of subsistence. In these cases, supplementary incomes earned by other family members were an essential means of household survival.

The bottom income strata and the classes associated with them can, thus, be identified not simply by occupational titles or by whether the job was formal or informal, but by the overall household strategy for obtaining an income. These were the urban poor for whom income pooling, sharing housing, food, and other resources, were essential means of urban survival. Family and friendship networks were also crucial to urban subsistence strategies by providing help with housing, food, and finding work. In the case of Mexico City, 90 per cent of individual migrants were preceded or followed by a family member (García, Muñoz, and Oliveira, 1979). Poor households might contain formally or informally employed workers, the self-employed, or domestic servants. They were less likely to have the spare resources to keep a child out of the labour-market, or a mother to devote herself solely to bringing up children and keeping up the household. Thus, the consequences of income inequality extend to inadequate standards of nutrition, health, and general welfare (Wood and Carvalho, 1988).

Poor households were not homogeneous in their social and economic characteristics. The greater use of available labour among households whose heads had low incomes led to considerable occupational heterogeneity among the poor. It was common for workers of different types to be present in the same household: workers in the manufacturing sector, service workers, white- and blue-collar workers, workers in the formal and informal sectors, and so on. Studies in Mexico City and Guadalajara indicated that in households headed by manual workers, the sons were usually also manual workers, whilst the daughters and wives—when they worked for a wage—worked in a variety of occupations (García, Muñoz, and Oliveira, 1981; González de la Rocha, 1986). Furthermore, households included both migrants and natives. For example, in Mexico City in 1970, the urban labour-force was mainly made up of migrant heads of families and their native children, and, in 1980, in Reynosa—one of the northern border cities of Mexico—migrants and natives were members of the same family and shared the same household (García, Muñoz, and Oliveira, 1981; Margulis and Tuirán, 1986).

Migration and ethnic differences need also to be considered as factors adding to inequality by the 1970s, though their impact varied between cities, and differed depending on whether the migrant was male or female. There is evidence for some cities that migrant selectivity declined after the 1960s, with those arriving in the cities possessing less education and skills relative to their populations of origin than had previous migrants. Migrants who arrived in the 1950s and 1960s in Monterrey—an industrial city to the north of Mexico City—were less skilled and were more likely to fill manual job positions than previous migrants (Balán, Browning, and Jelin, 1973: 146–7). Likewise, from the 1960s there was an increasing migration of unskilled rural workers to Mexico City who took up unskilled urban positions (Oliveira, 1975). However, migrants played an important part in the expansion of the industrial labour-force in Mexico City and of non-manual jobs (Oliveira and García, 1984). In Buenos Aires, migrants, both male and female, were as likely to be found working in manufacturing and construction as those born in the city (Marshall, 1978).

Usually, rural migrants are disproportionately found in unskilled manual jobs in construction, manufacturing, and services, whilst non-manual jobs—created by the rise of the modern services—which demanded relatively high levels of education fall to those born in the city or in other cities with good educational facilities. The city-born and those urban migrants with at least primary education worked, in general, in skilled or semi-skilled work in manufacturing, in service firms or in family workshops (Escobar, 1986).

Casual labour in personal services, in construction, and in manufacturing has often been provided by temporary rural–urban migration, emphasizing the rural or ethnic distinctiveness of such workers. Examples are Smith's (1984, 1989) account of Peruvian highland villagers picking strawberries and selling fruit in Lima, and Arizpe's (1982) account of Mazahua women in Mexico City. Where, as in Argentina, there was no peasant economy to provide temporary migrants, the unskilled migrants to Buenos Aires have come from neighbouring countries. This foreign immigration fulfilled similar functions to the temporary flows in Mexico (Balán, 1981, 1985). International migration—from Bolivia and Paraguay to Argentina, from Colombia to Venezuela—has created enclaves of poor and ethnically distinct migrants in several Latin American cities.

Though rural migrants did not, as we noted in the previous section, suffer substantial disadvantage when compared with the urban-born, this is not the case of those migrants who came from ethnic minorities that are clearly distinct from the bulk of the population. The *cabecitas negras* from the northern provinces of Argentina were looked down upon by European migrants and the city-born. Paraguayan and Bolivian migrants had by the 1970s become a distinct underclass in Buenos Aires.

In Brazil, despite its reputation for harmonious race relations, Blacks became the most disadvantaged among the urban population, though this inequality is lower in the most industrialized areas (Wood and Carvalho, 1988; Telles, 1994). Taking account both of levels of education, rural or urban origins, and region of the country, Blacks are paid less than Whites, live in the worst housing conditions, and suffer the worst health. The available evidence suggests a similar conclusion for the Indian groups that migrate to Latin American cities. In Guatemala, Indians are disadvantaged in all areas of urban stratification—income, housing, education, health—when compared to Whites (Pérez Sainz, 1990). In Lima, Aymara-and Quechua-speaking migrants are amongst the poorest in the city (Altamirano, 1984, 1988).

Gender remained a contributing factor to inequality in the 1970s. Female-headed households, with no male head present, have increased in numbers in the region and represent the most impoverished group of households (Jelin, 1978). However, households headed by a single female parent may be better off economically than households with male and female heads, particularly in those cases where the male head does not contribute all his income to household expenses and the female head is not in paid work. Female-headed households are, on average, smaller than those with male heads. The female head combined the tasks of earning a living with domestic ones and had also to maintain the networks that brought in non-monetary resources. In the 1970s, households with female heads accounted for about 17 per cent of households in Mexico City and Buenos Aires, 22 per cent in Recife (in the north-east of Brazil), and 20 per cent in Santiago, Chile (García, Muñoz, and Oliveira, 1982; Recchini de Lattes, 1977, 1983; Pantelides, 1976; García, Muñoz, and Oliveira, 1983). Almost 45 per cent of female-headed households in Belo Horizonte, Brazil, were below the poverty line, whilst the equivalent figure for two-parent households was approximately 28 per cent (Merrick and Schmink, 1978).

In a comparative study of two Brazilian cities (Recife and San José dos Campos), it was found that both men and women worked more when they belonged to a female-headed household. Since women earn, on average, lower incomes than men, the low income of the female head made more necessary the earnings of male and female children (García, Muñoz, and Oliveira, 1983). The fact that female-headed households need to use all their available resources to survive means that they have less flexibility in face of the worsening economic crisis (González de la Rocha, 1988*b*).

The Pattern of Occupational Mobility

Urbanization, the changes in urban occupational structure, and the relative decline of agriculture as a source of livelihood resulted in considerable social mobility, certainly up to the 1970s (CEPAL, 1989; Durston, 1986). The various patterns of social mobility in the region crystallize the differences and divergences in the social and occupational transformations that have been reviewed. In general, the four decades from 1940 to 1980 have been years of a high degree of social mobility because of the massive transfer of labour from agriculture to urban jobs (Filgueira and Geneletti, 1981). The countries that had the highest rates of total structural mobility (defined as the sum of the proportionate increase in non-manual jobs and the proportionate decline in agricultural employment in the period) were those with the highest rates of urbanization—in our examples, Brazil and Mexico. Countries which experienced urbanization earlier, for example, Argentina, showed less structural mobility than did Brazil and Mexico, both because they had fewer agricultural workers at the beginning of the period and because, at that time, they already had high levels of non-manual jobs (CEPAL, 1989: table 1–6).

The contrasts between Chile and Peru were also indicative of differences in the pace and nature of economic modernization. Chile had the higher level of structural mobility between 1960 and 1980, as a result of a sharp drop in the agricultural population and a large increase in lower non-manual jobs. However, if the high levels of unemployment in 1980 and the low levels in 1960 were taken into account, structural mobility in Chile would be low or non-existent. The drop in the proportion of the agricultural population in Peru was one of the lowest in the region, and structural mobility was concentrated in the lower non-manual strata with little increase in the non-agricultural manual strata.

The experience of social mobility was different according to gender and age. For men, the major form of mobility in the period up to 1980 was through the increase in the higher non-manual strata. For women, the traditional form of mobility—from the countryside to urban domestic service—was replaced by entry into the lower non-manual strata, such as shop assistants and clerks, but with lower incomes than those in skilled manual jobs that were still dominated by males.

Most mobility occurred between generations, with the 25–34 year age cohort being substantially more likely to have non-manual jobs than their counterparts twenty years earlier. There was, however, some evidence of mobility within a generation since the 25–34 year age cohorts of 1960 were more likely to have non-manual jobs, and less likely to have agricultural jobs in 1980 when they were 45–54 years old (CEPAL, 1989: Table 1–9).

Accompanying the high rates of social mobility among the young age cohorts were the rising levels of education in the region that benefited the younger age cohorts most. The increasing demand for non-manual workers in the 1960s and 1970s was met by an increase in the supply of people with the requisite levels of education, particularly the urban-born. Education became a more important avenue to occupational mobility than it had been in the previous period. The changes in educational levels between 1950 and the 1970s were dramatic ones, with the six countries reducing illiteracy by almost 50 per cent (ECLA, 1983, table 34, p. 102). Lack of comparability between censuses in educational classifications obscures the trend reported in Table 8.4, but, in general, there is a substantial increase between 1960 and 1980 in the proportion of the labour-force with more than primary education. Despite these gains, the educational levels of the economically active population remained low in several countries (Table 8.4). The exceptions were Argentina and Chile, with less than 30 per cent

TABLE 8.4 Educational Levels of Economically Active Population in Six Latin American Countries, 1960, 1970, 1980 (%)

Country	Census year	Number of years of education							
		None	1–3	4–6	7–9	10–12	13+	No info.	Total
Argentina	1960[a]	6.9	24.4	45.8	4.7	9.6	4.4	4.2	100.0
	1970[b]	0.0	15.8	20.3	36.7	13.1	5.9	8.2	100.0
	1980[c]	—	29.4[d]	48.4[e]	16.9[f]	—	5.3[g]	—	100.0
Brazil	1960[h]	41.6	30.6	19.2	1.9	3.0	3.2	0.5	100.0
	1970	36.0	27.6	22.9	6.1	4.9	2.3	0.1	100.0
	1980[h]	27.2[i]	24.7	28.1	10.1	6.8	3.1	0.1	100.0
Colombia	1960	—	—	—	—	—	—	—	
	1970[j]	21.6	31.1	27.8	10.0	5.6	2.8	1.1	100.0
	1980[k]	11.5	49.2[l]	—	29.8[m]	—	7.7[n]	1.8	100.0
Chile	1960[j]	14.1	21.3	35.2	12.3	10.0	2.3	4.8	100.0
	1970[j]	8.2	15.4	31.6	13.0	13.5	4.0	14.3	100.0
	1980[o]	4.9	11.3	25.9	14.9[p]	33.3[q]	9.7	—	100.0
Mexico	1960[j]	35.4	32.0	24.3	4.6	2.1	1.6	0.0	100.0
	1970[j]	27.1	30.3	29.7	5.9	3.7	3.3	0.0	100.0
	1980[y]	16.2	22.1	34.7	16.2	4.6	6.2	—	100.0
Peru	1960[r]	32.8	—52.2—		—11.6—		2.5[r]	0.9	100.0
	1970[o]	19.3	27.3	28.1	7.9	9.4	4.8	3.2	100.0
	1980[o]	12.4	24.2[t]	18.4[u]	13.1[v]	13.8[w]	12.1[x]	6.0	100.0

[a] Economically Active Population (EAP) 14 years and older.
[b] EAP, 10 years and older;
[c] EAP, 14 years and older;
[d] EAP without any year of schooling and incomplete primary;
[e] EAP with complete primary and incomplete secondary;
[f] EAP with complete secondary and incomplete higher;
[g] EAP with completed higher;
[h] EAP 10 years and older;
[i] Includes EAP with less than one year of studies;
[j] EAP of 12 years and older;
[k] Employed population of 12 years and older;
[l] EAP with some primary;
[m] EAP with some secondary;
[n] EAP with some higher;
[o] EAP, 15 years and older;
[p] EAP with 7 or 8 years of study;
[q] EAP with a minimum of 9 and a maximum of 12 years education;
[r] EAP, 6 years and older;
[s] University education;
[t] EAP, with incomplete primary;
[u] EAP, with complete primary;
[v] EAP, with incomplete secondary;
[w] EAP, with complete secondary;
[x] EAP, with higher education.
[y] Total population of 12 years and older, less those from 12–14 years attending primary school.

Sources: UNESCO, United Nations, *Statistical Yearbook for Latin America and the Caribbean, 1989*. For Mexico, 1980: *X Censo General de Población y Vivienda, 1980*, Resumen General, i, tables 11–15, INEGI, 1986.

of their economically active population with three or fewer years of education in 1980. The equivalent figure for Brazil was 52 per cent.

By contrast, in Peru the economically active population was 'over-educated' for the available non-manual jobs. De Gregori (1985) points to the frustrated mobility aspirations of secondary and university students as one factor in support for the guerrilla movement, *Sendero Luminoso*. A rapid increase in educational levels between 1970 and 1980 resulted in almost 26 per cent of the Peruvian economically active in 1980 having completed secondary education, most of whom would be young entrants to a labour-market which showed little increase in higher non-manual jobs between 1972 and 1982.

A distinctive characteristic of educational levels in Latin America has been their polarization (Tedesco, 1987). Alongside a persistently large percentage of those failing to finish primary school, there are growing numbers with secondary and university education. Though the trend is obscured, to some extent, by the lack of comparable classifications between the censuses, the increase of those with seven or more years of education in Table 8.4 is marked.[7] The rapid increase in non-manual jobs and in people with primary and higher levels of education meant a certain depreciation of these jobs and of educational qualifications. The increase in non-manual workers in Mexico meant a relative decline in their salaries with respect to skilled manual workers (Reyes Heroles, 1983). Whereas literacy was a sufficient qualification for most skilled manual jobs in the 1960s, primary education became essential by the 1970s. Similar increases in required educational qualifications occurred for non-manual jobs. Though part of the demand for higher qualifications arose from the requirements of new jobs based on advanced technology in the services or industry, even the same jobs required higher levels of education in 1980 than in 1960 (Bálan, 1969; Roberts, 1973*b*; CEPAL, 1989: 38–41).

Studies of occupational mobility that obtained lifetime and intergenerational mobility data enable us to give a more detailed account of the nature and extent of mobility than is possible using census data only (Bálan, Browning, and Jelin, 1973; Muñoz, Oliveira, and Stern, 1977).[8] They show that rates of social mobility have indeed been high in Latin America and comparable to those reported for the United States and the European countries since the Second World War. The analysis of life histories showed that migrants to the large cities competed reasonably successfully with urban natives to obtain the better jobs, particularly after a few years of urban residence. This was true, for example, of Balan, Browning, and Jelin's (1973) study of migration to Monterrey, Mexico, that included an analysis of migration from the small town of Cedral which had supplied Monterrey with migrants. However, their relative success was, in part, based on the fact that the migrants of the 1940s and 1950s were quite selective when compared to their populations of origin. These migrants often came from the provincial middle classes and had relatively high levels of education enabling them to take up the clerical and professional jobs that were opening up in the cities. Likewise, the emigration of craftworkers from villages and small towns provided a skilled working class for the cities.

While family status continued to be a significant factor in children's occupational attainment, its impact occurred mainly through its effect on children's education. As a determinant of children's educational attainment, the socio-economic status of the family was more important than place of origin (whether born in rural or urban areas) and continued to influence attainment even with the rapid expansion of educational opportunities in the 1960s (ibid., 1973).

There were, in fact, indications that the increase in private (usually religious) education at primary and secondary levels, in the 1960s and later, reflected the importance placed on education by middle-class parents and their dissatisfaction with overcrowded urban public schools (Roberts, 1973*b*). The 1960s and 1970s saw also an expansion of private universities, both religious and secular, catering for the increased demand for higher education and providing a more privileged educational environment for those with money than did the mass public universities (Levy, 1986). By 1980, education had become the single most important factor in obtaining higher status and better-paid occupations than their parents.

There was a sharp reduction in social mobility in the 1980s because of the declining importance of rural–urban migration, of the stagnation of the region's economies, and of policies aimed at reducing public expenditures that cut clerical employment. The mobility opportunities for the younger generation were, consequently, much less than for their parents. This greater rigidity in the social structure has been an important factor in urban politics, but operated differently according to country and city.

Conclusion

The patterns of urbanization and the transformations in urban social structure have been closely related to developments in the industrial sector that were linked to changes in the international division of labour. Before the 1940s, the various countries of Latin America were primarily exporters of primary products and had a weak development of the internal market. They were mainly rural countries. Non-agricultural employment was, above all, in commerce and in the crafts. Women had a very low participation in urban labour-markets. From the 1940s, import-substitution policies, first in basic goods, and then, at the end of the 1960s, in consumer durable, intermediate, and capital goods, resulted in rapid urbanization and a dramatic transformation in social stratification.

This transformation was based as much on demographic factors as on economic ones. Rural–urban migration was intense and the cities grew rapidly. The dynamism of manufacturing industry made possible the absorption of the increasing supply of workers, which, in the early period, did not grow as fast as it would in subsequent years when women began to enter the urban labour-market in increasing numbers and the new cohorts of workers were swollen by the delayed impact of national population growth. This early absorption made possible the consolidation of a working class—at times with a strong component of workers originating in rural areas—and the possibility of social mobility into non-manual work for those urban sectors that had higher levels of education. Despite the evident problems of social marginality with respect to housing and income, these years constituted a period in which the working classes shared, to a certain extent, in the benefits of development. At this time, there were also signs of the emergence of a welfare-oriented state, as the social bases for populist governments were created. The various forms of populism that appeared in Latin America in the early stages of import-substituting industrialization can be interpreted as based on national differences in the strength and composition of the urban social classes (Di Tella, 1990). Elites seeking to promote industrial development needed political support against entrenched commercial and agrarian interests. The extent and strength of the industrial working class determined, for instance, whether it retained a certain bargaining power while giving that support, or was organized and controlled from above (Weffort, 1973).

In the later stages of import substitution, the basis for even this limited working-class participation in government broke down. This was the period, in the 1960s and 1970s, of what has been called the politics of exclusion in Latin America. Military governments came to power espousing developmentalist and nationalist ideologies of rapid economic growth and seeking to curtail the demands for better wage and living conditions by both working and middle classes. Developmentalist ideology advocates top-down, state-directed development in which the state promotes economic growth at all cost, and thereby strengthens the nation so that, in the ideology, nationalism and developmentalism reinforce each other (Evans, 1989). Both working and middle classes had consolidated in this period, leading to increased demands made on the state—demands that were the more pressing because they were often occurring in the capital.

Urban economies had become more complex, increasingly crucial parts of national economies due to the growth of the internal market, and concentrated in one or two large cities in each country. Though this stage of import substitution was highly dependent on direct investment by multinational companies, the central role of the state in managing industrialization made it an inevitable target, and potential ally, of both multinational and national capital (Evans, 1979). At the same time, the rapid, unplanned growth of the cities and their poverty was fertile ground for neighbourhood-based social movements, such as those of squatters seeking to defend and enhance their settlements. Often, industrial workers and their unions took a lead in these struggles since they mainly lived in unplanned and poorly serviced settlements (Castells, 1983: 181–209; Nelson, 1979; Schuurman and Van Naerssen, 1989). Also, the new middle classes, especially government employees such as teachers, became increasingly effective in demanding higher wages and improved benefits as they became more numerous and better organized.

In the 1970s, a new set of factors began to produce considerable diversification based on the fragmentation and polarization of the different social sectors. The supply of workers increased rapidly—as a result of migration, natural increase, and higher female participation—while the manufacturing sector lost its capacity to absorb labour both as a result of technological change and because of the downturn in the Latin American economies. At the same time, there was a weakening in the developmentalist and nationalist orientation of the state as external indebtedness increased. This period was characterized by economic stagnation, a greater privatization of the economy, including the opening of 'strategic' sectors of the economy to foreign capital, and a reduction in the possibilities of social mobility. There was also, however, the rise of diverse forms of political participation around environmental, human rights, and neighbourhood issues that were weakly institutionalized (Jelin, 1992).

Austerity measures by governments led to urban riots throughout Latin America—the so-called IMF riots (Walton and Ragin, 1989). The new period also coincided with a resurgence of democracy throughout the region as authoritarian governments collapsed in face of economic crisis or, as in the case of Mexico, sought a new legitimacy by opening up the electoral process. In no country of the region, however, did democratic politics appear to be securely anchored through widespread participation in political parties or grass-roots social movements (Touraine, 1987). Underlying this political instability were the changes in urban stratification that we have reviewed.

These changes in urban stratification meant, at one and the same time, modernization, greater diversity, and social inequality. Modernization was clear in the expansion of new middle classes—professionals, managers, technicians, and office workers. These sectors provided the labour needed for an industrialization based on advanced technology and on transnational capital with its attendant services, for the expansion of social services, and for a range of personal services connected with the entertainment and tourist industry.

This modernization of the class structure created the appearance of uniformity among the countries of Latin America and a relative convergence with the class structure of advanced industrial countries. Though the Latin American middle classes represented a smaller proportion of the labour-force than in the industrialized world, they shared the perceptions, aspirations, and demands of their counterparts in Europe and the United States: consumerism, education as a means of social mobility, political interest in ecology, low taxes, and a preference for greater economic liberalism.

But, if by 'middle classes' we refer to groups with high levels of education and job and income security, then the situation in Latin America was very different from that of the industrialized countries. In Latin America, the social and economic bases of the middle classes were weaker than in the industrialized world because of their shallow historical roots and internal heterogeneity. The greater part of the Latin American middle classes were formed by occupations that had only recently been created. These occupations used a considerable amount of female labour, with low levels of education and low incomes (schoolteachers, bank and office workers). Likewise, a part of these non-manual workers came from working-class families. They were in the labour-market to obtain incomes that would allow their families to maintain minimum levels of welfare.

The economic crisis of the 1980s, with its high levels of inflation and sharp drop in real wages, contributed to the relative impoverishment of the middle classes. For these classes, modernization was halted and continuing social mobility less of a prospect. The impoverishment of the middle classes led them, at times, to seek to lower their expenditures by seeking cheaper housing in working-class areas (Portes, 1989; Rolnik, 1989; Valladares, 1989). The consequence was a greater social diversification at the neighbourhood level, diminishing spatial segregation in some cities. There were important differences in these respects between countries—this relative middle-class decline was more accentuated in Argentina than in Mexico or Brazil.

The working class of Latin America was never homogeneous and large-scale industry played only a relatively minor role in its formation. In this respect, there was no repetition of the historical experience that formed a working class in Britain and Germany during the nineteenth century and beginning of the twentieth. In Latin America, between 1930 and 1990, the working class was constituted by service and construction workers as much as by workers in manufacturing industry. Furthermore, the increase in the relative importance of wage labour ended by 1980. It was replaced by an increase in the proportions of the self-employed and non-waged workers.

This tendency led some commentators to emphasize the rise of an 'informal' economy in the region. Informal workers were not, however, a homogeneous sector since they included the very poor for whom self-employment was the only means of subsistence, small-scale entrepreneurs, and disguised wage-workers who worked at home or in a small workshop, but who were subcontracted by large national and multinational firms. Also, there were blurred boundaries between 'independent' workers and the casual workers who moved in and out of large-scale firms. Informalization was part of a secular change in the way in which labour was used and in the organization of labour-markets. As was the case in some advanced industrial countries, there was a move in Latin America towards more flexible forms of contract and more flexible use of labour, that resulted, at times, in greater instability, more part-time work, and fewer labour rights (Marshall, 1987; Michon, 1987; Roberts, 1989*b*). This tendency was based on the pressures to make use of labour in ways compatible with technological change in a highly integrated international market. In this situation, even informal employment had limited possibilities of expansion as was shown by increasing levels of open unemployment in several countries (Portes, 1989).

A further factor in the heterogeneity of the working classes was the centrality of the family for daily survival. The worsening in salaries made it necessary for several members of the household to enter the labour-market. The ever greater presence of households with several workers resulted in a greater occupational diversity within families. The job of the male head of household lost importance as a source of family income and as a source of identity for the members of the family. There were few examples in Latin America of the types of working-class community that were common in nineteenth- and early twentieth-century Europe in which popular culture was shaped by one predominant type of work.

These processes resulted in apparently contradictory tendencies in class formation. There was an increasing fragmentation of the working classes as fewer workers shared a similar position in the labour-market. Common work experience was less likely to be the key factor in social and political identity. Yet, family and community solidarity was important in times of crisis, and generated needs and interests that were common to broad sectors of the population. By 1990, the differences in urban context and in patterns of regional development were probably more important than in previous periods in shaping class relations and class differences. The social and economic significance of being a woman, of being a manual or non-manual worker, or of being self-employed depended on the city and country in which one lived. The heterogeneity of urban social structure and of social mobility in Latin America meant that there was no single pattern of social stratification to provide a common basis for politics in the region. The demands that governments faced from the various sectors of the urban population differed in kind and strength from country to country, affecting the leeway that governments had in adjusting to external economic pressures. In Mexico, these demands were used by government to promote free trade with the United States as a solution to Mexico's economic crisis and led to a certain democratization of the political process. In Peru, in contrast, the 'informalization' of the economy and social structure contributed to authoritarian politics. This heterogeneity is likely to have been one factor in the difficulty that the countries of the region had in the 1980s in making common alliance to negotiate with the advanced industrial countries over such issues as the external debt.

Notes

1. Metropolitan population concentration reached its high point at the beginning of the 1950s, reaching 40.6 per cent of the urban population in 1950, 39.5 per cent in 1960, and 39.4 per cent in 1970.
2. Urban primacy is usually measured by the ratio of the population of the largest city in an urban system to the next largest or next two largest. Urban systems have high primacy if the largest city is two or more times larger than the next largest. The 1990 census figures from Mexico indicate that Mexico City was five times larger than Guadalajara, the second city, as compared with six times larger in 1980.
3. We are grateful to Gustavo Garza for suggestions as to these types.
4. The undercount of population may have been particularly high for Mexico City, but even accepting the highest (and improbable) estimate of the city's 1990 population—19 million—gives an annual growth rate of 3 per cent from 1980 to 1990, indicating a slowing-down of the metropolitan area's growth, due to a slackening of the rate of in-migration and lower birth rates.
5. The PREALC studies of the 1970s tended to emphasize size, low levels of capital endowment, and market—and not state regulation—as the main factors in distinguishing informal and formal sectors.
6. Informal employment was defined as the sum of self-employment, unremunerated family employment, and employment in enterprises of less than five workers (thus including domestic service).
7. This is the case in the figures for Argentina and Colombia. The 1980 figures for Mexico are only available for total population, and not for the economically active population. This difference is likely to produce a slight upward bias by including those still in high school and beyond, but this is likely to be compensated by the downward bias of including those adults who are not part of the economically active population. The educational levels of this segment are, in the Mexican case, slightly lower than of those in the labour-market. CEPAL (1989: table I-11) provides figures for the changes in educational levels of the population of 15–24 years of age between 1960 and 1980 for Chile, Brazil, Panama, Peru, Ecuador, and Uruguay. This young population also shows substantial increases in the proportions with secondary and higher levels of education in all six countries.

8. The CEPAL method of calculating structural mobility consists in comparing equivalent cohorts taken from national censuses at different time periods. There is no guarantee, however, that a cohort identified in the first census will consist of the same people as its counterpart in the next census, and parents' and childrens' occupations cannot be linked.

References

Altamirano, T. (1984), *Presencia Andina en Lima Metropolitana* (Lima: Fondo Editorial, Pontificia Universidad Católica).

—— (1988), *Cultura Andina y Pobreza Urbana* (Lima: Fondo Editorial, Pontificia Universidad Católica).

Arias, P. (1988), 'La Pequeña Empresa en el Occidente Rural', *Estudios Sociologicos*, 6/17, 405–36.

Arizpe, Lourdes (1978), *Migración, Etnicismo y Cambio Económico* (México, DF: El Colegio de México).

—— (1982), 'Relay Migration and the Survival of the Peasant Household', in Helen Safa (ed.), *Towards a Political Economy of Urbanization in Third World Countries* (Delhi: Oxford University Press), 19–46.

—— (1985), *Campesinado y Migración* (México, DF: Secretaría de Educación Pública).

—— (1990), 'Fin de Epoca: Nuevas Opciones', in *México en el Umbral del Milenio* (México, DF: Centros de Estudios Sociológicos, El Colegio de México), 459–74.

Balán, Jorge (1969), 'Migrant-Native Socioeconomic Differences in Latin American Cities: A Structural Analysis', *Latin American Research Review*, 4/1 (Feb. 1969): 3–29.

—— (1981), 'Estructuras Agrarias y Migración en una Perspectiva Histórica: Estudios de Casos Latinoamericanos', *Revista Mexicana de Sociologia*, 43/1, 141–92.

—— (1985), *International Migration in the Southern Cone* (Buenos Aires: Centro de Estudios de Estado y Sociedad).

—— Browning, Harley, and Jelin, Elizabeth (1973), *Men in a Developing Society* (Austin, Tex. and London: ILAS/University of Texas Press).

Bean, Frank D., Edmonston, Barry, and Passel, Jeffrey S. (1990), *Undocumented Migration to the United States: IRCA Experience of the 1980s* (Santa Monica, Calif., and Washington, DC: Rand Corporation and the Urban Institute).

Benería, Lourdes (1991), 'Structural Adjustment, the Labour Market, and the Household: The Case of Mexico', in Guy Standing and Victor Tokman (eds.), *Toward Social Adjustment: Labour Market Issues in Structural Adjustment* (Geneva: International Labour Office), 161–83.

Blanco Sánchez, M. (1990), *Empleo en México: Evolución y Tendencias*, Ph.D. diss. in Social Sciences, El Colegio de México, México, DF.

Brambila, C. (1985), *Migración y Crecimiento Demográfico en la Región Centro*, Centro de Demografía e Desarrollo Urbano (México, DF: El Colegio de México).

Briones, Carlos (1991), 'Economía Informal en el Gran San Salvador', in J. P. Pérez Sainz and Rafael Menjívar (eds.), *Informalidad Urbana en Centroamérica: Entre la Acumulación y la Subsistencia* (San José, Facultad Latinoamericana de Ciencias Sociales and Caracas: Editorial Nueva Sociedad), 91–148.

Bruschini, C. (1989), *Tendéncias da Força de Trabalho Feminina Brasileira nos Anos Setenta a Ochenta*, Comparacoes Regionais, 1/89 (São Paulo: Departamento de Pesquisas Educacionais-Fundacao Carlos Chagas).

Butterworth, D. and Chance, J. K. (1981), *Latin American Urbanization* (Cambridge University Press).

Canak, William L. (1989), 'Debt, Austerity, and Latin America in the New International Division of Labor', In W. L. Canak (ed.), *Lost Promises: Debt, Austerity and Development in Latin America* (Boulder, Colo.: Westview Press), 9–30.

Cantú Gutiérrez, José Juan, and González, Rodolfo Luque (1990), 'Migración a la Zona Metropolitana de la Ciudad de México', *Demos (Carta Demográfica sobre México)* (México, DF: Coordinación de Humanidades, UNAM), 17–18.

Carrillo, Jorge (1989), 'The Restructuring of the Automobile Industry of Mexico: Adjustment Policies and Labor Implications', *Texas Papers on Mexico 89–07*. (Austin, Tex.: Mexican Center, University of Texas).

Carrillo Huerta, Mario (1990), *The Impact of Maquiladoras on Migration in Mexico*, Working Paper no. 51 (Washington, DC: Commission for the Study of International Migration and Cooperative Economic Development).

Castells, Manuel (1983), *The City and the Grassroots* (London: Edward Arnold).

CEPAL (1986), *América Latina: Las Mujeres y los Cambios Socio-ocupacionales 1960–1980*, Documento LC/R.504, División de Desarrollo Social (Santiago: United Nations).

—— (1989), *Transformación Ocupacional y Crísis Social en América Latina* (Santiago: United Nations).

Chamorro, Amália, Chávez, Mario, and Membreño, Marcos (1991), 'El Sector Informal en Nicaragua', in J. P. Pérez Sainz and Rafael Menjívar (eds.), *Informalidad Urbana en Centroamerica: Entre la Acumulación y la Subsistencia* (San José: Facultad Latinoamericana de Ciencias Sociales, and Caracas: Editorial Nueva Sociedad), 217–58.

Chant, Sylvia (1991), *Women and Survival in Mexican Cities: Perspectives on Gender, Labour Markets and Low-Income Households* (Manchester University Press).

Cornelius, Wayne (1991), 'Labor Migration to the United States: Development

Outcomes and Alternatives in Mexican Sending Communities', in S. Díaz-Briquets and S. Weintraub (eds.), *Regional and Sectoral Development in Mexico as Alternatives to Migration* (Washington, DC: Commission for the Study of International Migration and Cooperative Development), 89–131.

Corona, Rodolfo (1988), 'Movilidad Geográfica: Búsqueda de Bienestar', *Demos, 1. (Carta Demográfica sobre México)* (México, DF: Coordinación de Humanidades, UNAM), 7–8.

—— (1989), 'Aspectos Cuantitativos de la Migración Femenina Interestatatal en México, 1950–1980', in Jennifer Cooper, Teresita de Barbieri, Teresa Rendón, Estela Suárez, and Esperanza Tuñón (eds.), *Fuerza de Trabajo Femenina Urbana en México, Características y Tendencias*, (México, DF: Miguel Angel Porrúa-UNAM), 255–303.

—— (1991), 'Confiabilidad de los Resultados Preliminares del XI Censo General de Población y Vivienda de 1990', *Estudios Demográficos y Urbanos*, 16, 33–68.

Cortés, Fernando and Rubalcava, Rosa María (1991), *Autoexplotación Forzada y Equidad por Empobrecimiento* (Mexico, DF: El Colegio de México/Jornadas 120).

De Barbieri, Teresita (1984*a*), 'Incorporación de la mujer a la economía en América Latina', en Memoria del Congreso Latinoamericano de Población y Desarrollo (Mexico, DF: PISPAL-COLMEX-UNAM), 355–89.

—— (1984*b*), *Mujeres y Vida Cotidiana* (Mexico, DF: SEP-Fondo de Culture Económica).

—— and Oliveira, Orlandina de (1987), *La Presencia de las Mujeres en America Latina en una Década de Crisis* (Santo Domingo: Editora Bho).

De Gregori, Carlos (1985), *Sendero Luminoso: Los Hondos y Mortales Desencuentros*, Documento de Trabajo, Serie Antropología, no. 2 (Lima: Instituto de Estudios Peruanos).

Despres, Leo A. (1991), *Manaus: Social Life and Work in Brazil's Free Trade Zone* (Albany, NY: State University of New York).

Di Tella, Torcuato (1990), *Latin American Politics: A Theoretical Framework* (Austin, Tex.: University of Texas Press).

Durand, Jorge and Massey, Douglas (1992), 'Mexican Migration to the United States; A Critical Review', *Latin American Research Review*, 27/2: 3–42.

Durston, J. (1986), 'Transición estructural, movilidad ocupacional y crísis social en America Latina, 1960–1983', Documento de Trabajo, LC/R.547 (Santiago: CEPAL).

Echeverría, Rafael (1985), *Empleo Público en América Latina*, Colección Investigación sobre Empleo, 26, (Santiago: PREALC).

Economic Commission for Latin America (ECLA) (1983), *Statistical Yearbook for Latin America, 1983* (Santiago: ECLA).

Economic Commission for Latin America and the Caribbean (ECLAC) (1989), 'The dynamics of social deterioration in Latin America and the Caribbean in the 1980s', Reference Document LC/G.1557 (Santiago: ECLAC).

—— (1993), *Statistical Yearbook for Latin America and the Caribbean, 1992* (Santiago: United Nations).

Edel, Matthew and Hellman, Ronald G. (1989), *Cities in Crisis: The Urban Challenge in the Americas* (New York: Bildner Center for Western Hemisphere Studies).

Escobar, Agustín (1986), *Con el Sudor de tu Frente* (Guadalajara: El Colegio de Jalisco/CIESAS).

—— (1988), 'The Rise and Fall of an Urban Labor Market: Economic Crisis and the Fate of Small-scale Workshops in Guadalajara, Mexico', *Bulletin of Latin American Research*, 7/2: 183–205.

—— (1992), 'Occupational Mobility in Guadalajara: 1975–1982–1990', paper to Conference on the Sociodemographic Effects of the 1980s Economic Crisis in Mexico, Austin, Tex., April.

—— and Roberts, Bryan R. (1991), 'Urban Stratification, the Middle Classes, and Economic Change in Mexico', in Mercedes Gonzales de la Rocha and Agustín Escobar (eds.), *Social Responses to Mexico's Economic Crisis of the 1980s* (San Diego, Calif.: Center for US–Mexican Studies), 91–113.

Evans, Peter (1979), *Dependent Development: The Alliance of Multinational, State and Local Capital in Brazil* (Princeton University Press).

—— (1989), 'Predatory, Developmental, and Other Apparatuses: A Comparative Political Economy Perspective on the Third World State', *Sociological Forum*, 4/4: 561–87.

Fernández-Kelly, María Patricia (1983), *For We Are Sold, I and My People: Women and Industry in Mexico's Frontier* (Albany, NY: State University of New York).

Filgueira, C. and Geneletti, C. (1981), *Estratificación y Movilidad Ocupacional en América Latina*, Cuaderno de la CEPAL, no. 39 (Santiago: CEPAL).

García, Brígida (1988), *Desarrollo Económico y Absorción de Fuerza de Trabajo en México: 1950–1980* (México, DF: El Colegio de México).

—— and Oliveira, Orlandina de (1992), 'Trabajo Femenino y Cambios Económicos en México, 1976–1987', paper to Conference on the Sociodemographic Effects of the 1980s Economic Crisis in Mexico, Austin, Tex., April.

——, Muñoz, Humberto, and Oliveira, Orlandina de (1979), *Migración, Familia y Fuerza de Trabajo en la Ciudad De México*, Cuadernos del CES, no. 26 (México, DF: CES/El Colegio de México).

—— —— —— (1981), 'Migration, Family Context and Labour-force Participation in Mexico City', in Jorge Balán (ed.), *Why People Move* (Paris: UNESCO).

—— —— —— (1982), *Hogares y Trabajadores en la Ciudad de México* (Mexico, DF: El Colegio de México e Instituto de Investigaciones Sociales, UNAM).

—— —— —— (1983), *Familia y Mercado de Trabajo, un Estudio de dos Ciudades Brasileñas* (México, DF: El Colegio de México e Instituto de Investigaciones Sociales-UNAM).

García y Griego, Manuel (1990), 'Emigration as a Safety Net for Mexico's Labor Market: A Post-IRCA Approximation', in Georges Vernez (ed.), *Immigration and International Relations: Proceedings of a Conference on the International Effects of the 1986 Immigration Reform and Control Act* (Santa Monica, Calif.: Rand Institute, and Washington, DC: Urban Institute), 115–34.

Garza, Gustavo and Departamento del Distrito Federal (1987), *Atlas de la Ciudad de México* (México, DF: Departamento del Distrito Federal).

Gilbert, Alan and Ward, Peter (1985), *Housing, the State and the Poor: Policy and Practice in Three Latin American Cities* (Cambridge University Press).

—— and Varley, Ann (1990), 'Renting a Home in a Third World City: Choice or Constraint?', *International Journal of Urban and Regional Research*, 14: 89–108.

Goldani, Ana María (1977), 'Impacto de los Inmigrantes sobre la Estructura y el Crecimiento del Área Metropolitana', in Humberto Muñoz, Orlandina de Oliveira, and Claudio Stern, *Migración y Desigualdad Social en la Ciudad de México* (México, DF: El Colegio de México e Instituto de Investigaciones Sociales, UNAM), 129–37.

González de la Rocha, Mercedes (1986), *Los Recursos de la Pobreza: Familias de Bajos Ingresos en Guadalajara* (México: El Colegio de Jalisco/CIESAS y SPP).

—— (1987), 'Crisis, Economía Doméstica y Trabajo Femenino en Guadalajara', paper given at Coloquio de Estudios de la Mujer, Encuentro de talleres, México, DF, El Colegio de México-PIEM, March.

—— (1988*a*), 'Economic Crisis, Domestic Reorganization and Women's Work in Guadalajara, Mexico', *Bulletin of Latin American Research*, 7/2: 207–23.

—— (1988*b*), 'De por qué las Mujeres Aguantan Golpes y Cuernos: Un Análisis de Hogares sin Varón en Guadalajara', in Luisa Gabayet *et al.*, *Mujeres y Sociedad: Salario, Hogar y Acción Social en el Occidente de México* (Guadalajara: El Colegio de Jalisco/CIESAS del Occidente), 205–27.

—— and Escobar Latapí, Agustín (1988), 'Crisis and Adaptation: Households of Guadalajara', Texas Papers on Mexico, Pre-publication Working Papers of the Mexican Center, Institute of Latin American Studies, University of Texas at Austin.

—— ——, de la O, María, and Castellaños, Martínez (1990), 'Estrategias Versus Conflicto: Reflexiones Para el Estudio del Grupo Doméstico en Época de Crisis', in Guillermo de la Peña, Juan Manuel Durán, Agustín Escobar, and Javier García de Alba (eds.), *Crisis, Conflicto y Sobrevivencia: Estudios Sobre la Sociedad Urbana en México* (Guadalajara: Universidad de Guadalajara, and Tlalpan, México DF: CIESAS), 355–88.

Hardoy, Jorge E. (1975), 'Two Thousand Years of Latin American Urbanization', in Jorge Hardoy (ed.), *Urbanization in Latin America* (Garden City, NY: Anchor Books), 3–55.

Humphrey, John (1987), *Gender and Work in the Third World: Sexual Divisions in Brazilian Industry* (London and New York: Tavistock Publications).

Iglesias, Enrique (1983), 'La Crisis Económica Internacional y las Perspectivas de América Latina', in Centro Latinoamericano de Economía Humana, *América Latina y la Crisis Internacional*, Serie Estudios CLEH no. 29 (Montevideo), 53–76.

—— (1985), 'Balance Preliminar de la Economía de América Latina en 1984', *Comercio Exterior*, 35/2: 171–90.

Instituto Brasileiro de Geografia e Estatística (IBGE) (1992), *Censo Demográfico de 1991, Análises Preliminares*, i. Fundação Instituto Brasileiro de Geografía e Estatística.

Instituto Nacional de Estadística, Geografía e Informática (INEGI) (1977), *Encuesta Nacional de Empleo Urbano: Indicadores Trimestrales de Empleo* (México, DF: INEGI).

—— (1988), *Encuesta Nacional de Empleo Urbana: Indicadores Trimestrales de Empleo (Enero-Marzo de 1987)* (Aguascalientes: INEGI).

—— (1990), *Resultados Preliminares, XI Censo General de Población y Vivienda, 1990* (Aguascalientes: INEGI).

International Labour Office (1989), *World Labour Report, 1989* (Geneva: International Labour Office).

Jelin, Elizabeth (1978), *La Mujer y el Mercado de Trabajo Urbano* (Estudios CEDES, Buenos Aires: CEDES).

—— (1984), *Familia y Unidad Doméstica: Mundo Público y Vida Privada* (Estudios CEDES, Buenos Aires: CEDES).

—— (1992), *De la Clase al Movimiento: Procesos Sociales Urbanos en América Latina, 1930–1990* (Buenos Aires: CEDES).

Kowarick, Lucio (1977), *The Logic of Disorder: Capitalist Expansion in the Metropolitan Area of Greater São Paulo*, Discussion Paper no. 102, Institute of Development Studies (Brighton: University of Sussex).

—— (1979), *Espoliacáo Urbana* (Rio de Janeiro: Editorial Paz e Terra).

Lattes, Alfredo E. (1984), *Algunas Dimensiones Demográficos de la Urbanización Reciente y Futura en América Latina'*, Cuadernos del CENEP, no. 31 (Buenos Aires: Centro de Estudios de Población).

Lawton, R. (1978), 'Census Data for Urban Areas,' in R. Lawton (ed.), *The Census and Social Structure* (London: Frank Cass), 82–145.

Leff, Gloria (1976), *Las Migraciones Femeninas a la Ciudad de México*, Informe de investigación (Mexico, DF: El Colegio de México).

Levy, Daniel (1986), *Higher Education and the State in Latin America: Private Challenges to Public Dominance* (University of Chicago Press).

Lustig, Nora and Ros, Jaime (1986), 'Economic Crisis and Living Standards in Mexico: 1982–1985', document prepared for UNU/WINDER Conference on 'Stabilization and Adjustment Programs and Policies', August, Helsinki.

Margulis, Mario and Rodolfo Tuirán (1986), *Desarrollo y Población en la Frontera Norte: El Caso de Reynosa* (Mexico, DF: El Colegio de México).

Marshall, Adriana (1978), *El Mercado de Trabajo en el Capitalismo Periférico: El Caso de Argentina* (Santiago: PISPAL).

—— (1987), 'Non-Standard Employment Practices in Latin America', Discussion Paper OP/06/1987 (Geneva: International Institute for Labour Studies).

Massey, Douglas, Durand, Jorge, Alarcón, Rafael, and González, Humberto (1987), *Return to Aztlan: The Social Process of International Migration from Western Mexico* (Berkeley: University of California Press).

Merrick, T. W. and Schmink, M. (1978), 'Female-Headed Households and Urban Poverty in Brazil', paper presented at the workshop on 'Women in poverty: What do we know?', Belmont Conference Center, April.

Mesa-Lago, Carmelo (1978), *Social Security in Latin America: Pressure Groups, Stratification and Inequality* (Pittsburgh: University of Pittsburgh Press).

—— (1983), 'Social Security and Extreme Poverty in Latin America', *Journal of Development Economics*, 12: 83–110.

—— (1986), 'Social Security and Development in Latin America', *CEPAL Review*, 28: 135–50.

Michon, F. (1987), 'Time and Flexibility: Working Time in the Debate on Flexibility', *Labor and Society*, 12/1: 3–17.

Muñoz, Humberto, Oliveira, Orlandina de, and Stern, Claudio (1977), *Migración y Desigualdad Social en la Ciudad de México* (México, DF: Instituto de Investigaciones Sociales, Universidad Autónoma de México y El Colegio de México).

Negrete Salas, M. and Sánchez, Héctor Salazar (1986), 'Zonas Metropolitanas en México, 1980', *Estudios Demográficos y Urbanos* 1/1: 97–124.

Nelson, Joan M. (1979), *Access to Power: Politics and the Urban Poor in Developing Nations* (Princeton University Press).

Oliveira, Orlandina de (1975), 'Industrialization, Migration and Entry Labor Force Changes in Mexico City, 1930–1970', Ph.D. thesis, Department of Sociology, University of Texas at Austin.

—— (1989*a*), 'Empleo Femenino en México en Tiempos de Recesión Económica: Tendencias Recientes', in J. Cooper, T. de Barbieri, T. Rendon, E. Suarez, and E. Tuñon (comps.), *Fuerza de Trabajo Femenina Urbana en México* (Mexico, DF: Coordinacíon de Humanidades, UNAM/Porrúa), 29–66.

—— and García, Brígida (1984), 'Migración a Grandes Ciudades del Tercer Mundo: Algunas Implicaciones Sociodemográficas', *Estudios Sociológicos*, 2/4: 71–103.

—— —— (1988), 'Expansión del Trabajo Femenino y Transformación Social en México: 1950–1987', mimeo (Mexico, DF: Centro de Estudios Sociológicos, El Colegio de México).

—— —— (1990), 'Cambios en Fecundidad, Trabajo y Condición Femenina en México', paper presented at the 12th World Congress of Sociology, Madrid.

—— and Roberts, Bryan (1989), 'Los Antecedentes de la Crisis Urbana: Urbanización y Transformación Ocupacional en América: 1940–1980', in

D. Veiga and M. Lombardi (eds.), *Las Ciudades en Conflicto: Una Perspectiva Latinoamericana* (Montevideo: Ediciones de la Banda Oriental), 23–80.

—— —— (1994), 'Urban Growth and Urban Social Structure in Latin America, 1930–1990', in L. Bethell (ed.), *The Cambridge History of Latin America*, vi (Cambridge University Press), 253–324.

Pantelides, Edith (1976), *Estudios de la Población Femenina Económicamente Activa en América Latina, 1950–1970*, serie C, no. 16 (Santiago: CELADE).

Pedrero Nieto, Mercedes (1990), 'Evolución de la Participación Economica Femenina en los Ochenta', *Revista Mexicana de Sociología*, 52/1: 133–49.

de la Peña, Guillermo, Durán, Juan Manuel, Escobar, Agustin, and de Alba, Javier García (1990) (eds.), *Crisis, Conflicto y Sobrevivencia: Estudios Sobre la Sociedad Urbana en Mexico* (Guadalajara: Universidad de Guadalajara, and Tlalpan, Mexíco DF: CIESAS).

Pérez Sainz, J. P. (1990), *Ciudad, Subsistencia e Informalidad: Tres Estudios sobre el Area Metropolitana en Guatemala* (Guatemala: FLASCO-Guatemala).

—— and Menjívar, Rafael (1991), *Informalidad Urbana en Centroamerica: Entre la Acumulación y la Subsistencia* (San José: Facultad Latinoamericana de Ciencias Sociales and Caracas: Editorial Nueva Sociedad).

Portes, Alejandro (1985), 'Latin American Class Structures', *Latin American Research Review*, 20/3: 7–39.

—— (1989), 'Latin American Urbanization During the Years of the Crisis', *Latin American Research Review*, 24/3: 7–49.

—— and Schauffler, Richard (1993), 'Competing Perspectives on the Latin American Informal Sector', *Population and Development Review*, 19/1: 33–60.

—— Castells, Manuel, and Benton, Lauren A. (1989) (eds.), *The Informal Economy: Studies in Advanced and Less Developed Countries* (Baltimore: Johns Hopkins University Press).

PREALC (1982), *Mercado de Trabajo en Cifras 1950–1980* (Santiago: International Labour Office).

—— (1983), *Empleo y Salarios* (Santiago: International Labour Office).

—— (1987), 'Pobreza y Mercado de Trabajo en Cuatro Países: Costa Rica, Venezuela, Chile y Perú', *Documentos de Trabajo* (Santiago: International Labour Office).

—— (1988), 'La Evolución del Mercado Laboral Entre 1980 y 1987', Working Document Series, no. 328 (Santiago: International Labour Office).

Raczynski, Dagmar (1983), *La Población Migrante en los Mercados de Trabajo Urbanos: El Caso de Chile*, Notas Técnicas, no. 55 (Santiago: CIEPLAN).

—— and Serrano, Claudia (1984), *Mujer y Familia en un Sector Popular Urbano: Resultado de un Estudio de Caso* (Santiago: CIEPLAN).

Ramos, Silvina (1984), *Las Relaciones de Parentesco y Ayuda Mutua en los Sectores Populares Urbanos: Un Estudio de Caso* (Estudios CEDES, Buenos Aires: CEDES).

Recchini de Lattes, Zulma (1977), 'Empleo Femenino y Desarrollo Económico: Algunas Evidencias', *Desarrollo Económico*, 17/66: 301–17.

—— (1983), *Dinámica de la Fuerza de Trabajo Femenina en la Argentina* (Paris: UNESCO).

—— and Mychaszula, Sonia María (1993), 'Female Migration and Labour Force Participation in a Medium-sized City of a Highly Urbanized Country', in *Internal Migration of Women in Developing Countries: Proceedings of the United Nations Expert Meeting on the Feminization of Internal Migration* (New York: United Nations), 154–77.

Reyes Heroles, J. (1983), *Política Macroeconómica y Bienestar en México* (Mexico, DF: Fondo de Cultura Económica).

Roberts, Bryan R. (1973*a*), *Organizing Strangers: Poor Families in Guatemala City* (Austin, Tex. and London: University of Texas Press).

—— (1973*b*), 'Education, Urbanization and Social Change', in R. Brown (ed.), *Knowledge, Education and Cultural Change* (London: Tavistock), 141–62.

—— (1979), *Cities of Peasants: The Political Economy of Urbanization in the Third World* (Beverly Hills, Calif.: Sage).

—— (1989*a*), 'Employment Structure, Life-cycle and Life Chances: Formal and Informal Sectors in Guadalajara', in A. Portes, M. Castells, and L. Benton (eds.), *The Informal Economy in Comparative Perspective* (Baltimore: Johns Hopkins University Press), 41–59.

—— (1989*b*), 'The Other Working Class: Uncommitted Labor in Britain, Spain and Mexico', in M. Kohn (ed.), *Crossnational Research in Sociology* (Beverly Hills, Calif.: Sage), 352–72.

—— (1991), 'The Changing Nature of Informal Employment: The Case of Mexico', in Guy Standing and Victor Tokman (eds.), *Towards Social Adjustment: Labor Market Issues in Structural Adjustment* (Geneva: International Labor Office), 115–40.

—— (1993), 'The Dynamics of Informal Employment in Mexico', in Gregory Schoepfle and Jorge Pérez-Lopez (ed.), *Work Without Protections: Case Studies of the Informal Sector in Developing Countries* (Washington, DC: US Department of Labor, Bureau of International Labor Affairs), 101–25.

Rodriguez, Daniel (1987), Agricultural Modernization and Labor Markets in Latin America: The Case of Fruit Production in Central Chile, Ph.D. diss., University of Texas at Austin.

Roldán, Marta and Benería, Lourdes (1987), *The Crossroads of Class and Gender* (University of Chicago Press).

Rolnik, Raquel (1989), 'El Brasil Urbano de los Años 80: Un Retrato', in M. Lombardi and D. Veiga (eds.), *Las Ciudades en Conflicto* (Montevideo: CIESU/Ediciones de la Banda Oriental), 175–94.

Saldanha, Rosangela, Mais, Rosane, and Camargo, José Marangoni (1988), *Emprego e Salário no Sector Público Federal*, Textos para la Discussáo, 5 (Brasilia: Ministry of Employment/SES).

Schuurman, F. and Van Naerssen, T. (1989) (eds.), *Urban Social Movements in the Third World* (London: Routledge, Chapman and Hall).

Secretaría de Programación y Presupuesto (SPP) (1979), *La Ocupación Informal en Areas Urbanas* (Mexico, DF: SPP).

Selby, Henry A., Murphy, Arthur D., and Lorenzen, Stephen A. (1990), *The Mexican Urban Household Organizing for Self-Defense* (Austin, Tex.: University of Texas Press).

Singelmann, Joachim (1993), 'Levels and Trends of Female Internal Migration in Developing Countries, 1960–1980', in *Internal Migration of Women in Developing Countries: Proceedings of the United Nations Group Meeting on the Feminization of Internal Migration* (New York: United Nations), 77–93.

Smith, Gavin (1984), 'Confederations of Households: Extended Domestic Enterprises in City and Country', in Norman Long and Bryan R. Roberts, *Miners, Peasants, and Entrepreneurs: Regional Development in the Central Highlands of Peru* (Cambridge University Press), 217–34.

—— (1989), *Livelihood and Resistance: Peasants and the Politics of Land in Peru* (Berkeley: University of California Press).

Tapia Curiel, J. (1984), *El Estado Nutricional en los Niños de Dos Grupos Sociales de Guadalajara*, Cuadernos de Difusión Cientifica, Serie Salud Pública, no. 2 (Guadalajara: Universidad de Guadalajara).

Tedesco, Juan Carlos (1987), *El Desafio Educativo: Calidad y Democracia* (Buenos Aires: Grupo Editoral Latinoamericano).

Telles, E. (1988), 'The Consequences of Employment Structure in Brazil: Earnings, Socio-demographic Characteristics and Metropolitan Differences', Ph.D. diss., University of Texas at Austin.

—— (1994), 'Industrialization and Racial Inequality in Employment: The Brazilian Example', *American Sociological Review*, 59: 46–63.

Tironi, E. (1987), 'Pobladores e Integración Social', *Proposiciones*, 14: 64–84.

Tokman, Victor (1987), 'El Sector Informal: Quince Años Después', *El Trimestre Económico*, 215: 513–36.

—— (1991), 'The Informal Sector in Latin America: From Underground to Legality', in Standing and Tokman (eds.), *Towards Social Adjustment*, 141–57.

Touraine, Alain (1987), *Actores Sociales y Sistemas Políticos en América Latina* (Santiago: PREALC).

UNESCO (1983), *Statistical Yearbook* (Geneva: Unesco).

United Nations (1980), *Patterns of Urban and Rural Population Growth*, Population Studies 68 (Department of International Economic and Social Affairs, New York: United Nations).

—— (1989), Statistical Yearbook, Latin America and the Caribbean (Santiago: ECLAC).

Valdes, Teresa and Weinstein, Mariza (1993), *Mujeres que Sueñan: Las Organizaciones de Pobaldores en Chile, 1973–1989* (Santiago: Libros FLACSO).

Valladares, Licia (1989), 'Rio de Janeiro: La Visión de los Estudiosos de lo Urbano', in M. Lombardi and D. Veiga (eds.), *Las Ciudades en Conflicto* (Montevideo: CIESU/Ediciones de la Banda Oriental), 195–222.

Walton, John and Ragin, Charles (1989), 'Austerity and Dissent: Social Bases of Popular Struggle in Latin America', in W. L. Canak (ed.), *Lost Promises* (Boulder, Colo.: Westview Press), 216–32.

Ward, Peter (1990), *Mexico City: The Production and Reproduction of an Urban Environment* (London: Belhaven Press).

—— Jimenez, Edith, and Jones, Gareth (1993), 'Residential Land Price Changes in Mexican Cities and the Affordability of Land for Low-income Groups', *Urban Studies*, 30: 1521–43.

Weffort, Fransisco C. (1973), 'Clases Populares y Desarrollo Social', in F. C. Weffort and A. Quijano (eds.), *Populismo, Marginalización y Dependencia* (San José: Editorial Universitaria Centroamericana), 17–169.

Wood, Charles and Carvalho, José de (1988), *The Demography of Inequality in Brazil* (Cambridge University Press).

Yujnovsky, Oscar (1976), 'Urban Spatial Configuration and Land Use Policies in Latin America', in Alejandro Portes and Harley L. Browning (eds.), *Current Perspectives in Latin American Research* (Austin, Tex.: Institute of Latin American Studies, University of Texas at Austin), 17–42.

Name Index

Abernethy, D. 242
Abu-Lughod, J. 184–208, 185, 187, 194, 207
Achebe, C. 245
Adam, H. 239
Afonja, S. 233
Ahmad Baba 214
Akbar 22
Alatas, S. 164
Alsayyad, N. 207
Altamirano, T. 292
Alves, M. H. Moreira 11
Anaf, A. 164, 165
Aphilas 218
Arias, P. 278
Arizpe, L. 259, 261, 262, 284, 292
Ashoka (Asoka) 20
Aten, A. 146
Aurangzeb 23
Austen, R. A. 217
Awanohara, S. 137
Azis, J. J. 154

Babur 21
Baker, P. H. 238
Bakir, S. Z. 163
al-Bakrī 214
Balán, J. 260, 291, 292, 293, 296, 297
Balazs, E. 56
Banton, M. 228, 235, 237
Bapat, M. 125
Barlow, G. 56
Barnes, S. T. 240
Basham, A. L. 20
Bates, R. H. 66, 224
Bean, F. D. 260
Becker, C. M. 93, 243
Behrman, J. R. 244
Benería, L. 278, 284
Benton, L. A. 278
Bergquist, C. 11
Birks, J. S. 196
Blanco Sánchez, M. 274
Blank, G. 66
Booth, A. 151, 175
Boxer, C. R. 50
Bradshaw, Y. 12, 243
Brambila, C. 261
Briones, C. 280
Brown, B. B. 244
Brown, K. 207
Browning, H. 291, 296
Bruschini, C. 277
Busteed, H. E. 24
Buthelezi, G. M. 238
Butterworth, D. 258, 259

Camargo, J. M. 273
Canak, W. 255
Cantú Gutiérrez, J. J. 260
Carillo, J. 263
Carillo Huerta, M. 262
Carvalho, J. de 286, 288, 289, 290, 292
Castells, M. 11, 278, 299
Castles, L. 85, 139, 141, 142, 163
Celik, Z. 207
Cell, C. P. 65
Chamorro, A. 283
Chan, Kam Wing 62, 64
Chance, J. K. 258, 259
Chandler, D. P. 48
Chandler, T. 242
Chang, K. C. 32, 55
Chang, S. D. 34
Chant, S. 261
Charnock, Job 24
Chávez, M. 283
Chen, Jiyuan 85
Chen, Xiangming 60–90, 62, 67, 70, 77, 81
Ch'in Shih Huang Ti 33
Cobban, J. L. 147
Coedes, G. 48, 49
Cohen, A. 238, 244
Cohen, H. 146
Collier, W. R. 164
Connah, G. 215–16, 242
Copans, J. 245
Coquery-Vidrovitch, C. 242
Cornelius, W. 260
Corona, R. 260, 261, 268

Cortés, F. 282, 284, 288
Cotton, H. E. 24
Crenshaw, E. M. 12, 66, 74, 85

Dandekar, H. C. 56
Davis, D. 8
De Barbieri, T. 275, 276, 285
De Gonzalez, N. L. S. 165
De Gregori, C. 296
De Haan, F. 142
Demko, G. J. 176
Despres, L. A. 263
Diamond, L. 240
Di Tella, T. 298
Douglass, M. 174, 176, 177
Drake, P. W. 11
Durand, J. 260
Durston, J. 256, 293

Eberhard, W. 56
Echeverría, R. 273
Edel, M. 271
Edwardes, S. M. 56
Ekeh, P. 244
Epstein, A. L. 235
Escobar Latapí, A. 274, 276, 278, 282, 288, 291
Evans, P. 299
Evers, H. 168
Ezana 218

Fairbank, J. K. 56
Feldman, H. 56
Fernández-Kelly, M. P. 261
Figueira, C. 261
Findley, S. E. 12, 61, 64, 74, 243
Firman, T. 152
Fisher, C. A. 48, 49, 148
Fisher, H. B. 175
Flanagan, W. G. 229, 234, 243, 245
Fox, R. G. 55
Freeman-Grenville, G. S. P. 215
Freund, B. 245
Friedman, S. 240
Fuchs, R. J. 176

Gama, Vasco da 20, 23
Gao, Xiaoyuan 88
García, B. 274, 276, 277, 278, 289, 290, 291, 293
García y Griego, M. 260
Gardiner, P. 154
Garza, G. 267
Geneletti, C. 293
Germen, A. 207
Gernet, J. 56
Gilbert, A. 12, 61, 244, 269, 270
Gillion, K. L. 56
Ginsburg, N. 67, 70, 72, 154
Gluckman, M. 235
Gold, T. B. 83
Goldani, A. M. 261
Goldstein, A. 75
Goldstein, S. 74, 75
González, R. L. 260
González de la Rocha, M. 276, 282, 284, 285, 291, 293
Grunebaum, G. von 207
Gugler, J. 1–14, 6, 7, 12, 210–51, 223, 229, 231, 232, 234, 237, 243, 244, 245

Hageman, J. 153
Haggett, P. 152
Haile Selassie 214
Hakim, B. 207
Hall, D. G. E. 48
Hall, J. W. 45
Hambly, G. 22
Hamer, A. D. 175
Hansen, K. T. 233
Hardoy, J. E. 269
Harris, J. R. 244
al-Hathloul, S. 207
Hay, Jr., R. 207
Heeren, H. J. 163
Hellman, R. G. 271
Henthorn, W. E. 56
Herodotus 212
Hilton, A. 218
Hindson, D. 226
Hoddinott, J. 243
Hopkins, A. G. 242
Hoselitz, B. F. 66
Hu, Dapeng 70
Hu, Xin 74, 75, 76
Hugo, G. J. 132–83, 133, 134, 138, 144, 150, 152, 162, 163, 164, 167, 168, 171, 172, 174
Humphrey, J. 275
Hussain, Saddam 191

Ibn al-Mukhtār 214
Ibn Battuta 38
Ibrahim, S. E. 197
Iglesias, E. 286, 288

Jain, M. K. 105
Jamal, V. 224
Jamison, D. T. 80
Jelin, E. 275, 291, 292, 296, 297, 300
John of Marignolli 38
Johnson, D. G. 66
Jones, G. W. 174, 177

Kasarda, J. D. 12, 66, 74, 85
Kelkar, V. 99
Kerr, M. 205
Kipling, R. 25, 27
Kirkby, R. J. R. 64
Koentjaraningrat 163
Koestoes, R. H. 163
Koppel, B. 154
Kowarick, L. 270
Kumar, R. 99

Lanning, G. 56
Laquian, A. A. 176
Lattes, A. E. 260
Lawton, R. 257
Lee, K. B. 56
Leeds, A. 237
Lee-Smith, D. 243
Leff, G. 262
Leinbach, T. R. 163
Leung, Shu-yin 83
Levine, H. B. 12
Levine, M. W. 12
Levy, D. 297
Lewandowski, S. 56
Lewis, W. A. 119
Li, Mengbai 74, 75, 76
Li, Si-ming 76
Lim, E. 83
Lim, Lin Lean 164
Lin, Nan 83
Lipton, M. 66, 67
Liu, Zuefeng 74
Lofchie, M. F. 224
London, B. 61, 243
Lu, Han-long 83
Lucas, R. E. B. 243
Luckham, R. 245
Ludwar-Ene, G. 223, 229, 234, 243
Lustig, N. 288

Mabogunje, A. L. 215, 242
MacAndrews, C. 175
McBeth, J. 177
McCune, G. M. 56
McGee, T. G. 12, 138, 139, 140, 149, 151, 152, 154
Mais, R. 273
Mandela, N. 241
Mantra, I. B. 163, 168, 172
Margulis, M. 291
Marshall, A. 260, 278, 291, 301
Martin, J. R. 54
Martin, M. 62
Marx, K. 70
Massey, D. 260
Maxey, K. 226
Mauny, P. A. 214
Mayer, P. 235
Mehta, S. D. 56
Membreño, M. 283
Memon, P. A. 243
Menjívar, R. 280, 283
Merrick, T. W. 293
Mesa-Lago, C. 273
Michon, F. 301
Mills, E. S. 93
Milone, P. D. 143, 145, 147
Mohan, R. 92–131, 93, 96, 101, 108, 109, 128, 130
Molo, M. 163
Moodley, K. 239
Morris, M. D. 56
Morrison, A. R. 243
Moßbrucker, H. 12
Mote, F. W. 56
Muñoz, H. 289, 290, 291, 293, 296
Munro-Hay, S. 214
Murphey, R. 16–58, 24, 30, 42, 51, 53, 56
Murray, P. 64, 65
Mychaszula, S. M. 262

Nasser, Gamal Abdul 189
Negrete Salas, M. 267
Nelson, J. M. 299
Nelson, N. 233
Ngugi wa Thiong'o 211, 245
Noonan, R. 12, 243

O'Connor, A. M. 220, 242
Oliveira, O. de 252–314, 264, 274, 276, 277, 278, 284, 285, 289, 290, 291, 293, 296
Orleans, L. A. 62

Paden, J. N. 238
Pant, C. 93, 109, 130

Pantelides, E. 293
Parfitt, T. W. 245
Parish, W. L. 60–90, 62, 66, 67, 70, 80
Parkinson, C. N. 53
Pedrero, N. 277
Pérez Sainz, J. P. 280, 283, 292
Perkins, D. H. 66, 83
Polo, Marco 38
Portes, A. 12, 66, 262, 271, 275, 278, 280, 282, 286, 288, 289, 301, 302
Possehl, G. L. 55

Raczynski, D. 262, 276, 284
Raffles, T. S. 139, 141
Ragin, C. 300
Ramos, S. 276
Ranneft, J. M. 142, 143, 147, 148
Recchini de Lattes, Z. 262, 293
Reid, A. 138, 139
Renaud, B. 153
Reyes Heroles, J. 281, 296
Richardson, H. W. 66
Rietveld, P. 133
Riley, S. P. 245
Roberts, B. R. 13, 252–314, 258, 264, 270, 278, 280, 282, 288, 290, 296, 297, 301
Robinson, D. 242
Rodriguez, D. 266
Roldán, M. 278
Rolnik, R. 270, 271, 275, 301
Ros, J. 288
Rouch, J. 228
Rubalcava, R. M. 282, 284, 288

Saad, E. N. 214
Sabot, R. H. 227
al-Sa'di 215
Saksena, B. P. 22
Saldanha, R. 273
Sánchez, H. S. 267
Sansom, G. 44
Schatzberg, M. G. 237, 240
Schauffler, R. 282
Scheltema, A. M. P. A. 147
Schmink, M. 293
Schuurman, F. 299
Schwarz, A. 134
Sekhar, A. U. 113
Selby, H. A. 284, 289
Serrano, C. 276, 284
Shah Jahan 22
Shefner, J. 11
Shi, Ruohua 74
Shih Hwang-ti 33
Sibero, A. 175
Sinclair, C. A. 196
Singelmann, J. 261
Singer, M. 20
Sinha, K. N. 25
Skinner, G. W. 67, 146
Sklar, R. L. 238, 240
Smith, A. 54
Smith, D. A. 61, 243
Smith, G. 259, 292
Soeharto 151
Soekarno 149
Solinger, D. J. 74, 75, 76
Soto, H. de 85
Stark, O. 243
Steinberg, D. J. 49
Stern, C. 296
Storry, R. 44
Stren, R. E. 12, 224–5, 242
Sung, Shuwei 66
Suwarno, B. 163
Szelenyi, I. 64, 65

Tamerlane 21
Tapia Curiel, J. 284
Tedesco, J. C. 296
Telles, E. 280, 281, 292
Thottan, P. 101
Tian, Xueyuan 80
Tidrick, G. 85
Timur 21
Tironi, E. 284
Todaro, M. P. 244
Tokman, V. 278, 280
Tolley, G. S. 82
Touraine, A. 300
Tughuluq (Tughluq), Firuz 21
Tuirán, R. 291
Turok, B. 226

Valdes, T. 284
Valladares, L. 270, 271, 301
Van Naerssen, T. 299
Van Velsen, J. 229
Varley, A. 270
Vatikiotis, M. 173
Vries, E. de 146

Wadia, A. R. 56
Walton, J. 300
Ward, P. 269, 270, 271, 272, 275

Weber, M. 70
Weeks, J. 224
Weffort, F. C. 298
Weinstein, M. 284
Weisner, T. S. 5, 230
White, R. R. 224–5
Whyte, M. K. 70, 80
Wilks, I. 218
Wiseman, J. 241
Wolf, D. L. 164, 165
Wolfe, B. L. 244
Wolpert, S. 21
Wood, A. 83
Wood, C. 286, 288, 289, 290, 292
Wright, A. 38
Wright, G. 187

Xie, Bai-San 81, 82
Xu, Xueqiang 62, 68, 76

Yeh, Anthony Gar-On 68
Yujnovsky, O. 269
Yule, H. 38
Yunus, D. 163

Zelinsky, W. 148
Zhang, Yidi 87
Zhou, Yixing 70

Subject Index

(Cities and sites are listed by country)

administration 19–51, 67, 138, 145, 149, 151, 155, 168, 224, 256, 257, 273
  see also urban governance
Africa South of the Sahara 1–13, 28, 67, 210–51
  see also individual countries
Aksumite state 213, 214, 216, 218
Algeria 189, 190, 192, 193, 197–8, 202–3
  Algiers 184, 197, 203
Angola 219, 222
  Luanda 210, 213, 219
  São Salvador (Mbanza Kongo) 213, 218
  see also Kongo (kingdom)
Arab states 1–13, 184–208
  see also individual countries
architecture 220–1
Argentina 254–303
  Buenos Aires xx, 252, 256, 259, 260, 264, 265, 266, 270, 291, 292, 293
  Córdoba 252, 263
  Rosario 252
Ashanti 218, 219
Asia 16–58
  see also individual countries
Aymara 292

Bahrain 192, 193, 200, 203
Bangladesh 19–31
  Chittagong 16, 25
  Dacca xx, 16, 25
  see also India
Benin 222, 241, 244
  Abomey 213, 218
  kingdom, *see* Nigeria, Benin
Bolivia 260, 292
  La Paz 252
Botswana 13, 233, 243, 244
Brazil 254–303
  Belém 252
  Belo Horizonte 252, 265, 293
  Brasília 252
  Campinas 252
  Curitiba 252
  Fortaleza 252, 266
  Goiânia 252
  Manaus 252, 263–4
  Minas Gerais 266
  Paraná 266
  Pôrto Alegre 252, 265
  Recife 252, 265, 266, 293
  Rio de Janeiro xx, 252, 265, 270, 271
  Rio Grande do Sul 266
  Salvador 252, 265, 266
  San José dos Campos 293
  Santos 252
  São Paulo xx, 252, 256, 263, 264, 265, 266, 270, 271, 286
Brunei:
  Bandar Seri Begawan (Brunei) 17
Burkina Faso 222
Burma, *see* Myanmar
Burundi 222

Cambodia 47, 48
  Angkor 48, 49
Cameroon 222
  Douala 210
Caribbean 1–13
  see also individual countries
caste 9–10
Ceylon, *see* Sri Lanka
Chad 222, 233
child labour 284, 289, 290
Chile 254–303
  Putaendo 266
  Santiago xx, 252, 265, 266, 271, 284
  Valparaiso 265
China 1–13, 18, 19, 27, 28, 32–42, 42–4, 47, 49, 50, 60–90, 129
  Amoy 17, 41
  Anshan 60
  Anyang 19, 32–3
  Ao 32
  Baotou 60
  Beijing (Peking, Tiananmen, Yenching) xx, 9, 35, 39, 40, 41, 52, 65, 66, 67, 74, 85
  Canton, *see* Guangzhou

China (*cont.*):
  Ch'ang An 34, 35, 38, 39, 44
  Changchun 60
  Changsha 35, 41, 60
  Chengdu 35, 41, 60
  Chinan (Tsinan) 35
  Chongqing (Chungking) 35, 41, 60
  Chuanchow, *see* Quanzhou
  Dalian (Dairen) 17, 41, 52, 53, 60
  Datong 60
  Fushun 60
  Fuzhou (Foochow) 17, 35, 40, 60
  Guangzhou (Canton) 17, 35, 40, 41, 42, 52, 60, 67, 74
  Guiyang 60
  Handan 60
  Hangkow (Hankou), *see* Wuhan
  Hangzhou (Hangchow) 38–9, 40, 56, 60
  Harbin 60, 74
  Hong Kong xx, 17, 32, 41, 42, 52, 53, 54, 60, 76
  Jilin 60
  Jinan 60
  Kaifeng 38, 39, 40
  Kunming 60
  Lanzhou 60
  Luoyang (Loyang) 32–3, 34, 35, 39, 60
  Macao 17, 40
  Mukden, *see* Shenyang
  Nanchang 60
  Nanjing (Nanking) 35, 39, 40, 41, 42, 44, 56, 60
  Nanning 60
  Ningpo 35
  Peking, *see* Beijing
  Qingdao (Tsingtao) 17, 41, 53, 60
  Qiqihar 60
  Quanzhou (Chuanchow) 17
  Qufu 87
  Shanghai xx, 17, 40, 41, 42, 44, 53, 55, 56, 60, 66, 67, 70, 74, 75, 76, 83
  Shenyang (Mukden) xx, 35, 60, 70
  Shijiazhuang 60
  Sian, *see* Xian
  Suchow 17, 40
  Swatow 35
  Taiyuan 35, 60
  Tangshan 60
  Tiananmen, *see* Beijing
  Tianjin (Tientsin) xx, 17, 40, 52, 53, 60, 74, 83
  Tsinan, *see* Chinan
  Tsingtao, *see* Qingdao
  Wenchow 17
  Wuhan (Hangkow, Hankou) 17, 35, 40, 41, 52, 60
  Xian (Sian) 32, 33, 60
  Yenching, *see* Beijing
  Zengzhou 60
  see also Taiwan
class, *see* stratification
Colombia 254–303
  Barranquilla 252, 265
  Bogotá xx, 252, 265, 271
  Cali 252, 265
  Medellín 252, 256, 265
colonial rule 2, 23–31, 40–2, 43, 50–5, 139–49, 186, 219–21, 225
commercialization, *see* trade
commuting 134, 137, 152, 155, 162–3, 167, 168
Costa Rica 12, 283
  San José 283
Côte d'Ivoire 222, 245
  Abidjan 210
  Yamoussoukro 243
Cuba 12
  Havana 252
Cush, *see* Nubia

Dahomey (country), *see* Benin
Dahomey (kingdom) 218
democratization 11, 240–1, 300
dependency 154, 186–9, 194–5, 202, 206, 253–5
Dominican Republic 12
  Santo Domingo 252

Ecuador 303
  Guayaquil 252
  Quito 252
education 29, 75, 77–8, 85, 137–8, 149, 151, 164, 167, 172–3, 176, 216, 229, 231–2, 233, 274, 281, 291, 294–7
Egypt 189, 190–1, 192, 193, 195–7, 204–5, 218
  Alexandria 184
  Cairo xx, 184, 190–1, 196, 205
  Qasr Ibrim 212–3
El Salvador 280
  San Salvador 283
employment, *see* labour market
environment 53–5, 79, 177, 256

Eritrea:
  Adulis 214
Ethiopia 214, 217, 218, 222, 244
  Addis Ababa 210
  Aksum 212, 213, 214, 217
  see also Aksumite state, Eritrea
ethnicity 8, 9–10, 52, 141, 145–7, 151, 167, 168, 199–201, 234, 236–40, 292

family, *see* household; kinship

Gabon:
  Libreville 219
Gambia 233
gender, *see* women
Ghana (country) 222
  Accra 210
  Kumasi 213, 218
Ghana (kingdom) 214, 218, 228
Guatemala 12, 292
  Guatemala City 270, 283
Guinea:
  Conakry 210

Haiti 12, 252
Honduras:
  Tegucigalpa 283
Hong Kong, *see* China
household 5–6, 165, 230–1, 284–6, 289–93, 302
housing 63, 76, 77–9, 80, 81–2, 83, 84, 122–7, 173, 176, 177, 197, 225, 270, 274

Igbo (Ibo) 236–7
income, GNP, GDP 2, 3, 4, 12, 77, 84, 98, 102, 109–13, 134–5, 137, 151, 159–60, 191–2, 201, 222–3, 254–6, 259, 286–93
India 1–13, 19–31, 42, 47, 48, 49, 52, 67, 92–131
  Agra 21–2
  Ahmedabad 21, 22, 27, 31, 56, 92, 107, 112, 126
  Ajmer 23
  Allahabad 23, 31
  Amritsar 23
  Balasore 16
  Bangalore xx, 29, 92, 99, 107
  Bassain 16
  Benares, *see* Varanasi
  Berhampur 16
  Bhilai 99
  Bhiwandi 99
  Bhopal 92, 99
  Bhubaneshwar 99
  Bombay xx, 16, 23, 25–8, 29, 30, 52, 53, 92, 107, 112, 117
  Calcutta xx, 16, 23–5, 26, 28, 29, 31, 52, 53, 54, 72, 92, 107, 112, 125
  Calicut 16, 20, 23
  Cochin, *see* Kochi
  Coimbatore 92
  Cuttack 16
  Delhi xx, 21–2, 31, 52, 70, 92, 99, 107
  Diu 16
  Durgapur 99
  Faridabad 99
  Fatehpur Sikri 22
  Goa 16, 20, 23
  Haldia 99
  Hyderabad xx, 29, 92, 107, 125
  Indore 92
  Jaipur 23, 92
  Jamshedpur 27
  Jodhpur 23
  Kanpur 23, 28, 31, 92
  Kochi (Cochin) 16, 20, 92
  Lucknow 92
  Madras xx, 16, 23, 28–9, 52, 53, 56, 92, 107, 112, 126
  Madurai 20, 92
  Mangalore 16, 20
  Masulipatam 16
  Murshidabad 23
  Mysore 29
  Negapatam 16
  Nagpur 29, 92
  Pataliputra 19–20
  Patna 23, 31, 92
  Pondicherry 16
  Puhar 20
  Pulicat 16
  Pune (Poona) 92, 107, 125
  Salem 29
  Surat 16, 21, 22, 26, 92, 99
  Tamralipiti 20
  Tanjore 20
  Ulhasnagar 92
  Vadodara 92
  Varanasi (Benares) 20, 23, 92
  Vijayanagar 21
  Visakhapatnam 92

Indonesia 1–13, 47, 48, 51, 132–83
Aceh (Acheh), *see* Banda Aceh
Ambon (Amboina) 17, 49, 171
Balikpapan 17, 170
Banda Aceh (Aceh, Acheh) 16, 49, 139, 163
Bandung 132, 143–4, 152, 163, 168–9, 172
Banjarmasin (Bandjarmasin) 17, 140, 141, 143, 170
Banten 17, 139, 140, 145
Batam 168
Batavia, *see* Jakarta
Bekasi 169
Bengkulu (Bencoolen) 17
Bogor 169
Borobodur 48, 140
Cianjur 169
Cilacap 169
Cirebon 152, 169
Demak 139, 140
Den Pasar 162, 171
Depok 169
Garut 169
Jakarta (Sunda Kelapa, Batavia) xx, 17, 51, 53, 54, 70, 132, 139–45, 147–8, 152, 154–6, 163, 164, 165–72, 174–7
Jambi 170
Jember 169
Kediri 140, 169
Klaten 169
Kudus 169
Madiun 169
Magelang 169
Majapahit 140
Makassar, *see* Ujung Pandang
Malang 143, 152, 163, 169
Manado 170
Mataram (on Java) 140, 171
Mataram (on Lombok) 17
Medan 16, 51, 54, 132, 142–4, 145, 163, 168, 170
Melayu 140
Padang 16, 162, 170
Palembang 17, 48, 132, 140, 143, 162, 163, 170
Pasuruan 169
Pekalongan 143, 169
Pekanbaru 170
Pematang Siantar 170
Pontianak 170
Purwokerto 169
Salatiaga 169
Samarinda 170
Semarang 17, 51, 132, 139–44, 148, 152, 169
Singhasari 140
Sukabumi 169
Sunda Kelapa, *see* Jakarta
Surabaya (Surabaja) 17, 51, 132, 139–44, 148, 152, 162, 168–9, 172
Surakarta 142–4, 162, 169
Tanjung Karang 170
Tasikmalaya 169
Tegal 169
Ternate 17, 49
Ujung Pandang (Makassar) 17, 139, 140, 141, 143, 162, 163, 168, 170
Yogyakarta 142–4, 152, 163, 168–9
industrialization, industry, *see* manufacturing
infant mortality 3, 12, 137, 222–3, 286
informal sector 84, 85, 123, 125–7, 137, 162–3, 176, 224, 256, 271, 275–84, 289, 301–2
Iran 9
Teheran xx
Iraq 189, 190, 191, 192, 193, 197–8, 202–3
Baghdad xx, 184, 197, 198, 203
Basra 198
Ivory Coast, *see* Côte d'Ivoire

Jamaica 12
Japan 43–7, 52
Kamakura 44
Kanazawa 45, 46
Kobe 17, 46, 53
Kyoto (Heian) 44–5, 46
Nagasaki 17, 45
Nagoya 45, 46, 51
Naniwa 43
Nara 44
Okayama 70
Osaka xx, 43, 44, 45, 46, 51
Sakai 45
Tokyo (Kamakura, Edo) xx, 44, 45–7, 51
Yokohama 17, 46, 53
Jordan 190, 192, 193, 195–7, 204, 207
Amman 190, 191, 196, 197, 204

Kenya 211, 222, 225, 233, 242, 243, 245
Kisumu 219

Kenya (*cont.*):
  Mombasa 219
  Nairobi 210, 211, 233
kinship 8, 85–6, 229, 233–5, 237
Kongo (kingdom) 218
Korea 34, 42–3
  Inchon 17, 43
  Kaesong 43
  Kwangju 43
  Kyongju 43
  Pusan 17, 43
  Pyongyang (Lo-lang) 34, 42
  Seoul xx, 17, 43, 52
  Taegu 43
Kuwait 190, 192, 193, 197, 199, 200, 201
  Kuwait City 184, 207

labour market 63, 64–5, 67, 72–4, 77, 84–5, 96, 100–3, 119, 128–30, 134–7, 148, 162, 172–6, 231–3, 272–87
language 29, 215, 236, 237
Latin America 1–13, 243, 252–314
  see also individual countries
Lebanon 190, 192, 193, 195, 204, 207
  Beirut 184, 207
Liberia 218, 245
  Monrovia 219
Libya 190, 192, 193, 199, 200, 203, 204
  Tripoli 184

Malawi 222, 225
Malaysia (Malaya) 12, 49, 51, 152, 168
  Johor Baharu 17
  Kuala Lumpur 51
  Kuching 17
  Malacca 17, 49, 50
  Penang 16, 50
  Port Swettenham 51
Mali (country) 223
  Bamako 220
  Gao 213, 214, 217
  Jenné 213, 214, 217
  Timbuktu 212, 213, 214, 217
  Tindibi 218
  see also Ghana (kingdom); Songhai (empire)
Mali (empire) 214, 216
manufacturing 20, 23, 25, 27, 29, 31, 38, 41, 46, 64–6, 70–1, 72, 77, 84, 96–103, 109–10, 112–14, 119, 135, 137, 148, 149, 151, 154, 155, 162, 175, 193, 204, 220, 221–2, 228, 253–5, 257, 259, 261, 263, 264, 267–8, 277–8
Mauretania:
  Kumbi-Salih 213, 214
Mazahua 262
Mexico 12, 254–303
  Acapulco 267
  Aguascalientes 267
  Cedral 297
  Celaya 267
  Chihuahua 267
  Ciudad Juárez 267
  Cuautla 267
  Guadalajara 252, 266, 268, 274, 291
  Guadalupe 252
  Hermosillo 263, 267
  Irapuato 267
  Léon 267
  Matamoros 267
  Mérida 267
  Mexicali 267
  Mexico City xx, 252, 256, 261, 262, 266–9, 270, 284, 290, 291, 292, 293
  Monterrey 252, 266, 268, 270, 291, 297
  Naucalpan 252
  Nuevo Laredo 267
  Pachuca 267
  Puebla 252, 263, 267, 272
  Puerto Vallarta 261
  Querétaro 267, 272
  Reynosa 267, 291
  Saltillo 263, 267
  San Luis Potosí 267
  Tijuana 267, 268
  Toluca 263, 267, 272
  Torreón 267
migrants, migration 4, 52, 67, 72–6, 99–100, 111, 114, 134, 137, 141–2, 144–8, 152, 159–72, 174, 194–5, 196, 198–205, 221, 224, 256–62, 265–6, 296–7
  policy 63–5, 74–6, 84–5, 174–6, 225–8; *see also* migrants, migration policy
  strategies 5–8, 228–31, 258
  see also slavery
mineral resources 20, 159, 191, 193, 194, 197–205, 219, 267–8
modernization 55, 300
Morocco 189, 190, 192, 193, 194–5, 202

Morocco (*cont.*):
Casablanca 184
Fez 195
Rabat 184
Salé 207
Mozambique 223
Beira 219
Maputo 210
Myanmar 48, 49, 51
Akyab 16
Ava 49
Bassein 16
Mandalay 50
Moulmein 16, 49
Pagan 49
Pegu 16, 49
Prome 48
Rangoon 16, 50, 51, 52
Syriam 49
Tavoy 16
Toungoo 49

neighbourhood, *see* social organization
Nicaragua 9
Managua 283
Niger 223, 233
Nigeria 223, 236–7, 245
Abuja 243
Benin 213, 215
Ibadan 210, 213, 215, 238
Ife 212, 213, 215
Igbo-Ukwu 215
Kano 213, 217, 219, 238
Lagos xx, 210, 219, 238, 240
Sokoto 213, 217
Zaria 213, 217
Nubia 212–13, 216, 218

oil, *see* mineral resources
Oman 190, 192, 193

Pakistan 18, 19–31, 42
Karachi xx, 16, 29, 52, 53, 56
Lahore xx, 21, 22, 31
Lothal 20
Mohenjo Daro 19, 20
Multan 30
Peshawar 31
Rawalpindi 31
see also India
Palestine, Palestinians 189, 190–1, 193, 195, 197, 201, 204, 207
Panama 303
Papua New Guinea 12
Paraguay 260, 292
patronage 232, 238, 240
Peru 12, 254–303
Arequipa 265
Chiclayo 265
Chimbote 265
Huancayo 265
Lima xx, 252, 256, 264, 265, 284, 292
Trujillo 265
Philippines 47, 49–50, 51
Cebu 17, 51
Davao 51
Iloilo (Ilailo) 17, 51
Legaspi 17
Manila xx, 17, 50, 51, 53, 54
policy 8
political conflict 9–11, 85, 129–30, 236–40, 256, 271, 298–302
political economy 206
population growth 3, 4–5, 94–100, 138, 199, 221, 222–3, 224
primacy, *see* urban hierarchy
Puerto Rico:
San Juan 252

Qatar 190, 192, 193, 199, 200
Quechua 292

railways 24–5, 26, 27, 29, 31, 41, 43, 46, 51, 52, 53, 54, 219–20
regional comparisons:
across developing countries and regions 1–14, 61, 63–7, 70–2, 77, 80, 84–6, 186–9, 206
within countries 2, 67–8, 107–14, 129
religion 8, 9–10, 19, 20, 23, 26, 29, 30, 48, 138, 234
Rhodesia, *see* Zambia; Zimbabwe
Rwanda 223

Saudi Arabia 190, 192, 193, 194, 199, 200, 204, 205
Jeddah 184
Riyadh 184
Senegal 223, 242
Dakar 210, 219, 220
service sector 64–5, 70–1, 75, 77–80, 83, 84, 97–9, 100–3, 135, 137, 151, 259, 261, 267–8, 274–5, 277–8, 282, 283
see also administration

shipping 20–55, 141–2, 218–20, 268
Sierra Leone:
  Freetown 219, 237
Singapore 17, 50–1, 52, 53, 152, 168
slavery 139–41, 216–18, 244
social organization 7–8, 84, 233–6, 276, 284–6
  see also household; kinship
social security 80–1, 85–6, 231, 273–4, 279–81
Somalia 223, 245
  Mogadishu 213, 215
Songhai (empire) 214, 218
South Africa 220, 221, 223, 225–6, 233, 238, 240, 241, 243
  Cape Town 210, 213, 219
  Durban 210, 213
  East Rand 210
  Johannesburg 210, 219
  Kimberley 219
  Port Elizabeth 219
squatters 11, 125, 127–8, 195, 270, 271, 299
Sri Lanka 30–1
  Anuradhapura 30
  Colombo 16, 30–1, 53, 56
  Galle 16, 30
  Jaffna 30
  Kandy 30
  Mannar 30
  Polonnaruwa 30
  Trincomalee 16, 30
stratification 10, 239–40, 253–6, 278–303
Sudan 190, 192, 193–4, 202, 223
  Kerma 212, 213, 218
  Khartoum 184, 210
  Meroë 212, 213, 218
  Napata 212, 213, 218
Swahili 215
Swaziland 13, 233
Syria 190, 192, 193, 195, 204
  Aleppo 184
  Damascus 184

Taiwan:
  Chi-lung (Tamshui-Keelung) 17
  Kaohsiung 17
Tanzania 223, 227–8, 233, 242
  Dar es Salaam 210, 219
  Dodoma 243
  Kigoma 219
  Kilwa 213, 215
  Zanzibar 213, 215
Thailand 47–8, 51
  Ayuthia 49
  Bangkok xx, 16, 49, 50, 51, 52, 53, 70
  Nakom Patom 47
  Pattani 16, 49
  Phuket 16
theoretical perspectives viii–ix, 185–9, 205–6
  see also dependency; modernization; political economy; stratification; urban bias; world system
Third World 1, 18
tourism, *see* service sector
trade 20–55, 93, 97, 139, 141–2, 148, 149, 154, 155, 164, 214–20, 255, 256, 258, 267, 268
trade unions 10–11, 227–8, 239–41, 273, 279–80, 287, 289, 298–9
Tunisia 189, 190, 192, 193, 194–5, 202
  Tunis 184, 195
Turkey:
  Istanbul xx, 207

Uganda 219, 223, 245
United Arab Emirates 190, 192, 193, 199, 200
  Dubai 200
Upper Volta, *see* Burkina Faso
urban:
  definition of 61–3, 68–72, 95, 133–4, 154–5, 173
  bias 4, 63, 64, 66–7, 72, 84, 114, 151, 173, 224–5, 241, 259
  civilizations, origin of 215–17
  concentration, *see* urban hierarchy
  ecology 19, 22, 24, 26, 28, 36, 39, 49, 147, 194, 220–1, 269–72, 275
  farming 137, 224, 270
  functions, *see* administration; education; manufacturing; mineral resources; religion; service sector; shipping; trade
  governance 68–70, 115–22, 129–30, 175
  growth 3, 4–5, 64, 84, 94–9, 103–14, 121–2, 128–9, 142–4, 148, 149–52, 168–71, 196, 199, 221, 222–3, 224, 256–8, 262–9; *see also* urbanization, level of
  hierarchy 3, 5, 66–9, 72, 95, 103–7, 139–44, 152–4, 174–6, 194, 196–7, 222–3, 225, 257, 262–3, 265–9

Urban (*cont.*):
history 2, 16–56, 138–59, 212–25, 256–72
infrastructure, *see* urban services
land use 82–3, 177, 196; *see also* urban ecology
layout, *see* urban ecology
services 76, 77–80, 83, 84, 96, 98, 115, 118, 119, 127, 151, 173, 175, 176, 177, 199, 224–5, 256, 271, 284
sex ratios 3, 5–8, 167, 222–3, 229
space, *see* urban ecology
urbanization, level of 2, 18, 20, 31, 40, 46, 62–6, 93–103, 107–14, 128–30, 133–4, 139, 156–60, 190, 191–2, 198–200, 219, 220, 221, 222–3, 256–8, 264
see also regional comparisons; urban growth
Uruguay 282, 303
Montevideo 252, 271

Venezuela 292
Caracas 252
Maracaibo 252
Valencia 252
Vietnam 34, 47–8, 51
Da Nang (Tourane) 17, 47
Hanoi (Haiphong) 17, 34, 47, 51, 53
Hue 48
Saigon 17, 49, 51, 53, 54
Poulo Condore 17

women 234, 285
and social security 80
household heads 292–3
in labour market 65, 232–3, 275–8, 281, 289–91, 294
in manufacturing 7, 164
in migration 5–8, 163–5, 167, 168, 229–31, 258, 260–2
see also urban sex ratios
world system 185–7, 206

Xhosa 238

Yemen 190, 192, 193–4, 202, 218
Yoruba 215, 236–7

Zaïre 223, 215, 245
Kinshasa (Léopoldville) 210, 219
Lubumbashi (Elisabethville) 219
Shaba Province (Katanga) 227
Zambia 215, 223, 245
Copperbelt 219, 227
Kitwe 219
Zimbabwe 223
ruins 213, 215, 217, 225, 233
Zulu 238